UNDERSTANDING
CYBER
CONFLICT

D1127656

Other Titles of Interest from Georgetown University Press

Cyber Blockades
Alison Lawlor Russell

Cyberspace and National Security: Threats, Opportunities,
and Power in a Virtual World
Derek S. Reveron, Editor

Optimizing Cyber Deterrence: A Comprehensive Strategy
for Preventing Foreign Cyberattacks
Robert Mandel

UNDERSTANDING CYBER CONFLICT

14 ANALOGIES

GEORGE PERKOVICH
ARIEL E. LEVITE
EDITORS

U
167.5
.C92
U54
2017

WITHDRAWN

Georgetown University Press / Washington, DC

© 2017 Georgetown University Press. All rights reserved. No part of this book
may be reproduced or utilized in any form or by any means, electronic or mechanical, including
photocopying and recording, or by any information storage and retrieval
system, without permission in writing from the publisher.

The publisher is not responsible for third-party websites or their content. URL links
were active at time of publication.

Library of Congress Cataloging-in-Publication Data

Names: Perkovich, George, 1958- editor. | Levite, Ariel, editor.
Title: Understanding cyber conflict : fourteen analogies / George Perkovich
    and Ariel E. Levite, editors.
Description: Washington, DC : Georgetown University Press, 2017. | Includes
    bibliographical references and index.
Identifiers: LCCN 2017003096 (print) | LCCN 2017008146 (ebook) | ISBN
    9781626164987 (pb : alk. paper) | ISBN 9781626164970 (hc : alk. paper) |
    ISBN 9781626164994 (eb)
Subjects: LCSH: Cyberspace operations (Military science)
Classification: LCC U167.5.C92 U54 2017 (print) | LCC U167.5.C92 (ebook) |
    DDC 355.4--dc23
LC record available at https://lccn.loc.gov/2017003096

♾ This book is printed on acid-free paper meeting the requirements of the
American National Standard for Permanence in Paper for Printed Library Materials.

18 17   9 8 7 6 5 4 3 2 First printing

Printed in the United States of America

Cover design by Jen Huppert. Cover image by robertiez/iStockphoto.

To Emily O. Goldman

for building a bridge from scholarship to practice (and back)
and for doing much else to encourage this book

CONTENTS

ILLUSTRATIONS

Tables

Figures

ACKNOWLEDGMENTS

The editors would like to thank several people for making this volume possible. The first category, of course, is the authors of the chapters, each of whom was a pleasure to work with and a model of diligence. The volume would not have been conceived without the earlier work and the encouragement of Emily O. Goldman and John Arquilla.

Wyatt Hoffman, Steven Nyikos, Bert Thompson, and Tim Maurer—our colleagues at the Carnegie Endowment—provided the needful coordination, facilitation, and administration, as well as helpful critiques of chapters.

Eli Sugarman and Larry Kramer of the Hewlett Foundation saw the value of exploring the benefits and pitfalls of analogical thinking about cyber conflict. The foundation's support made this book possible.

Finally, we thank the National Cryptologic Museum at Fort Meade, Maryland, for hosting a daylong workshop where the authors critiqued each other's chapters. They also enjoyed a fascinating tour of the museum led by its inimitable curator, Patrick Weadon.

Introduction

GEORGE PERKOVICH AND ARIEL E. LEVITE

In extensive conversations with senior civilian and military cyber policymakers in the United States, the United Kingdom, France, Canada, Israel, Russia, and China, the editors of this volume heard repeatedly that these individuals and their counterparts in government frequently invoke historical analogies—aptly and inaptly—as they struggle to manage new technologies. The cyber domain is new to most senior officials. Cyber capabilities have unique properties. Experience with them in conflict thus far has been limited. Consequently, it is difficult to make confident judgments about their effects and escalatory potential. Moreover, the range of adversaries and behaviors that policymakers and experts must strive to dissuade, deter, or defeat in and through cyberspace is unprecedented: massive-scale thievery, political subversion, terrorism, covert operations, and open warfare. In such circumstances the human mind naturally pulls up analogies from the past to guide thinking and acting amid the new.

One of our interlocutors, in early 2014, recommended that we read *Cyber Analogies*, a collection of essays edited by Emily O. Goldman of US Cyber Command and John Arquilla of the US Naval Postgraduate School.[1] We took this advice and found that, indeed, those essays sharpened our thinking about differences and similarities between cyber and previous military technologies and episodes. Goldman and Arquilla encouraged us to extend the exploration of analogies, with an eye toward adding examples and perspectives that would be pertinent to readers beyond the United States. The result is the present volume, which includes four revised essays from their collection, plus ten chapters that we commissioned to explore additional analogies.

Human beings think, learn, and communicate through analogies. We use analogies—naturally, often without trying—to familiarize that which is new. As Richard E. Neustadt and Ernest R. May recorded in their classic study, *Thinking in Time*, policymakers and pundits regularly invoke analogies as they struggle to make sense of and affect new situations, often without adequate reflection.[2] This practice occurs now regarding the cyber world, which is evolving with an ever-quickening pace. For people who were born in this era, the benefits and risks that flow from the enhancement and distribution of information and communications technologies are more familiar than the earlier technologies, episodes, and policy challenges to which elders analogize. Young readers may know how hacktivists operate and how cyber attacks brought Estonia to a standstill in 2007, but they may be less familiar with the eighteenth- and nineteenth-century privateering at sea that resembles the challenges posed by proxy actors in cyberspace. Cybersecurity professionals may be convinced that the speed of offensive attacks will require automated defensive responses, but they may be unaware of how governments wrestled internally over the pre-delegation of authority to launch nuclear weapons under attack. Curricula today in courses on history, political science, international relations theory, and security studies still derive from pre–cyber era experiences; relatively few explore whether and how the cyber era may be similar or different. So, too, the strategies, policies, and institutions that governments use to manage dual-use technologies today generally predate the World Wide Web. Therefore, analogies across eras can be instructive for the young as well as for the not-so-young.

Variations in culture, ideology, and circumstances affect how audiences perceive and understand analogies. The authors of this volume are American, British, Israeli, and Swiss. The analogies to which they compare cyber technology and the challenges arising from it tend to be especially meaningful in their countries and, probably, in the West more broadly. We have tried throughout to keep the aperture wide enough to invite readers with different backgrounds to consider whether observations and analyses offered here do or do not apply more broadly. Moreover, readers from other locales and perspectives may gain insight from considering how these well-informed Western authors think about the given topic even if it differs from their perspective. In any case, the Cyber Policy Initiative of the Carnegie Endowment for International Peace hopes subsequently to build on the present volume and invite authors from other countries and perspectives to write about analogies that may be especially important to them.

Learning from analogies requires great care. Analogies can mislead as well as inform. Indeed, their educational value stems in no small part from identifying where, when, and how an analogy does not work well. Differences between technologies, effects, and historical, political, and strategic circumstances are as important to understand as similarities are. For example, today one must take particular care in analyzing which attributes of the nuclear era carry forward into the cyber era and which do not, and what the implications of confusion on this score could be.

Stanley Spangler, a professor of national security affairs at the US Naval War College, noted in 1991 that "virtually every postwar American president has been influenced by parallels drawn from the 1930s when Great Britain and France failed to react soon enough and strongly enough to halt [Adolf] Hitler."[3] President Lyndon Johnson, for example, declared, "Surrender in Vietnam [would not] bring peace, because we learned from Hitler at Munich that success only feeds the appetite of aggression."[4] Ironically, in ensuing decades the Vietnam War itself became a frequently used analogy in American debates over military intervention in other distant lands. In the 2015 debate over the Joint Comprehensive Plan of Action, which was negotiated to resolve the crisis over Iran's nuclear program, critics made countless references to the Munich Pact of 1938 and Neville Chamberlain, while proponents invoked the need to avoid repeating the 2003 war in Iraq. The Iran debate of 2015, like the Vietnam debate, demonstrated the risk that analogies can be a flawed substitute for actual knowledge of the past and the present and for critical thinking about both. Nevertheless, people ineluctably employ analogies to conceptualize and manage new circumstances. Thus, it is necessary and salutary to examine analogies carefully and to search for what is apt and inapt in them.

We have organized the essays (and analogies) in this volume into three groups. The first section, "What Are Cyber Weapons Like?," examines the characteristics of cyber capabilities and how their use for intelligence gathering, signaling, and precision strikes compares with pertinent earlier technologies for such missions. The second section, "What Might Cyber Wars Be Like?," explores how insights from episodes of political warfare, preventive force, and all-out war since the early nineteenth century could apply or not apply to cyber conflict in the twenty-first century. The final section, "What Are Preventing and Managing Cyber Conflict Like?," suggests what insights that states seeking to civilize cyberspace might draw from earlier experiences in managing threatening actors and technologies. We introduce the essays here accordingly.

What Are Cyber Weapons Like?

The cyber domain—and its associated hardware, software, and human resources issues—is constantly growing and evolving. Information and communications technologies can serve manifold peaceful and coercive purposes in addition to providing legal and illegal means of generating wealth. In the context of inter-state conflict alone, hundreds of analogies could be drawn and analyzed between cyber weapons and their predecessors. Capabilities and plans exist and are being developed further to use cyber assets in large-scale, combined-arms military campaigns. Cyber operations could be conducted to cause massive disruption and, indirectly, significant human casualties. A literature is already emerging on these larger-scale capabilities and scenarios.[5] Essays in the second and third sections of this volume explore whether and how technologies and practices central to World Wars I and II and the management of nuclear deterrence offer insights to the conduct and prevention of cyber warfare.

Here, in this first section, we focus on analogues to less destructive capabilities. In an era when all-out warfare among major powers may be deterred by nuclear weapons, among other factors, and global dependence on networked information and telecommunications technologies creates unprecedented vulnerabilities, the instruments of stealth, speed, and precision that can be controlled from great distances will be particularly salient as states compete to influence each other in the coming years. These applications pertain to intelligence gathering, covert operations, "political warfare," and relatively low-intensity, precise offensive actions. Such activities are especially germane to operations in the gray zone between declared war and peace, when large numbers of boots on the ground are not envisioned but exercising covert influence and coercive power is deemed expedient or necessary.

"What we call cyber is intelligence in an important sense," Michael Warner writes in the first chapter. "Intelligence activities and cyberspace operations can look quite similar." Warner, the US Cyber Command's historian, describes how cyber capabilities have been applied rather straightforwardly to serve the functions of spying and counter-spying that human agents have performed for millennia. "The main difference," he notes, "is the scale that can be exploited" by cyber techniques. Similarly, the use of cyber capabilities to conduct covert operations and to inform the planning and conduct of military operations builds on methods developed through the advent of the telegraph and radio in the nineteenth and twentieth centuries. The similarities here extend to the importance of cryptography and counter-cryptography to facilitate offensive and defensive missions. A key difference in the cyber era is that previously "the devices that secured and transmitted information did not also store it." Today, however, past, current, and future data are vulnerable to spies and eavesdroppers in unprecedented ways. This raises several questions that Warner examines: Will cyber espionage be more likely to cause conflict than traditional spying has done? What can responsible states do to gain the benefits of more fulsome intelligence collection while minimizing the risks to international stability and their own reputations, as well as to the brand value of companies whose products they exploit?

"No one has ever been killed by a cyber capability," write Lt. Gen. Robert Schmidle, Michael Sulmeyer, and Ben Buchanan in their chapter, "Nonlethal Weapons and Cyber Capabilities." Schmidle, the deputy commander of US Cyber Command from 2010 to 2012, and Sulmeyer, formerly the director for plans and operations for cyber policy at the Office of the Secretary of Defense, have been deeply involved in US military cyber policymaking. Buchanan is a postdoctoral fellow at the Cyber Security Project in Harvard University's Belfer Center for Science and International Affairs. Their chapter analogizes cyber capabilities to nonlethal weapons that the United States and other states have developed for decades. The Department of Defense defines *nonlethal weapons*—such as pepper spray, spike strips to puncture tires of vehicles, rubber bullets, flash bangs, electronic jamming devices, and lasers—as "weapons, devices, and munitions that are explicitly designed and primarily employed to incapacitate targeted personnel or materiel immediately, while minimizing fatalities, permanent injury to personnel, and undesired damage to property in the target area of environment."[6] In a first-of-its-kind analysis, the authors compare and contrast potential utilities of nonlethal weapons and cyber capabilities in four ways: their ability to incapacitate, the reduced collateral damage they inflict, the reversibility of their effects, and their ability to deter. Schmidle, Sulmeyer, and Buchanan also address an interesting paradox: Why have US defense officials been particularly reluctant to approve the use of nonlethal capabilities, and can this reluctance be expected to continue, in the United States and in other states?

Moving up the ladder of coercive power, James M. Acton, a physicist and the codirector of the Nuclear Policy Program at the Carnegie Endowment for International Peace, explores the analogy between precision-guided munitions (PGMs) and cyber weapons. The development of PGMs—"guided gravity bombs and cruise missiles, in particular—has had profound implications for warfare," Acton begins. "Such weapons tend to cause much less collateral damage than their unguided predecessors do, and because they can remain effective when used from a distance, they can also reduce casualties sustained by the attacker. Thus, PGMs have altered national-level decision-making by lowering the political threshold for the use of force and by slowing the likely loss of public support during a sustained military campaign."

Cyber weapons may extend the militarily, politically, and morally attractive logic and functionality of PGMs. Cyber weapons offer the potential of "exquisite precision" in terms of targets and effects, although this potential may be very difficult for many actors to achieve in practice. They involve "minimal risk to the lives of the service personnel who 'deliver' them" and are "likely to cause fewer civilian casualties than even the most carefully designed and executed kinetic attack." As a result of these attributes, cyber weapons "could further lower the threshold for the use of force." At the same time, the effective use of cyber weapons requires sophisticated intelligence, surveillance and reconnaissance, and time-sensitive battle damage assessment. As with PGMs, it also remains questionable whether cyber weapons can accomplish larger, strategic political-military objectives. From all this the fundamental question arises of whether cyber weapons will augment deterrence of military conflict or make conflict more likely.

Drones, or unmanned aircraft used to surveil and precisely strike targets on the ground, have been celebrated and reviled since their use by the United States became an open secret in the mid-2000s. Armed drones are a form of PGM. What has made them more controversial, and perhaps more analogous to cyber weapons, are both the secrecy that for a long time shrouded the decision-making surrounding their use and the perception that their operators' immunity from physical harm lowers inhibitions on their use. David E. Sanger, the *New York Times*' chief Washington correspondent and author of *Confront and Conceal: Obama's Secret Wars and Surprising Use of American Power*, explores this analogy.

Sanger begins by recounting how outgoing-president George W. Bush told President-elect Barack Obama "there were two programs he would be foolish to abandon"—the drone program and a super-secret program called Operation Olympic Games, which was designing an offensive cyber operation to disable centrifuges in Iran's nuclear enrichment facility at Natanz. In the years that followed, Obama famously (or infamously to some) intensified the use of attack drones and authorized what became known as the Stuxnet attack on Iran. As Donald Trump stamps his imprint on US policy, he will need to grapple with the moral, legal, and strategic issues that these two types of weapons raise. What targets in what locations and under what circumstances are legitimate not only for the United States but for others too? What degree of confidence can realistically be attained that effects of cyber attacks (and drone strikes) will be limited to legitimate targets and will not cause unintended harm, or "collateral damage"? Many observers argue that drone strikes have incited escalatory revenge. Can cyber capabilities enhance deterrence of terrorism and other forms of aggression without this counterproductive effect? Sanger unpacks these issues by comparing and contrasting the nature and effects of drone and cyber attacks, and by drawing on the experience with drones, he considers how secrecy regarding cyber techniques and operations may affect prospects of governing them nationally and internationally.

What Might Cyber Conflicts Be Like?

The present conflicts in Ukraine, the Islamic State of Iraq and Syria operating in both states, and the cyber-abetted interference in the 2016 US presidential campaign may characterize prevalent challenges to peace and security in the twenty-first century, at least in cyberspace. At the same time, of course, the recent escalation in tensions between Russia and the West, and between China and its US-backed neighbors in East Asia, underscores the enduring importance of historical major power conflicts in continuing to shape perceptions and political discourse in the East and the West. Thus, the chapters in this section explore analogies from a wide span of history to draw implications for a range of confrontations and conflict contingencies that cyber-capable states may face and in which cyber operations may play a role.

In his chapter, Stephen Blank, of the American Foreign Policy Council, describes how Russia's contemporary use of offensive cyber operations against Estonia (2007), Georgia (2008), and Ukraine (2014–15) is not merely analogous but also a direct continuance of the strategy and practice of Soviet subversion of neighboring states. He writes, "Tactics and strategies developed and employed during the Soviet period have served as a foundation for establishing new strategies that incorporate some of the century-old Leninist repertoire and new trends like IW [information warfare], as defined by Moscow, for the conduct of continuous political warfare against hostile targets."

In describing the conduct of IW and cyber attacks in Estonia, Georgia, and Ukraine, Blank reports that Russia's aim was to "instill a feeling of constant political and economic insecurity among the target state's population" while testing whether and how European security institutions and the United States would respond. In Georgia and Ukraine, attackers believed to be linked to the Russian state penetrated and placed malware in electricity supply systems. When the Georgian conflict ended early, without Western intervention, no decision to execute destructive cyber attacks was made. In Ukraine, nationalists sabotaged electricity supply lines to (Russia-annexed) Crimea in November 2015 and cut off power there. Russian retaliation, prepared well in advance, was executed four weeks later in the form of a sophisticated, measured cyber attack that shut down three regional electric power distribution companies. Thus, as Blank details, cyber capabilities provide Russian actors with a spectrum of relatively inexpensive and risk-mitigating coercive instruments to impose Russian interests on adversaries below the threshold of violence that would prompt military escalation, especially by Western powers. "Russia has already engaged its adversaries in information warfare," Blank concludes, "thus, its adversaries must understand and learn from it for their own security."

Moving up the ladder of force, many assessments posit that offensive cyber operations would optimally be undertaken secretly, before armed warfare has commenced, to impair an opponent's capacity to fight or to create facts on the ground that could motivate an opponent to stand down. In "An Ounce of (Virtual) Prevention?," John Arquilla, the chair of defense analysis at the US Naval Postgraduate School, considers how the use of preventive force in the Napoleonic Wars and leading up to World War I may hold insights for the cyber era. Arquilla describes how the British navy in 1801 and 1807 conducted attacks on the Danish fleet, the coastal artillery emplacements, and the city of Copenhagen to prevent Denmark from colluding with Napoleon in closing the Baltic Sea to British trade. While the British attacks accomplished their tactical and strategic objectives, the exercise of preventive force also motivated Germany in the late nineteenth and early twentieth centuries to build up its navy to deny Britain the option of preventive force. Fast forwarding a hundred years, Arquilla analogizes that the Stuxnet cyber attack conducted by the United States and Israel against Iran's centrifuge program not only successfully slowed Iran's acquisition of enriched uranium but also may have spurred Iran and future potential nuclear proliferators to take defensive measures that will make counter-proliferation more difficult in the future. Ultimately, Arquilla concludes, twenty-first-century states are likely to see cyber techniques and operations as useful for preventive force—including against terrorist groups—and will therefore compete offensively and defensively in this type of conflict.

Francis J. Gavin, an international historian and director of the Henry A. Kissinger Center for Global Affairs at the School of Advanced International Studies at Johns Hopkins University, addresses the issue of war instigation from a different angle, assessing whether and how the technology of railroads drove Germany, France, the United Kingdom, and Russia into World War I. Early historiography on the war posited that the European great powers' reliance on railways to transport military forces to their borders placed a premium on deploying their forces before their adversaries did. Ambiguities about the purpose of mobilization—either offensive or defensive—exacerbated crisis dynamics. Moreover, the logistics of railway mobilization made it difficult to pause or reverse once it started. Consequently, according to early historiography, once mobilization began, it acquired too much momentum to be stopped in the amount of time that the complicated diplomacy to prevent war would have required.

Modern historians have corrected the overly simplistic determinism of the railway narrative, yet, as Gavin notes, this work has not prevented the notion of technological determinism from influencing conceptions of cyber warfare. Nor should it necessarily. Indeed, the military implications of major, globally infused dual-use technologies can and should be analyzed independently. Comparing their similarities and differences with prior technologies can be helpful in this regard.

In Gavin's view, rail and especially cyber technologies are more facilitating technologies than they are instruments for killing adversaries, destroying their military assets, and occupying their territory. Both rail and cyber technology quickly spread around much of the world because they were vital to national and international economies, even as they also serve military purposes. The economic indispensability of these technologies complicates efforts to control their military or other coercive uses. Both technologies condense the effects of space and time, making the world smaller and faster, which, in turn, dramatically increase the pressures on decision-making during a crisis.

Yet, as Gavin analyzes, differences between cyber technology and railways may be most instructive. In any case, looking from 1914 to the future of potential cyber conflict, a portentous question is whether states in tense regions possess the "institutional capacities . . . to deal with massively increased amounts of information coming from a variety of different sources and in an environment where cyber attacks might be oriented toward degrading and blinding" decision-making capabilities.

World War I offers another analogy to potential cyber warfare in the twenty-first century, as the British historian Nicholas A. Lambert considers in "Brits-Krieg: The Strategy of Economic Warfare." Lambert, the Class of 1957 Chair in Naval Heritage (2016–17) at the US Naval Academy, fascinatingly describes how the advent of the telegraph and undersea cables enabled an unprecedented, global movement of goods, money, knowledge, and information that transformed international commerce. In earlier eras, traders purchased and stockpiled large amounts of goods. In the newly globalized system, traders relied on processes such as just-in-time delivery, credit-based purchase, and transfer of goods, all underpinned by new information technology. Britain was the hub of much of this global trade and finance. Realizing this, a few strategists in the Admiralty began in 1901 to consider how, in a time of war, Britain could leverage its dominant naval and commercial position to halt global trade and thereby cause a quick and devastating economic shock to an adversary's economy and society, in this case Germany's. Unlike the interdiction of ships and the preventive and attrition bombing of military-economic assets, "the British aim" would be "far higher: . . . delivering an incapacitating 'knock-down' blow that would obviate the need for less intense but more prolonged types of war."

In the cyber era, an analogous act would be to use "cyber means as a weapon of mass destruction or disruption, targeting an enemy's economic confidence as well as its infrastructure, with the aim of causing enemy civilians to put political pressure on their government." For example, a sophisticated actor could corrupt the integrity of data and the processing algorithms in one or more major financial institutions in ways that would profoundly undermine the confidence on which modern international commerce depends. Yet, as Lambert recounts, the United Kingdom's application of economic warfare at the onset of war in 1914 was so effective that it ultimately backfired and had to be abandoned. Trade plummeted, and with it went the well-being of British traders, financiers, and labor. "As the scale of the economic devastation [in the United Kingdom] became increasingly apparent, domestic interest groups became ever more vocal in clamoring for relief and lobbying for special exceptions, and neutrals [countries] howled in outrage at collateral damage to their interests." Soon, "political commitment to the strategy began to crumble; more and more exceptions to the published rules were granted, thereby further undermining the effectives of economic warfare." In October 1914, the government aborted the strategy. Readers can easily imagine how in the globalized, digitally intertwined world of today, a strategy to cause massive economic disruption through cyber attack could pose similar challenges. Not only would the intended object of the attack suffer enormously but so too would the attacking state if its labor force, employers, and treasury were dependent on global trade and finance. Lambert's conclusion details some of these possible challenges and ways of anticipating them.

Pearl Harbor presents the most frequently deployed analogy to cyber warfare, at least in US discourse. In October 2012 Secretary of Defense Leon Panetta warned of a possible "cyber Pearl Harbor," saying a malicious actor could launch devastating cyber attacks to "paralyze and shock the nation and create a new, profound sense of vulnerability."[7] Since then, cyber Pearl Harbor has become a recurring motif for officials, journalists, and experts warning of the dangers of a massive surprise cyber attack, especially in the United States. The image invoked is of a bolt-from-the-blue attack that catches defenders by surprise. Yet Emily O. Goldman, the director of the US Cyber Command–National Security Agency Combined Action Group, and Michael Warner clarify in chapter 9 that Pearl Harbor was not a surprise. "The United States was exercising coercive power to contest Japan's occupation of China and other Asian states, and Washington expected war. Pearl Harbor was a logical, if misguided, result of Imperial Japan's long-term strategy to expand its Pacific empire and blunt the United States' effort to stop it." Faulty American analysis and communication of intelligence data, and mistaken assumptions that the adversary (Japan) would calculate the risks of attacking as American personnel did, produced the sense of surprise. This observation makes what happened at Pearl Harbor even more salient for the United States and perhaps others today. Insofar as weaker actors embroiled in confrontations with powerful states may calculate, correctly or incorrectly, that a surprise cyber attack could temporarily weaken their adversary's political resolve and military capability, they may see such an attack as the least bad alternative. By creating a fait accompli, with relatively few casualties on both sides, they could shift the burden of escalation to the stronger party to choose war rather than compromise. Goldman and Warner conclude that the United States and other states whose militaries, economies, and societies are extremely reliant on cyber capabilities should both increase their vigilance and create resilience in their military cyber networks. Unlike the case of Pearl Harbor, the vectors of attack could be located not only in military networks but also through privately owned and managed networks. This possibility greatly complicates the challenge of detecting, defending against, and responding to attack.

What Are Preventing and Managing Cyber Conflict Like?

Capabilities to conduct cyber information warfare, criminal activities (including terrorism), covert operations, and preventive military force are spreading faster than the international community's capacity to establish agreed rules for managing them. This is normal; all major disruptive technologies have emerged and created challenges that states have then struggled for years and decades to regulate. These management struggles have been waged first on a national basis and then later, if at all, internationally. Cyber capabilities may emerge and evolve faster, and spread more extensively and quickly, than have antecedents such as nuclear power plants and weapons, air transportation, radio, and so on. Moreover, cyber capabilities are less geographically bounded than preceding technologies are. Nevertheless, the inherent interests of states and societies dictate that norms and rules for managing these new capabilities must be proposed, negotiated, and ultimately agreed on, even if their enforcement will be imperfect. Otherwise, the dangers and costs of threatening activities will be too severe for most states and societies to bear.

States have already begun to address the complexities of regulating the underlying technologies of cyberspace, including the Internet's infrastructure. The struggle to establish rules for cyber capabilities and activities is intertwined with a broader, ongoing struggle over the governance of the Internet and the nature of sovereignty in cyberspace. This plays out in various formal bodies, such as the International Telecommunications Union, nongovernmental organizations including the Internet Corporation for Assigned Names and Numbers, and multistakeholder groups including the Internet Governance Forum. More tentatively, informal and formal efforts at various levels have begun to develop norms for the use of cyber weapons and the conduct of cyber conflict. Most notably they come from such groups as the G20 (or Group of Twenty), the United Nations Group of Governmental Experts on Developments in the Field of Information and Telecommunications in the Context of International Security, and the participants in the *Tallinn Manual on the International Law Applicable to Cyber Warfare* project. Clear, internationally agreed-on rules remain elusive, but unilateral and multilateral initiatives can begin to reduce the risks of unrestrained cyber conflict. These efforts can be enlightened by past experiences in managing threats to national and international security.

In the first essay in this section, Steven E. Miller, the director of the International Security Program at Harvard University's Belfer Center for Science and International Affairs, compares essential features of the nuclear era with those emerging in the cyber era. Miller notes that the nuclear age emerged publicly in 1945 with ferocious suddenness as nuclear weapons were detonated over Hiroshima and Nagasaki. The technology was born through secrecy, was militarized, and was tightly controlled—first by one government, the United States, and then by another, the Soviet Union. Civilian applications of the technology came later and never lived up to the advertisements of its progenitors. In contrast, cyber technology, notwithstanding its origination in the US defense establishment, quickly and widely took root and spread through commercial channels. Countless, often unpredicted, civilian applications of the technology have fueled economic growth and affected the lives of billions of people who have become dependent on them. Thus, the nature, purposes, and stakeholders associated with cyber technology are profoundly different than those associated with nuclear weapons and with civilian applications of nuclear technology. In this context, Miller considers whether and how the four central "pillars" of the nuclear order—"deterrence, damage limitation, arms control, and nonproliferation"—may be useful or not in managing cyber threats.

Miller's essay provides a segue to the next three essays in this section, which explore key facets of the defensive challenge. John Arquilla, in "From Pearl Harbor to the 'Harbor Lights,'" leads off the discussion of analogues to defending against cyber attack and conflict. Arquilla illuminates some of the sometimes surprising difficulties in reducing the vulnerabilities of civilian and defense networks. He recounts how the United States for three months *after* Pearl Harbor failed to "turn off" the lights in the country's coastal cities and harbors at night. As a result, German U-boats easily identified the eastern coastline, lurked off open anchorages and undefended harbors, and inflicted enormous casualties and destruction. Once the order to darken the coasts was implemented, along with other defensive measures, the German navy significantly reduced its operations in US waters. Arquilla likens the US failure to dim the harbor lights to the ongoing, inadequate government and private sector policies and actions to make their computers, networks, and data less accessible to attackers, and he suggests ways to redress these liabilities.

One of the growing policy conundrums in cyberspace is whether and how states and legitimate non-state entities should be permitted to actively defend themselves against intrusion and attack. Passive defenses such as encryption, firewalls, authentication mechanisms, and the like do not carry risks of international crisis. But some "active" cyber defenses that in some cases could harm another country raise serious risks and challenges. Intervention in an adversary's networks or computers that causes serious economic harm to an innocent entity in another country or that (unintentionally) impedes another state's national intelligence collection and defenses, could make the active defender liable to economic and criminal penalties or worse. Dorothy E. Denning and Bradley J. Strawser, professors at the US Naval Postgraduate School, explore the ethical and legal issues arising from active defense by analogizing air defense to active cyber defense. They focus mainly on state-conducted defensive actions while recognizing that such cyber actions by businesses and other legitimate non-state actors, although entirely plausible, pose additional complications.

To set up the analogy, Denning and Strawser describe a range of active defenses deployed against air and missile threats. Among them are aircraft, which the United States and the United Kingdom have deployed since September 11, 2001, to defend against hijacked aircraft; missile defense weapons, such as the Patriot surface-to-air system used in the Gulf War in 1991; other rocket and missile defense weapons, such as Israel's Iron Dome; and electronic warfare.

The authors then summarize some possible forms of active cyber defense and ask several questions about each to assess their ethical implications.

The development of missile-carried nuclear weapons in the 1950s confronted American (and Soviet and UK) authorities with an existential problem—that is, how to preserve political control over these forces when evolving technology and threats narrowed the time to respond to a nuclear attack. President Dwight D. Eisenhower's response in 1959 was to grant military commanders the authority to use nuclear weapons under carefully prescribed conditions. Peter Feaver, a professor of political science and public policy at Duke University, and Kenneth Geers, an ambassador with the North Atlantic Treaty Organization's Cooperative Cyber Defence Centre of excellence and a senior fellow at the Atlantic Council, reflect on how this challenge and the US response to it may be analogous to challenges posed by potential cyber warfare.

Three features of nuclear war motivated the adoption of nuclear pre-delegation: "the speed with which a nuclear attack could occur, the surprise that could be achieved, and the specialized nature of the technology (that meant only certain cadres could receive sufficient training to be battle competent)." While cyber war does not pose the civilization-ending threat that global thermonuclear war does, it may impose similar challenges on the management of cyber weapons (offensive and defensive). Feaver and Geers expertly unpack these challenges and the possible solutions to them.

The final chapter in this section explores a different and necessary way of reducing cyber threats—curtailing the operations of hostile private actors that operate as proxies of states or with state toleration. The analogy here is to naval privateering between the thirteenth and nineteenth centuries. Written by Florian Egloff of the Cyber Studies Programme at the University of Oxford, "Cybersecurity and the Age of Privateering" chronicles how governments commissioned privately owned vessels in wartime to operate against their adversaries' trade and in peacetime to attack merchants' ships in reprisal for harms attributed to a nation and to capture goods of equal value.

Analogies to the cyber domain abound here. Several states recently have used or allowed hackers and criminal organizations to conduct cybercrime and cyber-enabled espionage against adversarial states and economic interests. This practice is analogous to privateering and piracy. Meanwhile, if a state lacks the capacity to defend the cyber domain and obtain redress for harmful cyber activities, then the users are largely left to protect themselves. Naturally, private companies, like the earlier naval merchants, are now debating with governments the advisability of issuing letters of marque that would allow companies to counterattack against cyberespionage and theft. Of course, as Egloff discusses, the myriad state and non-state actors and interests at play in the cyber domain, and the pace of technological change, mean that ordering this space will be exceptionally difficult and will take considerable time. He offers a thought-provoking framework for understanding differences and similarities in the naval and cyber domains and how this understanding could inform efforts to secure cyberspace.

Each of these chapters is valuable and instructive in its own right. Together, as we describe in the conclusion, they suggest insights into the challenges that cyber capabilities and operations pose to individual states and the international community. We expect that this work will stimulate readers to think of additional analogies that could augment their understanding of cyber capabilities and operations, as well as policies to manage them in ways that reduce conflict and enhance international well-being. It would be especially welcome if scholars, journalists, and officials from non-Western countries were to elucidate analogies from their own technological and historical experiences to the cyber era, for the unprecedented benefits of cyber technology are the relative ease and affordability of its global dissemination. To realize its benefits, and to minimize the technology's destructive potential, the widest possible range of societies and states must learn to steward it wisely. The authors here seek to contribute to this outcome and encourage others to do the same.

Notes

1. Emily O. Goldman and John Arquilla, eds., *Cyber Analogies* (Monterey, CA: Naval Postgraduate School, 2014), 5.

2. Richard E. Neustadt and Ernest R. May, *Thinking in Time: The Uses of History for Decision-Makers* (New York: Free Press, 1986).

3. Stanley E. Spangler, *Force and Accommodation in World Politics* (Maxwell Air Force Base, AL: Air University Press, 1991), 52.

4. Ibid., 62.

5. See, for example, Joseph Nye, "Nuclear Lessons for Cyber Security?," *Strategic Studies Quarterly* 5, no. 4 (2011); Andrew Krepinevich, *Cyber Warfare: A "Nuclear Option"?* (Washington, DC: Center for Strategic and Budgetary Assessments, 2012), http://csbaonline .org/research/publications/cyber_warfare_a_nuclear_option; and Richard Clarke and Robert Knake, *Cyber War: The Next Threat to National Security and What to Do about It* (New York: HarperCollins, 2012).

6. Ashton B. Carter, "DOD Executive Agent for Non-Lethal Weapons (NLW), and NLW Policy," Number 3000.03E (Washington, DC: Department of Defense, April 25, 2013), 12, http://www.dtic.mil/whs/directives/corres/pdf/300003p.pdf.

7. Elisabeth Bumiller and Thom Shanker, "Panetta Warns of Dire Threat of Cyberattacks on U.S.," *New York Times*, October 11, 2012, http://www.nytimes.com/2012/10/12 /world/panetta-warns-of-dire-threat-of-cyberattack.html.

PART I

What Are Cyber Weapons Like?

1 Intelligence in Cyber— and Cyber in Intelligence

MICHAEL WARNER

Cyber technologies and techniques in some respects originated in the intelligence profession. Examining cyberspace operations in the light of the history and practice of technology helps illuminate both topics.[1] Intelligence activities and cyberspace operations can look quite similar; what we call cyber is intelligence in an important sense. The resemblances between the two fields are not coincidental. Understanding them opens new possibilities for exploring the applicability of intelligence concepts to a growing understanding of cyberspace.

To appreciate the evolutionary connections between these fields, it is necessary to define the multiple functions that intelligence performs. Intelligence guides decisions by providing insight to leaders and commanders, of course, but its definition is broader still. The field has always included espionage and counterespionage, and today it includes technical collection as well. Such clandestine activities are but a short step from covert operations, which fall under the ambit of intelligence organizations in many states. Finally, intelligence, with its partner activities of surveillance and reconnaissance, has become a key component of today's real-time, networked warfare. This chapter explores these functions of intelligence and how cyber capabilities resemble or differ from the capabilities that earlier technologies provided, as well as how cyberspace capabilities and operations pose new policy dilemmas. It does so from a US perspective, but the phenomena and issues discussed here are probably pertinent to other countries too.

Spy versus Spy

Intelligence has evolved over the last century, giving rise to two overlapping but not congruent definitions of the field. US military doctrine views *intelligence* as information that a commander finds vital in making a decision, plus the sources, methods, and processes used to produce that information. Not all information is intelligence, of course. Only information on the adversary and the conditions under which the commander's force might have to fight is considered *intelligence*.[2] One should note, however, that this concept of intelligence is relatively new. Indeed, it was formally stated in such terms only in the 1920s.[3] Spying, however, dates to the dawn of history; ancient texts from around the world mention spies and their exploits on behalf of rulers and commanders. The emergence of modern intelligence from classic spy craft resembles a millennia-wide "before" and "after" picture of the subject.

The Chinese sage whom we call Sun Tzu composed one of the earliest reflections on intelligence sometime around 300 BC. His classic *The Art of War* was hardly the first written reflection on this topic, although earlier authors (as far as we know) did not match Sun Tzu's insight and brevity in his thirteenth and final chapter, "On the Use of Spies." He described a lonely and deadly craft that occasionally became very important. A *spy*, in Sun Tzu's telling, might collect secrets, spread disinformation or bad counsel in the enemy's camp, or even assassinate enemy officials. He thus combined a range of activities far broader than merely passing information to his commander. A spy could potentially become a fulcrum of history, providing information or taking direct action to ensure the downfall of a dynasty and a shifting of the mandate of heaven.

Such considerations have relevance today, even for those who no longer see a change in regimes as cosmically important. Spy craft did not evolve much in the two millennia between Sun Tzu's day and the Industrial Revolution, so we can take his ideas as fairly representative of the field up until roughly the age of Napoleon Bonaparte. Indeed, while campaigning, Napoleon ran his spy network from his tent, filing agents' reports in pigeon holes in his camp desk. Even with the spread of intelligence collection by remote and then automated means in the twentieth century, individual spies retained importance for intelligence consumers and systems. Well-placed insiders could and did nullify expensive suites of technical collection assets during the Cold War, and more recently "insider threats" (even if not spies per se) precipitated media leaks that have significantly complicated international relations.

Spies have been eclipsed by technical collection, of course, but security and counterintelligence offices continue to focus significant resources on finding (and deterring) enemy agents. Leaders and their advisers intuit the danger that any human penetration poses to technological advantages, military operations, and diplomatic ties. The mere possibility of a spy can disrupt an intelligence bureau or even an alliance; the genuine article can do grave harm and cause effects that reverberate for years. Entire disciplines of the security field (e.g., background checks, compartmentation, and so on) grew up around the imperative to minimize and mitigate the damage that spies could inflict. Counterintelligence, of course, emerged precisely to guard against spies in a more active manner. The most effective counterintelligence operations (like Britain's Double-Cross system in World War II) managed to take control of not only enemy spies but the perceptions of their spymasters as well. They fooled the latter into believing their espionage network was still collecting valuable secrets, which naturally turned out to be misleading "chicken feed."[4]

Cyberspace operations have obvious parallels to traditional human espionage. An *implant*, for example, can sit in a computer for weeks, months, or years, collecting secrets great and small. The finding of such an implant, like catching a spy, evokes mingled satisfaction and fear. Not finding one, moreover, might not inspire confidence. It could mean there was no intruder to catch. Alternatively, it might mean that one looked in the wrong place.

In strategic terms, catching a spy or finding an implant is not exactly a casus belli, although running a spy (or placing an implant) is obviously a provocation. States have tacitly established protocols for handling espionage flaps. Typically the actual spy stands trial, while his or her foreign case officers are declared personae non gratae and expelled. Foreign intelligence officers (like Russia's Anna Chapman, whom the Federal Bureau of Investigation [FBI] caught in 2010) are jailed in a glare of publicity. Soon, however, when the media's attention has wandered elsewhere, the spies are quietly exchanged for individuals in their homeland's prisons. We have not developed such protocols for handing disconnected computer implants back to their originators, but one suspects that similar understandings around cyber espionage will emerge over time.

How much cyber espionage is there? That depends on how broadly we define *espionage* as the acquisition of data in ways unbeknownst to its "owner." At the risk of stating the obvious, entire sectors of the world economy now rest on the ability of corporations to aggregate and sell information about the online habits of consumers. Few computer users worry about such aggregation. They implicitly permit much (though by no means all) by pressing "Accept" after scrolling through the fine print in lengthy end-user agreements. This chapter must leave such matters to abler minds, though certainly a fair amount of illegal or at least unethical mischief is directed against the software sold to consumers to facilitate the harvesting and sale of their data.[5] Going from such mischief to actively cyber spying on unsuspecting people is a short step. Today anyone with a network connection can be a victim of espionage mounted from nearly anywhere. A cottage industry has grown up around efforts to find and expose such cyber espionage schemes. From the instances uncovered so far, anyone possessing modest resources and sufficient motivation can readily download highly intrusive, capable, stealthy suites of surveillance tools.[6] The publicly available evidence—not to mention the complaints by many governments and the myriad allegations based on leaked documents—should lead any fair-minded observer to conclude that many examples of cyber espionage were perpetrated by state actors.

The counterintelligence parallel with cyberspace operations seems to be developing another analogous aspect as well. The most ruthless counterintelligence services since at least the czars' Okhrana have planted agent provocateurs among groups they deemed to be subversive. Their role was not only to report from within but to incite rash or premature action that would expose and discredit the groups. A whole literary subgenre explored the dramatic possibilities such plots entailed; think of Joseph Conrad's *The Secret Agent* (1907) or G. K. Chesterton's *The Man Who Was Thursday* (1908). Such agents were not just the stuff of fiction—Vladimir Lenin devoted his landmark essay "What Is to Be Done?" (1902) to countering them—and they spread fear and distrust among revolutionaries across Europe before World War I.

Attentive watchers of the cyber news will see an echo of these operations. Security services like the FBI seem to be learning how to persuade cyber criminals to switch allegiance while maintaining contact with their online cohorts (on secretly monitored connections, of course). Once the authorities identify the network and record enough evidence against its members to warrant prosecution, the nations involved in the investigation mount simultaneous raids— sometimes across multiple continents—to round up the suspects.[7] Court filings soon expose the mole in the network, of course, but by then the person has been whisked to safety and perhaps even living under a new, state-provided identity.[8] The hacker world today is turning paranoid, worried that many of the anonymous contacts in the dark web have switched sides and started providing evidence. This spreading distrust represents a direct application of counter-intelligence tradecraft to cyberspace.[9]

In sum, espionage and counterespionage operations made the jump from the proverbial dark alleys to cyberspace virtually intact. What is new is old. How readily both of these ancient crafts adapted their techniques to the new cyber domain is astonishing. The main difference between their traditional operations and their cyber counterparts is the scale that can be exploited in the latter.

Common Roots

The history of intelligence provides still another template for understanding cyber operations. Intelligence connected itself to communications technology in the early twentieth century, with profound implications for itself and for diplomacy, security, and privacy. The modern era of communications began with the improvement of the telegraph, allowing quantities of messages and data to be transferred across global distances in near-real time. Wireless telegraphy and then radio broadcasting accelerated this trend, creating mass audiences and markets, as well as new military requirements for not only the equipment to transmit and receive such communications but also the cryptographic support to secure them and the messages they relayed. Intelligence, of course, grew in parallel with what Stephen Biddle terms the "new system" of military operations, in which real-time communications allowed generals to synchronize combined-arms actions involving infantry and artillery, and soon armor, aircraft, and ultimately guided weapons as well.[10] This revolution in military affairs began with the battlefield use of radio in World War I and accelerated across the remainder of the twentieth century. Over the last generation, modern militaries have become dependent on sensors, networks, bandwidth, and surveillance. This dependence is encapsulated in the ubiquity (at least in military affairs) of the term "C4ISR," meaning command, control, communications, computers, intelligence, surveillance, and reconnaissance.

The parallel growth of advanced, technologically enabled intelligence alongside the new system was not coincidental; rather, it was (and is) organic. These two trends share a common root in the widespread impulse across the industrialized powers to gain real-time control of military forces at a distance while monitoring and frustrating adversaries who seek to control their own assets and forces. This sea change took place quite suddenly and dramatically during World War I, in which vast armies, navies, and soon air forces had to communicate securely in real time or lose to adversaries who did. To cite but two examples, the Russian disaster at Tannenberg in August 1914 showed the occasionally strategic consequences of lapses in communications security, while the Royal Navy's exploitation of German naval systems demonstrated what operational possibilities could be opened by a sustained cryptologic campaign against poor security practices and vulnerable technology.[11] The shift to technical collection and analysis of machine-generated data revolutionized the intelligence business, transforming it seemingly overnight from an ancient craft into an industrial enterprise.

Every Western military sought to learn communications security lessons from World War I. Modern codes and encryption had arisen with the printing press in the Renaissance, but they took off anew with the telegraph revolution in the nineteenth century and especially with the wireless in the twentieth century. The difference between private cryptography and governmental and military systems, of course, was the sensitivity of the information they carried and hence the length of time (in hours, days, months, or decades) that the information's owner would want eavesdroppers to have to devote to decrypting the intercepted messages. Despite the higher stakes for official uses, however, the quality of cryptographic support to both private and government messages for centuries remained roughly equivalent—in other words, not very good. That began to change with governments' quests for reliable enciphering machines for tactical communications, such as the Swiss-made Enigma, which was marketed to commercial firms but was soon adopted and improved by the German military in the late 1920s. These machines had become widespread by World War II, at least among the major combatants in that conflict. In 1939 the use of coded communications had also prompted several states (most notably Britain) to mount concerted efforts to divine the secrets of those enciphering machines and the codes they protected. Enlisting their American allies, soon the British applied a new technology to the problem—the digital computer.

The Anglo-American signals intelligence alliance after World War II hastened the evolution of computers and of America's computer industry in the 1950s. The enduring Anglo-American partnership henceforth kept its team members, particularly the National Security Agency (NSA), up to date with the evolution of computers, their concentration in networks, and the progress of a new field, computer security.

From the beginning, the NSA's expertise in securing digital communications and networks influenced the concepts for and debates over securing computers and the data they stored and shared.[12] Such effects quickly became embroiled in debates over encryption, particularly regarding the extent of the US government's role in fostering high-grade cryptography. For decades the point had been moot, as the best cryptographic solutions were treated as military secrets (which in a sense they were) and their export was banned. With the de facto merging of telecommunications devices and computers by the 1970s, however, a new dilemma arose—that is, how to secure digital data for governmental agencies, banks, and other institutions that shared sensitive communications and files but did not need export-controlled, military-grade ciphers. The initial answer was the Data Encryption Standard (DES), which the National Bureau of Standards proposed in 1975 after its development by IBM and vetting by the NSA. Various observers soon found weaknesses with the DES algorithm, however. Some alleged that the US government had exploited its role in creating DES to leave "backdoors" in the standard that would allow government officials routine (or at least emergency) access to private data.[13] For their part, the relevant agencies and even a congressional investigation insisted the government had done no such thing.[14] The controversy over DES created a template that has been followed ever since—for instance, in the debates during Bill Clinton's administration about the proposed "Clipper Chip" in the 1990s and the 2015–16 contretemps between Apple, Inc., and the FBI concerning the data residing on a smartphone used by one of the San Bernardino killers.[15] Then as now, various government officials' insistence on some official method of bypassing encryption standards for urgent national security and law enforcement purposes alarmed those who feared that US intelligence had already compromised the standards.[16]

This chapter cannot hope to resolve the policy issues over encryption or allay suspicions about the US government's motives and actions. The author supports strong encryption for everyone and would like all governments to resist the urge to install backdoors in any cryptographic systems. The point of this chapter, however, seeks to add perspective by noting today virtually anyone can routinely use encryption that, historically speaking, is fantastically effective. Nevertheless, governments, hacktivists, and organized criminals have found various ways around that wonderful encryption. Most observers would surely agree that encryption has never been better, yet those observers might nonetheless concede that never have so many users lost exclusive control of so much of their data.[17]

The burgeoning computer security field has an additional connection to intelligence that has been largely overlooked. In certain ways the concepts of computer security grew directly from the painful education in counterintelligence and security practices that US intelligence agencies gained during and after World War II. There was nothing like operating behind the Iron Curtain for making an organization interested in end-to-end security measures. This is precisely why the Central Intelligence Agency (CIA) established a comprehensive "automated data processing" (ADP) security regime that congressional investigators publicly praised forty years ago! Committee staffers surveying federal computer security in 1976 applauded the CIA for its thorough approach, which worked "on the assumption that not only is there potential for compromise in any ADP system[,] it is likely that an attempt will be made to effect that compromise." Though agency officials declined to offer their computer security regime as a template, the committee's study nevertheless suggested "trying to apply certain ADP security techniques which had evolved at CIA to other Federal programs where the issue may not be national security but at stake were considerations of nearly equal consequence, such as individual privacy data and . . . financial transactions leading to disbursements of large amounts of public funds."[18]

The spread of computers had heralded something novel for both communications and intelligence. Hitherto the devices that secured and transmitted information did not also store it. Computers did, at least as soon as they were given built-in memory. Thus, the level of care taken to transmit messages securely now must extend over the entire life cycle of that data and even to the machines that touch that data. Not only is your current data vulnerable to spies and eavesdroppers, it is now at risk forever in cyberspace. This raised the security bar tremendously for average users as well for governments. Consequently, the NSA since 2009 on its public website has urged "customers" to make prudent preparations now for the day when their encrypted data will be vulnerable to attack by quantum computers.[19] Permanency of data not only has broadened the practice of intelligence (as hinted above) but also has drawn a line of demarcation between some traditional, passive forms of intelligence collection and the new digital methods.

Everything Goes Digital

The early development of radio suggests yet another aspect to the analogy between intelligence and cyberspace. Certain security and policy issues relating to computers and networks strongly resemble those associated with radio as that earlier medium evolved and spread in the first decades of the twentieth century. Indeed, many of the terms we routinely use to describe the workings of cyberspace—"network," "bandwidth," "wireless," and others—came from radio terminology. As noted, both radio broadcasts and computer data can be intercepted in midstream and analyzed in various ways to deduce information on one's opponents, even if one cannot read the content of the intercepted messages. Both radio and computer communications therefore must be used with care so as not to disclose too much information to opponents.

Furthermore, the kinship between intelligence and deception exactly parallels the relationship between radio (and computer) operations and the field of electronic warfare (EW). Radio was weaponized in World War II, and EW has been a standard feature of modern conflict ever since. An opponent's employment of both radio and computer networks can be denied by jamming or flooding of one form or another. And, of course, those who intercept radio transmissions or computer data can be actively deceived by a clever originator. These intelligence dimensions of the new cyber realm (i.e., the principles of attack, defense, and exploitation) are readily apparent; indeed, they guided the US Department of Defense's thinking on information warfare for the first couple decades of this doctrine's existence. EW is thus one of the taproots of cyberspace operations, at least in the United States, as military thinking about command, control, and communications countermeasures in the early 1980s led directly to the earliest policy pronouncements on information warfare in 1992.[20]

Historians should not forget a related point: substantial impetus for the computer industry's maturation derived from the US military's drive to make weapons smarter and to share data to and from the battlefield. This vast field again falls beyond the scope of this exploration of the intelligence analogy, but it is important to note certain additional links between the evolution of computers and the realm of intelligence support to battlefield commanders. Smart weapons emerged in the early 1970s, motivated in part by the Pentagon's desire to increase the precision of bombs in Vietnam (thereby reducing the danger to aircrews and minimizing morally and politically harmful collateral damage to civilians). The weapons were "smart" not only because they could be guided to their targets but also because they depended on intelligence about those targets (e.g., precise locations) and on copious and timely data flows to increase their accuracy and lethality. Such data flows eventually demanded quantum leaps in bandwidth, processing power, and networking architecture. The US military thus helped drive improvements to digital communications to increase their resilience and volume, and in the 1970s it began setting standards for the security of these burgeoning systems and the data they carried. All these developments spurred research in the computer industry and provided growing markets for innovations that initially seemed to lack consumer markets.

Government links with industry had a direct, strategic focus as well. That nexus brought the intelligence and computer sectors together at the dawn of cyberspace. As historian Jonathan Winkler has shown, the US government has jealously guarded a national interest in the progress of international telecommunications, beginning in World War I and continuing unabated to our day.[21] Among many examples, Ronald Reagan's administration in 1984 took note of the de facto blending of the computer and telecommunications fields and found this trend had significant implications for US security. President Reagan accordingly issued a top-secret directive giving the NSA responsibility for setting standards to protect sensitive but unclassified data in all US federal government computers. Though Congress soon overturned Reagan's measure, the mere fact a president had ordered such a step demonstrated the growing overlap between the intelligence and computer security worlds.[22] Washington has quietly secured the strategic high ground in the nation's communications sector, using intelligence both to guard and to exploit that advantage. The importance of that access for the nation's intelligence function needs no reiteration here.

Cyber operations grew out of and still resemble EW, as noted earlier, with one key difference. Traditional EW aimed to guide, target, or protect weapons systems but remained an activity extrinsic to those weapons as such. Cyberspace, in contrast, includes many of those weapons. The "Internet of things" arrived early in modern military arsenals. Their interconnectivity not only makes them smart but also potentially leaves them vulnerable as an adversary could theoretically find ways to make those systems hinder rather than help operations. Here is another way in which cyberspace operations both learn from and affect intelligence activities.

What Is New in Cyberspace?

So far the parallels described between intelligence activities and cyberspace operations are not merely hypothetical but are already working themselves out in practice around the world. Other parallels can be envisioned as well, at least in listing the possible warning signs that might accompany their emergence over the foreseeable future.

The most obvious and oft-discussed association between intelligence activities and cyberspace operations is the confusion they can cause among those on their receiving end. Human espionage can look quite like subversion or worse as authors such as Sun Tzu and the Indian sage Kautilya noted thousands of years ago. They urged commanders and princes to have their spies assassinate rival chiefs.[23] Active intelligence operations, like cyberspace collection campaigns, are by definition quiet but potentially provocative. They can appear similar to preparations for war, and from time to time they have increased tensions between states. But has anyone gone to war over an intelligence operation that was exposed or blew up in a crisis?

Here the parallel with intelligence can be informative. People have gone to war having bad intelligence that was either misconceived or spoofed by the adversary (see the Iraq War in 2003). Wars have started over assassinations, to be sure, but an *assassination* is by definition a successful operation designed to provoke hostilities and is not the inadvertent cause of them. Outside of these unrepresentative examples, the list gets thin. As noted previously, states by and large do not fight over blown technical collection activities. History yields many such examples of states catching spies or finding wiretaps, telephone bugs, and so on, without those states declaring war in response. The net result of blown intelligence activities is typically the loss (or turning) of the source, sometimes with a well-publicized protest, an expulsion of diplomats, or an execution or two. Even military reconnaissance is not usually dangerously provocative, as a single aircraft or patrol boat can hardly be mistaken for an invasion force. Overflights of the Soviet Union in the 1950s did not provoke a strategic military response by Moscow (apart from the Soviets' downing reconnaissance aircraft such as Francis Gary Powers's U2 in 1960). Similarly, aggressive US overflights of Cuba in the Missile Crisis (1962) agitated local air defense and concentrated minds in Moscow and Havana, but they did not prompt Soviet strikes on the United States.

Cyberspace operations gone awry, like intelligence revelations, so far have not provoked wars. The net effect in cyberspace is typically the quiet purging of an implant, the updating of an operating system, or the closing of a port, combined with perhaps a diplomatic complaint, possibly via the press. The reason for this lack of panic and escalation might have been explained (in another context) by the Atlantic Council's Jason Healey. As he notes, cyberspace operations rarely if ever proceed in isolation. That two states are at odds over some issue certainly assists in the attribution of contemporaneous cyber attacks to one or both of them.[24] Although Healey does not explicitly flip this coin, his argument also hints that policymakers virtually always know some context behind the events that places noncrisis cyber developments in perspective, usually by showing that the state allegedly perpetrating the cyber transgression is not currently deploying for war.

Can cyber operations cause instability and even escalate a crisis? Of course, they might, if perhaps only because no one can definitively prove that they will not. What we can say is that no cyberspace operation to date has made a crisis spiral into war. Indeed, the United States has experienced more than its share of cyber penetrations and cyber attacks, yet it has never come close to initiating hostilities over a cyber incident. As far as we know, no one else has either. Some observers might cite Solar Sunrise, the Department of Defense's name for a 1998 cyber intrusion that originally looked as if Iraq had penetrated US military networks (and which turned out to be an Israeli hacker working with two American teenagers). Solar Sunrise did indeed unfold amid a diplomatic crisis with Iraq, leading American observers to suspect Iraqi complicity, yet it also happened at a time when Defense Department defenses and cyber decision-making were still nascent. The diplomatic net result of Solar Sunrise was nothing. Calmer heads prevailed, and the United States did not strike Iraq over the misattributed intrusion. What Solar Sunrise proves about crisis instability and escalation is anyone's guess. Nevertheless, every year since 1998, cyber attacks have been misattributed, but so far such mistakes have not caused any wars. One wonders how many years it takes to notice a pattern here.

One note of caution while listing the parallels between intelligence activities and cyberspace operations is that the intelligence-cyber analogy helps to illuminate cyberspace operations but not cyberspace as a war-fighting domain. The analogy also seems stretched when one ranks the relative scales of intelligence activities and cyberspace operations; the former tend to be minute, and the latter look comparatively vast. Other analogies in this volume can help explain such aspects of cyberspace and the events that happen there. Let us then close with the observation that the hitherto tight parallelism between intelligence activities and cyberspace operations could well witness a divergence of potentially strategic consequence. One sees such signs in the lingering reputational damage to the United States and American firms caused by the media's revelations over the last few years. It is difficult to measure the effects, which are primarily commercial and consist of missed opportunities as much as actual expenses. Much anecdotal evidence points to forfeited sales for American products, and Washington has certainly (for the time being) lost control over the global narrative regarding Internet security and privacy. This development adds a new element rarely if ever seen in traditional espionage cases, and we would be wise to remain sensitive to how it unfolds.

Conclusion

We hardly need an analogy to compare cyberspace operations with intelligence activities, as one exaggerates only a little to say they are mostly the same thing. A biologist might likewise say the same about dinosaurs and birds, for the latter developed from the former with no evolutionary "seam" to distinguish the two types of animals (indeed, they are both members of the *dinosauria* clade). We also know from Sun Tzu that intelligence is concomitant with force; intelligence guides and sharpens force, making it more secret, subtle, and sometimes more effective. Further, force follows people and wealth; thus, wherever they are, aggressors will try to use force to control those people and to take the wealth.

Cyberspace operations can and do work along the same lines, for the same purposes, and for the same leaders. The steadily growing scale of intelligence activities expanded dramatically with the global diffusion of cyberspace, allowing formerly state-monopolized means and capabilities to be used by almost anyone with an Internet connection. That same diffusion of intelligence tools in cyberspace also made virtually everyone a potential collector of intelligence or a potential intelligence target. The lines between spying and attacking have always been blurry in intelligence activities as well as in cyberspace operations. Both are inherently fragile and provocative. While neither is necessarily dangerously destabilizing in international relations, we must learn to perform cyberspace operations as we learned to perform intelligence activities—that is, with professional skill, with strict compliance with the law, and with careful oversight and accountability.

Notes

Michael Warner serves as the command historian for US Cyber Command. The opinions in this chapter are his own and do not necessarily reflect official positions of the command, the Department of Defense, or any US government entity.

1. The Joint Chiefs of Staff define *cyberspace operations* as "the employment of cyberspace capabilities where the primary purpose is to achieve objectives in or through cyberspace." Joints Chiefs of Staff, Joint Publication 3–12 (R), *Cyberspace Operations* (February 5, 2013), v, http://www.dtic.mil/doctrine/new_pubs/jp3_12R.pdf.

2. See Joint Chiefs of Staff, Joint Publication JP 1–02, *Department of Defense Dictionary of Military and Associated Terms* (November 8, 2010 [as amended through October 15, 2015], http://www.dtic.mil/doctrine/new_pubs/jp1_02.pdf), which defines *intelligence* as "the product resulting from the collection, processing, integration, evaluation, analysis, and interpretation of available information concerning foreign nations, hostile or potentially hostile forces or elements, or areas of actual or potential operations" (and the product of this activity and the organization performing it).

3. Michael Warner, "Intelligence as Risk Shifting," in *Intelligence Theory: Key Questions and Debates*, ed. Peter R. Gill, Stephen Marrin, and Mark Phythian (London: Routledge, 2008), 26–29.

4. J. C. Masterman broke this story in *The Double-Cross System in the War of 1939 to 1945* (New Haven, CT: Yale University Press, 1972).

5. *See* Josh Chin, "Malware Creeps into Apple Apps," *Wall Street Journal*, September 21, 2015.

6. For example, Citizen Lab at the University of Toronto's Munk School of Global Affairs has done yeoman service tracking spies in cyberspace for nearly a decade. See the nonprofit lab's assessment of FinFisher's surveillance software used in more than thirty states: Bill Marczak et al., "Pay No Attention to the Server behind the Proxy: Mapping

FinFisher's Continuing Proliferation" (October 15, 2015), https://citizenlab.org/2015/10/mapping-finfishers-continuing-proliferation/. Several of the larger antivirus and Internet security companies have fielded their own research arms to find and publicize state-based and criminal espionage.

7. A case in point is the FBI's takedown of the hacktivist group Lulz Security, or LulzSec, by turning one of its leaders in 2011. The hacker in question was Hector Xavier Monsegur, called "Sabu" by other members of LulzSec. The bureau had initially arrested Monsegur in 2011 and made mass arrests of LulzSec members on March 6, 2012. See Mark Mazzetti, "F.B.I. Informant Is Tied to Cyberattacks Abroad," *New York Times*, April 24, 2014.

8. See the FBI's unsealed affidavits here: "LulzSec Indictment Documents," *The Guardian*, March 6, 2012, http://www.theguardian.com/technology/interactive/2012/mar/06/lulzsec-indictment-documents-prosecution-complaints.

9. Saul O'Keeffe, "Hacking Underworld Riddled with Secret FBI Informants," ITProPortal, July 24, 2015, http://www.itproportal.com/2015/24/07/hacking-underworld-riddled-secret-FBI-informants/.

10. Stephen Biddle, *Military Power: Explaining Victory and Defeat in Modern Battle* (Princeton: Princeton University Press, 2004), 28.

11. At Tannenberg an outnumbered German army defeated two Russian armies in detail after overhearing their plans broadcast en clair. The Russians apparently lacked compatible codebooks. The Royal Navy turned the tables on German ships and naval aircraft by geo-locating their transmissions and monitoring their stereotyped messages, which upon analysis revealed patterns that clearly indicated upcoming operations.

12. I treat this in more detail in "Notes on the Evolution of Computer Security Policy in the US Government, 1965–2003," *IEEE Annals of the History of Computing* 37, no. 2 (April–June 2015).

13. Gina Bari Kolata, "Computer Encryption and the National Security Agency Connection," *Science* 197 (July 29, 1977): 438.

14. US Senate, Select Committee on Intelligence, "Involvement of NSA in the Development of the Data Encryption Standard," 95th Cong., 2d sess., April 1978.

15. The San Bernardino, California, attack occurred in December 2015, when two individuals, Syed Rizwan Farook and Tashfeen Malik, killed fourteen civilians and injured twenty-two others in a shooting at the Inland Regional Center. The two assailants were killed in a gunfight with police. After police recovered Farook's cell phone, the FBI asked Apple to unlock the device, as the bureau believed that information related to the attack was on the phone. This request launched a nationwide debate regarding whether Apple should unlock the device. The dispute ended when the FBI purchased a vulnerability to access the device for more than $1 million. For more information, see Adam Nagourney, Ian Lovett, and Richard Pérez-Peña, "San Bernardino Shooting Kills at Least 14; Two Suspects Are Dead," *New York Times*, December 2, 2015, http://www.nytimes.com/2015/12/03/us/san-bernardino-shooting.html; and "FBI Paid More than $1M for San Bernardino iPhone 'Hack,'" *CBS News*, April 21, 2016, http://www.cbsnews.com/news/fbi-paid-more-than-1-million-for-san-bernardino-iphone-hack-james-comey/.

16. Witness, for example, recent allegations over dual elliptic curve deterministic random bit generator, or Dual_EC_DRBG encryption, as well as FBI director James Comey's public warnings that his bureau is "going dark" because it cannot unlock the encryption in perpetrators' smartphones. The NSA insists it uses publicly available encryption suites for its own data. "NSA relies on the encryption and standards we advocate for and advocate for the encryption standards that we use," Anne Neuberger, then director of the agency's Commercial Solutions Center, told a radio audience in 2013. "[W]hat we

recommend for inclusion in those cryptographic standards, we use ourselves in protecting classified and unclassified national security systems." See "Threat Information Sharing Builds Better Cyber Standards, Expert Says," Federal News Radio Custom Media, October 3, 2013, 5:05 p.m., http://federalnewsradio.com/technology/2013/10/threat-information-sharing-builds-better-cyber-standards-expert-says/.

17. "Purdue's Gene Spafford was correct, but early, when he likened network security in the absence of host security to hiring an armored car to deliver gold bars from a person living in a cardboard box to someone sleeping on a park bench." See Daniel E. Geer Jr., "Cybersecurity and National Policy," *Harvard National Security Journal* 1 (April 7, 2010).

18. US Senate, Committee on Government Operations, "Staff Study of Computer Security in Federal Programs," 95th Cong., 1st sess., February 1977, 135–37, http://babel.hathitrust.org/cgi/pt?id=mdp.39015077942954;page=root;view=image;size=100;seq=3.

19. The NSA "will initiate a transition to quantum resistant algorithms in the not too distant future. Based on experience in deploying Suite B [encryption algorithms], we have determined to start planning and communicating early about the upcoming transition to quantum resistant algorithms. Our ultimate goal is to provide cost effective security against a potential quantum computer." See NSA, "Cryptography Today," January 15, 2009, https://www.iad.gov/iad/programs/iad-initiatives/cnsa-suite.cfm.

20. For more on this, see my recent article, "Notes on Military Doctrine for Cyberspace Operations in the United States, 1992–2014," *Cyber Defense Review* (Army Cyber Institute), August 27, 2015, http://www.cyberdefensereview.org/2015/08/27/notes-on-military-doctrine-for-cyberspace/.

21. Jonathan Reed Winkler, *Nexus: Strategic Communications and American Security in World War I* (Cambridge, MA: Harvard University Press, 2008).

22. Warner, "Notes on the Evolution," 10–12.

23. Sun Tzu, *The Art of War*, trans. Samuel B. Griffith (New York: Oxford University Press, 1971 [1963]), ch. 13. See also books 1, 2, and 13 of Kautilya's *The Arthashastra*, trans. L. N. Rangarajan (New Delhi: Penguin Books India, 1992).

24. Jason Healey, ed., *A Fierce Domain: Conflict in Cyberspace, 1986 to 2012* (Washington, DC: Cyber Conflict Studies Association, 2013), 265–72.

2 Nonlethal Weapons and Cyber Capabilities

LT. GEN. ROBERT E. SCHMIDLE JR. (USMC, RET.),
MICHAEL SULMEYER, AND BEN BUCHANAN

Scholars have considered many analogies for cyber capabilities, grappling with how these capabilities may shape the future of conflict.[1] One recurring theme in this literature is the comparison of cyber capabilities to powerful, strategic capabilities with the potential to cause significant death and destruction.[2] This theme is understandable. Reports of malware that can penetrate air-gapped networks and cause physical effects can easily stimulate worst-case thinking. Moreover, relative silence from senior government leaders about cyber capabilities can fuel speculation that nations are amassing devastating arsenals of malware.[3] Increasing connectivity from consumer products to critical infrastructure control systems creates the prospect of widespread vulnerability across societies.[4] Analogies to different methods of state-to-state coercion are therefore quite common.

However, no one has ever been killed by a cyber capability. With this in mind, perhaps another set of analogies for cyber capabilities—not destructive, strategic capabilities but those that are nonlethal—should be considered. The US Department of Defense for decades has developed a range of nonlethal weapons for its forces, yet to our knowledge, scant academic work to date has considered how nonlethal weapons might provide some additional conceptual insight into cyber capabilities.

In this chapter, we examine nonlethal weapons and cyber capabilities and suggest that for conceptual purposes it may be useful to analogize between them across four areas: their ability to incapacitate, the reduced collateral damage they inflict, the reversibility of their effects, and their ability to deter. In so doing, we show the usefulness and the limits of analogizing cyber capabilities to nonlethal weapons. Ultimately, we conclude that these four areas of convergence between nonlethal weapons and cyber capabilities make for a novel conceptual analogy that would serve policymakers well as they consider future employment of cyber capabilities.

In our conclusion, however, we highlight one important limitation of this approach: Department of Defense leaders have faced difficulty in gaining approval to use nonlethal capabilities. We briefly explore reasons why nonlethal weapons have so seldom been authorized and offer some observations as to why cyber capabilities may be easier to employ in the future. We base this distinction on the fact that most nonlethal weapons target opposing personnel, whereas most cyber capabilities target opposing matériel.

Before commencing our analysis, we offer one preliminary note about terminology. Already we have noted that we examine cyberspace "capabilities" as opposed to cyber "weapons." The distinction is not pedantic. When we write of nonlethal "weapons," the intent of these tools is in clearer focus—to inflict bodily harm or physical damage.[5] However, the cyber tools discussed in this chapter are not always weaponized *ex ante*. Instead, they offer certain capabilities: some that may be used offensively, some in self-defense, and still others for penetration testing. Because code is not inherently weaponized, we use the term "capabilities" to cover the full of range of what technologies in cyberspace have to offer.

Characteristics of Nonlethal Weapons

To more fully understand the proposed analogy between nonlethal weapons and cyber capabilities, we must first understand the basics of nonlethal weapons. The Department of Defense defines *nonlethal weapons* as "weapons, devices, and munitions that are explicitly designed and primarily employed to incapacitate targeted personnel or materiel immediately, while minimizing fatalities, permanent injury to personnel, and undesired damage to property in the target area or environment."[6] Cyber capabilities are excluded from this definition. Nonlethal weapons can provide operating forces with options to de-escalate situations, minimize casualties, and reduce collateral damage. By providing commanders with these additional options, nonlethal capabilities can be of unique value, sometimes proving to be more appropriate than their lethal counterparts.

Nonlethal weapons are often divided into two categories depending on their direct target. First, many nonlethal weapons are identified as serving a "counter-personnel" role because they target the human body itself. A notable example is oleoresin capsicum spray, which is more commonly known as pepper spray. When sprayed at a target, the chemical compounds in the spray act as an irritant to the eyes, causing tears, pain, and temporary blindness. This effect makes it more difficult for the target to engage in combat or other threatening activities.

The second category of nonlethal weapons targets machines, not people. An example of this sort of capability is the so-called spike strip. Derived from the older caltrop—which was used as a counter-personnel, counter-animal, and counter-vehicle weapon—the spike strip comprises long, upward-facing metal barbs linked together in a long chain. Each barb is sufficient to puncture the tires of many vehicles; so, when laid across a roadway, the spike strip can slow or stop vehicle movement until the tires have been replaced. Many spike strips are designed to gradually let the air out of affected vehicles' tires, minimizing the harm done to passengers and reducing the risk of collateral damage.

Across both counter-personnel and counter-machine nonlethal weapons, four characteristics are evident. First, their primary purpose is to incapacitate their targets. Second, they do so with minimal collateral damage, and, third, in a way that is often temporary or reversible. Finally, nonlethal weapons can serve as a limited deterrent in tactical situations. These characteristics are key points of comparison in making the analogy to cyber capabilities.

Operational History of US Nonlethal Weapons

One can trace the origins of nonlethal weapons in warfare to the development of modern chemistry, which began in the eighteenth century. By the mid-nineteenth century, consideration was given to using chemical weapons in the Crimean and US Civil Wars.[7] To be sure, chemical weapons would eventually become quite deadly, but initially the intent behind their use was not to kill but to force the enemy to disperse. Militaries apparently did not embrace using chemicals in warfare until World War I, when the German army launched the first chemical weapons attack on April 22, 1915, near Ypres.[8] As the United States entered the war, it institutionalized its chemical munitions research and development into a Chemical Warfare Service with the US Army.[9] Among the chemical weapons developed during the war, multiple armies used tear gas, which remains a nonlethal weapon in today's law enforcement and military arsenals.[10]

At the war's conclusion, the US Army rapidly demobilized its chemical weapons corps and seemed poised to all but abandon research into this class of weaponry.[11] The army's experts secured employment in civilian jobs, and surplus material was either sold or transferred to other parts of the government.[12] Thus concluded the US Army's initial efforts to explore how gas could be used as a chemical, nonlethal weapon.[13] Thereafter, the 1925 Geneva Protocol prohibited the use of chemical weapons in war.[14]

Even without this protocol, it seems unlikely that tear gas–related chemical agents would have been as effective in World War II, at least in the European theater. The rise and increasing adoption of motorized and mechanized forces neutralized the utility of chemical agents to disperse forces from fixed positions.[15] However, militaries used smoke as a tactical, nonlethal enabler during World War II, often to obscure their own positions rather than to force the enemy to reposition.[16] Variants included white phosphorus, smoke pots, oil smoke generators, aircraft-delivered smoke tanks, and even colored smoke munitions for signaling.[17]

Development of chemical agents continued after World War II. The use of herbicides and other agents during the Vietnam War, while not deemed to violate the 1925 Geneva Protocol, proved to be sufficiently controversial and damaging that President Gerald Ford issued an executive order renouncing the first use of herbicides and riot control agents in war.[18]

Other technologies emerged that offered militaries options between "don't shoot" and "shoot to kill." The United Kingdom used rubber and plastic bullets in Northern Ireland in the 1970s. Indeed, by one account the British military fired 55,834 rubber bullets between 1970 and 1975.[19] During Desert Storm, the United States fired cruise missiles filled with carbon fiber that disrupted Iraq's power stations.[20] In March 1991 Secretary of Defense Dick Cheney asked his lieutenants

Paul Wolfowitz and Zalmay Khalilzad to lead a Non-Lethal Warfare Study, but it is unclear what, if anything, came of this examination.[21]

Just how useful nonlethal weapons could be was perhaps most clearly demonstrated during the US Marine Corps' presence in Somalia in the mid-1990s. Their commander, Lt. Gen. Anthony Zinni, in a 1994 hearing spoke of the virtues of nonlethal weapons. "Non-traditional operations," he said, "often involve police-like actions that would be best dealt with by non-lethal means. Crowd control, demonstrations, petty theft, acts of urban violence in populated areas, are examples of situations that could best be handled all or in part by non-lethal weapons. ... These non-lethal means also permit forces to demonstrate resolve or provide a show-of-force without endangering lives."[22]

A year later, Zinni's Marines provided cover when several thousand United Nations (UN) forces withdrew from Somalia. The former had trained to use a variety of nonlethal weapons, including pepper spray, flash bangs, and road spikes.[23] To control hostile crowds, they were equipped with foam guns and sticky guns, as well as hard sponge projectiles.[24] The Marines also warned the local populace that they possessed these nonlethal weapons. Ultimately, the mission to secure the extraction of the UN forces was successful. No Marines were killed.[25] Zinni noted afterward, "Our experience in Somalia with non-lethal weapons offered ample testimony to the tremendous flexibility they offer to warriors on the field of battle."[26]

Later in the 1990s, the Defense Department attempted to institutionalize research and development for a broader array of nonlethal weapons.[27] Yet few capabilities were available to support US forces after they invaded and occupied Iraq in 2003. A 2004 Council on Foreign Relations task force on nonlethal weapons found that these weapons "could have helped to reduce the damage done by widespread looting and sabotage."[28] Its report was one of the last major studies of the US military's use of nonlethal weapons. There is little evidence that prioritization or resources have changed since then.

With this history of experimentation but not integration in mind, we return to the analysis of how four qualities of nonlethal weapons, especially those that are counter-matériel, make for a conceptually useful analogy to cyber capabilities.

Incapacitation

Nonlethal weapons incapacitate their targets by attacking critical parts of the targeted machine, such as tires on a vehicle, and disabling them. Cyber attacks can work in the same way, attacking critical parts of a computer system and either overwhelming them or disabling them. Information security professionals have long argued that a cyber operation can do harm in one of three ways.[29] First, it can target the confidentiality of data in a computer system, stealing sensitive data and perhaps making it public. Second, it can target the integrity of a computer system by inputting malicious commands that adversely (and clandestinely) affect its functionality or by corrupting important data. Third, it can target the availability of a computer system, disabling access to it at a critical time.

An example of the incapacitation function is the cyber operation that accompanied the purported Israeli air strike on Syria in 2007. The cyber operation corrupted the integrity of the Syrian air defenses. While operators of the Syrian air defense system believed their radar was functional and that it presented them with an accurate display of the area, in fact the radar systems did not show the Israeli jets entering Syrian airspace.[30]

A more common example of incapacitation via cyber operation is known as a *denial of service* attack, which targets the availability of important computer services by overwhelming them with data. An ocean of incoming data prevents the targeted systems from responding to legitimate requests. Finally, some capabilities achieve an incapacitating effect by targeting both the integrity and the availability of a target. For example, the 2014 attack on the Sands Casino in Las Vegas targeted the integrity of critical computer code and adversely impacted the availability of the overall system. When this critical code was erased or corrupted, the affected computers did not function.[31]

By definition, cyber capabilities target machines. As a result, it is more difficult, but not impossible, to imagine a cyber capability that is directly counter-personnel. One possible lethal capability is code that manipulates a vital medical device, such as a pacemaker. Indeed, in 2007 Vice President Cheney had the wireless functionality on his pacemaker disabled out of fear that it could be attacked.[32] More broadly, weaknesses in the Internet of things could allow malicious code to incapacitate critical devices at critical times, leading to the possibility of targeted attacks with a direct effect on personnel.[33] Even if cyber capabilities are not lethal now, if these sorts of attacks become more achievable, they might be more lethal in the future as well.

Whether an attack has lethal effects or not, electronic systems targeted by cyber capabilities might in some instances be so important to an individual that incapacitating the system could have debilitating counter-personnel effects. For example, targeting cellular phone networks or other communications systems can affect an individual's ability to coordinate illegal, hostile, or otherwise dangerous behavior. It could also perhaps be argued that targeting confidential systems, such as the theft of data from personnel databases, has an effect on personnel and could be used for blackmail. In this last case, however, the delay between operation and effect is substantially longer than is the case for most nonlethal weapons. Thus, on the matter of incapacitation, the analogy is strongest between counter-matériel nonlethal weapons and cyber capabilities that attack the integrity and availability of targeted systems.

Minimization of Collateral Damage

Similar to nonlethal weapons, some cyber capabilities can be deployed to minimize collateral damage. When it comes to malicious computer code, this sort of minimization can take one or both of two forms—first, preventing the spread of computer code beyond the target and, second, minimizing the harm the code causes to nontarget systems if it does in fact spread.

On the first point, intermediate systems are commonly breached in a cyber operation as stepping-stones to reach the target. This is especially true if direct access to the target is denied. For example, as a means of getting malicious code into a facility that is not connected to the Internet and is thus harder for an attacker to access, the authors of Stuxnet reportedly targeted a number of Iranian contractors who were servicing the country's nuclear program.[34] But such intermediate infections can be difficult to control in cases where the capability's propagation mechanism, or the code it uses to spread from machine to machine, is automatic. In the Stuxnet case, the code spread beyond the original authors' intent, reaching other systems and eventually coming to the attention of the information security community.[35]

Second, authors of malicious code have shown some capability to minimize the harm such code can do, even if it spreads. For example, the authors of Stuxnet, Gauss, and other malicious code placed targeting guidance in the code.[36] This targeting guidance prevented the code from launching its most significant and damaging payloads unless the malware arrived at the correct target. While reports indicate these mechanisms were not perfect at preventing all ill effects, they automatically and substantially constrained the damage done by the malicious code once it spread.[37] It is worth noting, however, that adding such constraints requires a great deal of information about the particulars of the target system, information that will likely need time and previous operations to collect.[38]

Another important area of overlap between nonlethal weapons and cyber capabilities is related to minimizing collateral damage. Policy guidance offered by the US Defense Department does not prioritize cyber capabilities or nonlethal weapons over potentially more destructive kinetic ones. While cyber capabilities, at some point in the future, might offer a commander the ability to achieve military-relevant effects with only a minimal risk for collateral damage or loss of life, the complexity of computer networks at present greatly complicates the confidence a commander can have in the ability to achieve precise effects exactly when desired. Battle damage assessment is subject to similar limitations. In some instances, therefore, a commander would reasonably prefer non-cyber capabilities over a vast arsenal of cyber capabilities if the former could give greater odds for the success of an operation.

Based on these examples, given enough effort, time, and operator ability, sophisticated cyber capabilities present some prospects for minimizing collateral damage to systems besides the target. However, it is hard to generalize this point and argue that this central characteristic of nonlethal weapons can be a characteristic of all cyber capabilities. In addition, failures to prevent collateral damage do occur. Especially with capabilities as new and complex as cyber ones, the unintended consequences of particular capabilities may cause additional or unexpected damage. On the matter of collateral damage, then, the analogy is as much aspirational as operational. Some cyber capabilities are narrowly targeted and may be wielded carefully by sophisticated actors, but certainly not all of them are.

Reversibility

The analogy functions similarly when it comes to reversibility, for some cyber capabilities, but not all, are reversible. We identify four categories of reversibility: capabilities that are not reversible; capabilities that are reversible after some reasonably constant period of time, depending on environmental conditions; capabilities that are reversible at the discretion of the operator; and capabilities and their effects that could be reversible by the target but require some time, material, or effort to do so.

Various nonlethal weapons fall into each of the four categories. For example, in the first category, some kinds of nonlethal munitions do harm to the body that, though not fatal, cannot be undone; however, they are comparatively rare. For example, a rubber bullet could possibly cause some harm to the body that is not easily undone. In the second category, flash bang grenades and tear gas cause paralysis for a time, but their effects eventually dissipate. In the third category, operators can turn on and off electronic jamming, lasers, or sonic capabilities. And in the last category, the spike strips discussed earlier require the target to acquire new tires.

Cyber capabilities exist in three of the four categories. In the first category, some sabotage attacks are difficult to reverse easily, especially if they destroy critical material or data. Stuxnet is an example, though it was substantially more destructive than nonlethal weapons are. We do not know of any cyber capabilities that fall into the second category, which sees effects dissipate over time, depending on environmental conditions.

Other cyber capabilities, such as ransomware, fall into the third category because they paralyze systems until an operator directs otherwise. When ransomware affects a system's capability, important data is encrypted in such a way that the legitimate user cannot access it until the criminal operating the ransomware decrypts it—usually for a fee. Capabilities that have an intentionally intermittent or time-bound effect would also fall into this third category. Still other cyber capabilities, such as some wiping operations, are best placed in the fourth category, as the target may be able programmatically to reverse it but would require a substantial amount of effort or time to do so. For example, a target might possibly recover data from "wiped" hard drives, depending on how the wiping attack was done, but it is beyond the capabilities of most ordinary users.

It is worth noting that some capabilities exist in both the third and fourth categories. For example, denial of service attacks—which overwhelm a target with meaningless data—can be not only turned off by an operator but also thwarted by the target's taking certain countermeasures.

From this analysis, we conclude that the qualities of reversibility that are most often intended when using nonlethal weapons are often similar to the most

frequent kinds of cyber capabilities employed today. With some rare but important exceptions, such as attacks that destroy physical infrastructure, the damage
caused by even some data-destroying cyber capabilities is often reversible in
that computers and systems can be repaired with sufficient time and resources.
However, as a practical matter, most victims of such attacks may find replacing
rather than repairing their malfunctioning systems is more prudent. Given that
the majority of contemporary compromises of confidentiality, integrity, and
availability of data are perpetrated through reversible means (like denial of service), we feel the analogy to nonlethal weapons has value in this area of analysis.

Deterrence

Analogizing nonlethal weapons and cyber capabilities in the area of deterrence is
possible but not as straightforward as the preceding three areas of analysis.
Deterrence is an important but reasonably narrow concept when it comes to nonlethal weapons. For cyber capabilities, questions of deterrence are more complex,
as applications converge with and diverge from the concept's use in nonlethal
weapons. Much has been written about deterrence of cyber capabilities as well as
about using these capabilities for deterrence; thus, we briefly provide an outline
of the underpinning of deterrence and examine how the analogy applies.[39]

When considering deterrence, the initial questions to consider are, whom do
we wish to deter from doing what, and what would we like them do to instead?
Any discussion about deterrence must be tailored around this "deter whom from
doing what" foundation. During the Cold War, the term "nuclear deterrence"
was often shorthand for "deterring the Soviet Union from launching a nuclear-
armed attack."[40] But this case of deterrence can obscure the fact that other kinds
of deterrence exist. While the Cold War case mostly involved deterrence of a
specific actor, some deterrents are general and apply to large groups of actors.

Similarly, while nuclear deterrence is absolute—that is, seeking to stop any
use of an atomic weapon—other deterrents are restrictive and seek to minimize
the effects and occurrence of an unwanted activity as much as possible while
acknowledging implicitly that some will occur.[41] Deterring crime is an example
that is both general and restrictive: police do not always know which individual
in society is a would-be criminal, and they also recognize that despite measures
to deter its occurrence, some amount of crime is inevitable.

The two traditional methods of deterrence are cost imposition and denial.[42]
Deterrence by *cost imposition* operates via a (tacit or explicit) credible threat of
retaliation to such a degree that the attacking state would find commencing the
unwanted activity prohibitively costly. Deterrence by *denial* operates by convincing an adversary that even if it does not fear cost imposition, the benefits it
seeks will be checked due to effective defenses. Together the two can make certain actions unappealing. Deterrence by denial can reduce the chances of success, while fear of retaliation can make certain actions prohibitively costly.

Nonlethal weapons can function, depending on the capability, as deterrents
by denial or by threatening cost imposition. Many counter-matériel and counter-

personnel capabilities impose comparatively minimal costs on an adversary but can reduce or deny the adversary's capability to carry out an unwanted action.

A tactical example from Somalia demonstrates that nonlethal lasers functioned as a means of threatening total retaliation, signaling to potential adversaries that they had been identified and would be neutralized if they attacked US forces. That is, a laser beam shined on a target warned that a bullet could follow.

Some cyber capabilities also can work as deterrents by denial or deterrents by cost imposition, depending on the capability. China's Great Firewall is an example of deterrence by denial. The system, which actively intercepts unwanted Internet activity in Chinese networks and prevents it from connecting to blocked servers, aims not only to prevent but also to deter actions that the Chinese government deems undesirable. It is a scalable and general deterrent across the broader population rather than a narrowly crafted one for a small group of actors. Still, it is restrictive rather than absolute, as the Chinese surely know that some individuals find their way around the firewall.

China's so-called Great Cannon is an example of deterrence by cost imposition. In 2015 members of the popular code repository and software development site GitHub, to which anyone can upload code or text, began uploading *New York Times* articles and other content the Chinese viewed as subversive. In response, while leveraging their position of privilege on the Chinese Internet that is made possible by the Great Firewall, Chinese actors launched a massive denial of service attack and took GitHub offline for a time. By imposing costs on GitHub, the Chinese carried out a form of deterrence by cost imposition to GitHub and similar sites, though they ultimately ceased the attack without changing GitHub's behavior.[43]

Cyber capabilities, in some circumstances, can send a signal threatening greater non-cyber cost imposition. For example, a nation may reveal a cyber operation to another state as a means of showing that it can access the latter's strategically important networks. While it is unclear if Stuxnet was intended to have such a psychological effect, apparently the program introduced doubt into the minds of Iranian engineers, and the worm's revelation potentially impacted later nuclear negotiations.[44] In other cases, cyber capabilities—such as the capacity to send a message to anyone entering a certain area—can directly carry a warning. In 2014 protestors in Kiev received text messages of this sort.[45]

Nonlethal weapons and cyber capabilities are similar in that deployment of some forms of each can enable various kinds of deterrence. But a key difference emerges: nonlethal weapons, because they are more limited in their potential damage, are seldom the objects of deterrence. While hypothetically possible, it seems impractical for one entity to devote resources to deter another's employment of nonlethal weapons. The stakes are usually just too low. The threat of nonlethal weapons against American troops is not sufficiently serious to warrant either issuing powerful threats to impose costs or creating sufficient defenses to deny an adversary's benefit.

However, it is somewhat easier to conceive of situations where the United States might wish to deter another entity's use of nonlethal weapons by implementing denial. For example, if US forces embarked on a stabilization mission where the local population had demonstrated a desire or capability to employ nonlethal weapons, the United States might wish to demonstrate powerful defenses that easily blunt the effectiveness of those weapons.

Cyber capabilities, because they are potentially more destructive or—in the case of data theft—strategically damaging without being destructive, are different in kind. Nonlethal weapon deterrence yields a one-way question: How can nonlethal weapons be useful for deterrence? Cyber deterrence yields a two-way question: How can cyber capabilities be useful for establishing deterrence generally, and how can an adversary's use of cyber capabilities be deterred but not necessarily with cyber means?

As a result, the analogy between the two is attenuated. When asking how to deter the use of cyber capabilities by others, it is important not to limit oneself to thinking of one's own cyber capabilities. All elements of national power, including political clout, economic sanctions, kinetic retaliation, and cyber defenses, should be included in the deterrence discussion. Offensive cyber capabilities may be part of this calculus, but many are likely too subtle or too limited to fully act as a deterrent on their own. The analogy to nonlethal weapons here points to the need for a broader discussion of cyber deterrence.

Conclusion

A clear theme runs through this analysis: areas of overlap in both characteristics and in function exist between nonlethal weapons and cyber capabilities. These areas of overlap strengthen the case for the proposed analogy and point to the possibility that lessons learned about nonlethal weapons may be usefully applied to cyber capabilities. In short, recognizing how new and different cyber capabilities are, we need not consider them with an entirely blank slate. Analogizing to nonlethal weapons can be a valuable approach.

With that said, some cyber capabilities do not fit the analogy particularly closely—for example, those capabilities that do not seek to incapacitate a target (instead, they might steal data from it), those capabilities that do not seek to minimize collateral damage, and those capabilities that are irreversibly destructive. For discussions of these kinds of capabilities, nonlethal weapons are less obviously useful.

Another, more practical kind of limitation to this analogy concerns the employment of these weapons and capabilities. For reasons that remain largely elusive to the authors, the use of nonlethal weapons by US military forces has been restricted. Several military officers have informally observed and stressed that gaining authorization to employ lethal force was often easier than that for nonlethal force despite the latter's promise of lower collateral damage and only temporary effects. The question that remains unanswered in our research is, why are nonlethal weapons not better integrated and employed?

Further research into this question may be aided by bringing in literature on path dependency and the "stickiness" of entrenched traditions—or, in this case, the greater familiarity of employing kinetic, conventional weapons. Additional research may also tell us more about the inflexibility of military targeting procedures, which may have been designed to weigh specific variables in the context of a kinetic action but may be insufficiently malleable to more completely consider the authorization of nonlethal capabilities. These questions are important to examine, as they may tell us about the willingness and the process to employ cyber capabilities in the future.

Any answer to this question is going to depend on the type of nonlethal weapon in question and on the nature of the international legal regime that restricts those weapons. For example, the United States does not use riot-control agents in combat due to its commitments under the Chemical Weapons Convention.[46] Nor does the United States employ lasers to blind individuals, in compliance with the terms of the 1980 Convention on Certain Conventional Weapons.[47]

Further exploration of why nonlethal weapons have been deployed so seldom by US forces should be considered separately in more detail. For our purposes, it is worth noting that the lack of explicit international law around the employment of cyber capabilities may enable commanders to deploy possible future tactical capabilities with more freedom than they have with nonlethal weapons. In addition, cyber capabilities are—at least up until now—counter-matériel capabilities. Nonlethal weapons are both counter-matériel and counter-personnel. As such, the focus of cyber capabilities on counter-matériel missions may eventually give leaders less cause to eschew authorizing their tactical employment. This distinction may change, however, as wearable and related technologies create a new attack vector and open the possibility of cyber capabilities becoming counter-personnel capabilities.

Regardless of the reasons that inform the US military's decision to employ nonlethal weapons in only a limited fashion, practitioners would be wise not to take the nonlethal-cyber analogy too far for fear that, for whatever reason, cyber capabilities might become another instrument of power that is unwieldable even when they are the most appropriate tools available.

Indeed, despite very real concerns about a coming conflict in cyberspace, some of the most promising features of cyber capabilities are also common with other nonlethal weapons: their effects need not be permanent and could possibly be so narrowly tailored that collateral damage is all but eliminated. As with any other instrument of military power, cyber capabilities should be used only as a last resort. But when military coercion is required to secure US interests, cyber capabilities—like nonlethal weapons—may offer US military commanders the opportunity to do so in ways that greatly reduce the incidence of death and destruction on all sides of a future conflict.

Notes

The views expressed in this publication are those of the authors and do not necessarily reflect the official policy or position of the Department of Defense or the US government.

1. Joseph Nye, "Nuclear Lessons for Cyber Security?," *Strategic Studies Quarterly* 5, no. 4 (Winter 2001); Joseph Nye, "Cyber Power" (Boston: Belfer Center for Science and

International Affairs, May 2010), http://belfercenter.ksg.harvard.edu/files/cyber
-power.pdf; and Emily Goldman and John Arquilla, eds., *Cyber Analogies* (Monterey, CA:
Naval Postgraduate School, 2014).

2. Michael S. Goodman, "Applying the Historical Lessons of Surprise Attack to the
Cyber Domain: The Example of the United Kingdom," in Goldman and Arquilla, *Cyber
Analogies*; Nye, "Cyber Power"; Joel Brenner, *Glass Houses* (New York: Penguin, 2014); and
Richard A. Clarke and Robert Knake, *Cyber War: The Next Threat to National Security and
What to Do about It* (New York: HarperCollins, 2010).

3. Michael Daniel, "Heartbleed: Understanding When We Disclose Cyber Vulnerabili-
ties," The White House Blog, April 28, 2014, http://www.whitehouse.gov/blog/2014/04
/28/heartbleed-understanding-when-we-disclose-cyber-vulnerabilities; and Richard
Clarke et al., "The NSA Report: Liberty and Security in a Changing World," President's
Review Group on Intelligence and Communications Technologies (Princeton: Princeton
University Press, 2013).

4. Rolf Weber, "Internet of Things: New Security and Privacy Challenges," *Computer
Law & Security Review* 26, no. 1 (2010); and Andy Greenberg, "Hackers Remotely Kill a Jeep
on the Highway—with Me in It," *Wired*, July 21, 2015, http://www.wired.com/2015/07
/hackers-remotely-kill-jeep-highway/.

5. The *Oxford English Dictionary* states that a weapon is "a thing designed or used for
inflicting bodily harm or physical damage."

6. Ashton B. Carter, "DOD Executive Agent for Non-Lethal Weapons (NLW), and NLW
Policy," Number 3000.03E (Washington, DC: Department of Defense, April 25, 2013), 12,
http://www.dtic.mil/whs/directives/corres/pdf/300003p.pdf.

7. Leo P. Brophy, Wyndham D. Miles, and Rexmond C. Cochrane, *The Chemical Warfare Service: From Laboratory to Field*, United States Army in World War II (Washington, DC:
Center of Military History, US Army, 1988).

8. Ibid., 2.

9. Ibid., 12.

10. Ibid., 70.

11. Ibid., 24.

12. Ibid., 25.

13. To characterize the overall effect of chemical weapons in World War I as nonlethal would be misleading, as the United Nations Office for Disarmament Affairs notes
that chemical weapons employed in that conflict eventually killed more than 100,000
individuals. See United Nations Office for Disarmament Affairs, "Chemical Weapons,"
https://www.un.org/disarmament/wmd/chemical/.

14. United Nations Office for Disarmament Affairs, "Protocol for the Prohibition of the
Use in War of Asphyxiating, Poisonous or Other Gases, and of Bacteriological Methods of
Warfare," 1925, https://www.un.org/disarmament/wmd/bio/1925-geneva-protocol/.

15. Brophy, Miles, and Cochrane, *Chemical Warfare Service*, 72–73.

16. Smoke is a nonlethal weapon, according to the definition used in this chapter, as
it does not directly inflict bodily harm or physical damage. This reference is merely to
note the evolution of how chemicals came to be used in World War II. Brophy, Miles, and
Cochrane, *Chemical Warfare Service*, 197.

17. Ibid., 197–225.

18. Gerald Ford, "Executive Order 11850—Renunciation of Certain Uses in War of
Chemical Herbicides and Riot Control Agents," *Federal Register*, April 8, 1975, http://www
.archives.gov/federal-register/codification/executive-order/11850.html.

19. Andrew Sanders and Ian S. Wood, *Time of Troubles: Britain's War in Northern Ireland* (Edinburgh: Edinburgh University Press, 2012), 127.

20. Richard Pike, *Phantom Boys: True Tales from Aircrew of the McDonnell Douglas F-4 Fighter-Bomber* (London: Grub Street Books, 2015), 105.

21. Barton Reppert, "Force without Fatalities," *Government Executive*, May 1, 2001, http://www.govexec.com/magazine/magazine-defense/2001/05/force-without -fatalities/8992/.

22. Senate Armed Services Committee, *Nomination of Maj. Gen. Anthony C. Zinni, USMC, for Appointment to the Grade of Lieutenant General and to Be the Commanding General, 1st Marine Expeditionary Force*, 103rd Cong., 2nd sess., June 16, 1994, Hrg. 103–873, 32.

23. Richard L. Scott, "Conflict without Casualties: Non-Lethal Weapons in Irregular Warfare" (thesis, Naval Postgraduate School, 2007), 6–7.

24. Nick Lewer and Steven Schofield, *Non-Lethal Weapons: A Fatal Attraction?* (London: Zed Books, 1997), 20.

25. For more on this episode, see F. M. Lorenz, "Non-Lethal Force: The Slippery Slope to War?," *Parameters*, 1996, 52–62.

26. Reppert, "Force without Fatalities."

27. Graham T. Allison and Paul X. Kelley, "Nonlethal Weapons and Capabilities" (Washington, DC: Council on Foreign Relations, 2004), 13–18.

28. Ibid., 1.

29. For one of the earliest articulations of this idea, see David D. Clark and David R. Wilson, "A Comparison of Commercial and Military Computer Security Policies," *Proceedings of the 1987 IEEE Symposium on Research in Security and Privacy*, 1987, 184–94.

30. John Leyden, "Israel Suspected of 'Hacking' Syrian Air Defences," *The Register*, October 4, 2007, http://www.theregister.co.uk/2007/10/04/radar_hack_raid/.

31. Ben Elgin and Michael Riley, "Now at the Sands Casino: An Iranian Hacker in Every Server," *Bloomberg*, December 11, 2014, http://www.bloomberg.com/bw /articles/2014-12-11/iranian-hackers-hit-sheldon-adelsons-sands-casino-in-las-vegas.

32. Dan Kloeffler and Alexis Shaw, "Dick Cheney Feared Assassination via Medical Device Hacking: 'I Was Aware of the Danger,'" ABC News, October 19, 2013, http://abc news.go.com/US/vice-president-dick-cheney-feared-pacemaker-hacking/story ?id=20621434.

33. Weber, "Internet of Things"; and Greenberg, "Hackers Remotely Kill."

34. Kim Zetter, *Countdown to Zero Day: Stuxnet and the Launch of the World's First Digital Weapon* (New York: Crown, 2014), 97.

35. Ibid.

36. Ibid.; and Kasperky Lab, "Gauss: Abnormal Distribution," SecureList, August 9, 2012, https://securelist.com/analysis/publications/36620/gauss-abnormal-distribution/.

37. Rachel King, "Stuxnet Infected Chevron's IT Network," *Wall Street Journal*, November 8, 2012, http://blogs.wsj.com/cio/2012/11/08/stuxnet-infected-chevrons-it-network/.

38. Zetter, *Countdown to Zero Day*.

39. For a sampling of perspectives, see Will Goodman, "Cyber Deterrence: Tougher in Theory Than in Practice?," *Strategic Studies Quarterly*, Fall 2010, 102–35; Murat Dogrul, Adil Aslan, and Eyyup Celik, "Developing an International Cooperation on Cyber Defense and Deterrence against Cyber Terrorism," in *2011 3rd International Conference on Cyber Conflict*, ed. C. Czosseck, E. Tyugu, and T. Wingfield (Tallinn, Estonia: Cooperative Cyber Defence Centre of Excellence, 2011), 43; and Amit Sharma, "Cyber Wars: A Paradigm Shift from Means to Ends," *Strategic Analysis* 34, no. 1 (2010).

40. To be sure, the United States also sought to deter the Chinese from undertaking similar activity, but in large part the goal was to influence the perceptions of the Soviet leadership that it should not undertake a nuclear-armed attack.

41. For a fuller explication of the two variables here—absolute versus general deterrence and specific versus restrictive deterrence—and an application to cyber operations, see Ben Buchanan, "Cyber Deterrence Isn't MAD; It's Mosaic," *Georgetown Journal of International Affairs*, International Engagement on Cyber IV, 15, no. 2 (2014): 130–40.

42. See also how the concept of entanglement relates to the calculation of an action's costs. Joseph Nye, "Can China Be Deterred in Cyber Space?," Foreign Policy Association (blog), April 6, 2016, http://foreignpolicyblogs.com/2016/04/06/can-china-be-deterred-in-cyber-space/.

43. Bill Marczak et al., "China's Great Cannon," Research Brief (Citizen Lab and Munk School of Global Affairs, University of Toronto, April 10, 2015), https://citizenlab.org/2015/04/chinas-great-cannon/.

44. David Sanger and William Broad, "Unstated Factor in Iran Talks: Threat of Nuclear Tampering," *New York Times*, March 21, 2015, http://www.nytimes.com/2015/03/22/world/middleeast/unstated-factor-in-iran-talks-threat-of-nuclear-tampering.html.

45. Heather Murphy, "Ominous Text Message Sent to Protesters in Kiev Sends Chills around the Internet," *New York Times*, January 22, 2014, http://thelede.blogs.nytimes.com/2014/01/22/ominous-text-message-sent-to-protesters-in-kiev-sends-chills-around-the-internet/.

46. Allison and Kelley, "Nonlethal Weapons and Capabilities."

47. Ibid., 53.

3 Cyber Weapons and Precision-Guided Munitions

JAMES M. ACTON

The development of precision-guided munitions (PGMs)—guided gravity bombs and cruise missiles, in particular—has had profound implications for warfare. Such weapons tend to cause much less collateral damage than their unguided predecessors do, and because they can remain effective when used from a distance, they can also reduce casualties sustained by the attacker. Thus, PGMs have altered national-level decision-making by lowering the political threshold for the use of force and by slowing the likely loss of public support during a sustained military campaign. PGMs have also increased the tactical effectiveness of military operations. They have dramatically improved force exchange ratios (at least against an adversary without these weapons) by reducing the likely number of weapons required to destroy individual targets. In doing so, they have eased logistical requirements and increased the pace at which military operations can be conducted.

Following the 1991 Gulf War, which provided the first high-profile demonstration of the effectiveness PGMs, these weapons were widely seen—both in the United States and abroad—as revolutionary (or, at least, as the technological component of revolutionary military changes).[1] Almost twenty-five years later, a number of analysts have argued that cyber weapons are effecting another revolution in military affairs.[2] This controversial claim is inspired, at least in part, by the analogy between PGMs and cyber weapons.

The similarities between PGMs and sophisticated cyber weapons are striking.[3] Cyber weapons also offer the *potential* of exquisite precision because, if well-designed, they may affect only specific targets and inflict carefully tailored effects.[4] Information technology (IT) is ubiquitous in military operations. As a result, the use of cyberspace for military purposes can confer potential tactical advantages to an attacker, including by further improving force exchange ratios, while placing few, if any, additional demands on the logistical network needed to supply frontline forces. Moreover, the use of cyber weapons involves minimal risk to the lives of the service personnel who "deliver" them and, in general, is likely to cause fewer civilian casualties than even the most carefully designed and executed kinetic attack.[5] As a result, they could further lower the threshold for the use of force. Overall, in fact, states' reasons for wanting cyber weapons are very similar to their reasons for wanting PGMs.

For all the benefits of cyber weapons, they undoubtedly have limitations too. The possibility that cyber weapons can be employed in highly discriminating ways does not imply they must be; like PGMs, cyber weapons can be used in indiscriminate ways. Indeed, many publicly known cyber attacks to date have had distinctly imprecise effects on the target system (for example, by destroying entire computers) and have caused collateral damage against undetermined numbers of other systems and users. That said, there is also reason to suppose that the public record is not representative of cutting-edge cyber capabilities, since more discriminate attacks are easier to hide.

Setting aside the technical and operational challenges of achieving precision in practice, this chapter seeks to exploit the analogy with PGMs to understand some of the other potential limitations of cyber weapons and how militaries might respond to them either by mitigating them or by capitalizing on them. The focus is on three challenges to the effective employment of PGMs and their cyber analogies. The first two challenges—intelligence, surveillance, and reconnaissance (ISR) and battle damage assessment (BDA)—relate to the effectiveness of enabling capabilities. The third challenge is the difficulty of achieving the political objectives for which a war is fought using only standoff attacks.

The Need for Effective Intelligence, Surveillance, and Reconnaissance

An important distinction is drawn in the physical sciences between precision and accuracy. The claim that the population of the United States is 62,571,389 is very precise, but it is not remotely accurate. Similarly, PGMs are almost invariably precise—in the sense that they almost always hit their aim points (or at least very nearby)—but because their intended targets may not always be located at those aim points, PGMs are not always accurate.

To ensure that PGMs are accurate, the location of the intended target must be known both correctly and precisely.[6] The ISR capabilities used to locate targets are, therefore, every bit as important as a weapon's guidance and navigation system. The development of various technologies for acquiring overhead images has made the process of locating stationary, aboveground targets much easier, but it has not guaranteed success. For example, during the bombing of Yugoslavia in 1999, US military planners knew the street address of the Yugoslav Federal Directorate for Supply and Procurement was Bulevar Umetnosti 2.[7] However, because of a combination of human error and out-of-date maps and databases, these planners incorrectly identified the building that corresponded to this address. As a result, although the weapons used in the subsequent strike did indeed hit their intended aim points, they destroyed not a legitimate military target but the Chinese Embassy.

Identifying the location of other types of targets—mobile and underground targets, in particular—is a much tougher problem. The challenge was illustrated during the 1991 Gulf War by the "great Scud hunt," in which Coalition forces attempted to destroy Iraq's Scud missiles and their mobile launchers. Coalition aircraft flew about 1,460 sorties against Scud-related targets—about 215 against the mobile launchers themselves—without scoring a single confirmed kill.[8] The *Gulf War Airpower Survey* attributes the cause of this failure to inadequate ISR and, in particular, "the fundamental sensor limitations of Coalition aircraft."[9] These limitations were compounded by effective Iraqi tactics, such as the use of decoys, which complicated the task of an already inadequate ISR system. Since 1991 significant improvements in ISR (as well as in tactics) have been central to enhancing—at least to some extent—the ability of advanced militaries to destroy dispersed mobile forces, as evidenced by Israel's moderately successful campaign to hunt down Hezbollah's mobile rocket launchers in the 2006 Lebanon War.[10]

Intelligence collection is a similarly important enabling capability for cyber attacks. It contributes to identifying how to penetrate the target IT system, to understanding the system sufficiently well to create a weapon payload with the desired effect, and to ensuring that the payload's effects are limited to the target network.

IT systems are most commonly penetrated as the result of human error. An attacker, for example, might send phishing emails containing a link that, if clicked on, causes malware to be installed. Such attacks are much more likely to be successful if the attacker exploits intelligence about targeted users' names, contacts, and behavioral characteristics—an approach known as "spear phishing." For example, a 2015 report by the cybersecurity firm FireEye details several recent spear-phishing attacks against Southeast Asian governments involved in territorial disputes with China.[11] These attacks appeared to exploit relatively detailed intelligence about targeted users. Much more detailed intelligence can be required to penetrate more sophisticated defenses. For example, to penetrate IT systems at Iran's Natanz enrichment plant, which are surrounded by an air gap, the perpetrators of the Stuxnet attack—believed to be the United States and Israel—reportedly first infected computers belonging to contractors. Personnel employed by these contractors then inadvertently transmitted the virus to Iran's enrichment control system on USB flash drives (other infection strategies were apparently employed too).[12] This approach could have been developed only with detailed knowledge about the organizational structure of Iran's enrichment program. Of course, not all infection strategies rely on user error, but most (if not all) others usually require detailed intelligence about the target, such as knowledge of "zero-day" vulnerabilities—that is, software or hardware flaws that are unpatched because they are unknown to the vendor.

Developing a payload that has the desired effects often requires equally—if not more—detailed intelligence. Stuxnet is a paradigmatic example. The code aimed to destroy Iranian centrifuges by reprogramming the enrichment plant's control system so it altered their rotation speed while simultaneously sending falsely reassuring signals to operators. The development of Stuxnet was reportedly preceded by a huge intelligence-gathering operation on the Natanz facility, which itself relied, at least in part, on cyber espionage.[13] The Stuxnet code was then tested on actual P-1 centrifuges (which are very similar to the IR-1 centrifuges operated by Iran). In one sense, Stuxnet—an exceptionally complicated and sophisticated virus—is something of an extreme example. However, it may well be representative of the challenges associated with developing cyber weapons that can have real-world effects similar to those of extremely precise kinetic weapons.[14] Indeed, that the Stuxnet code also migrated into nontarget machines underscores the practical challenges of achieving precision, while the fact that the code did not activate and thus disrupt the functioning of these machines demonstrates the possibility and importance of sophisticated target reconnaissance and malware engineering.

There are, of course, important differences in intelligence collection for cyber and PGM strikes. Usually, one major purpose of intelligence collection in planning a kinetic strike is to identify the exact location of the target; by contrast, the physical location of an enemy IT system is rarely a concern in planning a cyber attack.

The consequences of intelligence failures are also potentially dissimilar. Poor intelligence about the target of a kinetic attack—as the 1999 bombing of the Chinese Embassy in Belgrade typifies—can lead to high costs in the form of civilian deaths, diplomatic fallout, and reputational damage. For two reasons, the consequences of poor intelligence for a cyber attack are likely to be less significant than for a kinetic attack. First, the distinct chance is that a cyber attack based on poor intelligence will have no effect whatsoever. To be sure, this outcome is not guaranteed; poor intelligence can lead to the cyber equivalent of collateral damage. A 2008 cyber attack by the United States against a terrorist website in Saudi Arabia, for example, is reported to have disrupted more than three hundred other servers because the target IT system was insufficiently understood.[15] However, good programing can presumably minimize the risks of collateral damage, and even if it cannot, collateral damage restricted to cyberspace is likely to be less costly than collateral damage in physical space. Second, cyber attacks are more plausibly deniable than kinetic attacks are. As a result, the reputational cost of launching a cyber attack that causes collateral damage is likely to be less as well.

That said, it is also possible that cyber attacks will be held to a higher standard than kinetic strikes and thus raise the cost of intelligence failures, even if cyber collateral damage is indeed comparatively modest. In fact, precisely because the development of PGMs has changed expectations about what constitutes acceptable collateral damage, advanced states are now held to a much higher standard in assessing whether the application of kinetic force has been proportionate and whether sufficient care has been taken to discriminate between military and civilian targets. Given the potential for cyber weapons to be even "cleaner" than PGMs, cyber operations may be held to a still higher benchmark—at least where they are conducted by states with the capability to develop discriminating weapons.[16]

In any case, there are some interesting analogies about collecting intelligence for cyber operations and for kinetic strikes. One particular challenge of acquiring intelligence for cyber attacks is the inherent mutability of IT systems. For example, security protocols and antivirus software can be improved, zero-day vulnerabilities can be discovered and (usually) patched, software can be updated, and hardware can be replaced. As a result, a cyber weapon cannot remain effective indefinitely, and predicting how long it will remain potent is impossible. In this way, a particularly apt analogy from the physical world is the challenge of gathering intelligence for targeting a mobile asset. Locating a mobile target while it happens to be stationary makes striking it much easier, but given the difficulty of predicting when the target will next move, the window of opportunity for conducting the attack may be of an inherently unpredictable duration.

Given the challenges of targeting mobile assets, many nations have responded to the development of PGMs by increasing the mobility of their military forces (even though mobile systems are almost inevitably more lightly armored than their stationary equivalents and hence easier to destroy if their location is discovered). The analogous approach to cyber defense is to focus resources not only on hardening the IT system—that is, identifying and patching vulnerabilities—but also on regularly modifying an IT system simply for the sake of changing it, a strategy that has been termed "polymorphic cyber defense."[17] This approach attempts to render an attacker's knowledge of the target system obsolete almost as soon as it is obtained. One of the leaders in this field called its technology "Moving Target Defense," making the analogy to the physical world absolutely explicit.[18]

The primary challenge to polymorphic cyber defenses is probably the risk of introducing bugs that could prevent a system from performing as it should. The scale of this risk presumably depends on how much of the system and which parts of it are changed and on the size of the conceptual space of the allowed changes. Thus, there may well be a potential trade-off between greater security and reduced usability. Where states perceive the sweet spot to be may determine the prospects of polymorphic cyber defenses for military applications.

In the physical world, one approach to overcoming the challenge posed by mobility is to reduce the time between detection and engagement. To this end, sensors and weapons have been integrated into the same platform and, in some systems, given the capability to engage autonomously. Israel's Harpy unmanned combat aerial vehicle, for example, is designed to loiter and detect enemy air defense radars (which are frequently mobile) and to attack them automatically.[19] An analogous cyber weapon would have the capability to detect and exploit vulnerabilities autonomously.[20] This author is not qualified to speculate on whether such an "intelligent" cyber weapon could be developed, but the Defense Advanced Research Projects Agency is sponsoring research, including the Grand Cyber Challenge, into cyber defenses that completely autonomously could "identify weaknesses instantly and counter attacks in real time."[21] Such efforts may be dual use: research in detecting cyber vulnerabilities of friendly IT systems and enhancing their defenses could contribute to the development of offensive cyber weapons that can discover enemy IT vulnerabilities.

Beyond mobility, numerous other countermeasures to PGMs have been employed, including air defenses, hardening, deception, interference with navigation and command and control, and human shields. These countermeasures provide fertile ground for further extending the analogy between defenses to PGMs and defenses to cyber weapons (and take it far beyond interference with ISR capabilities), as a few examples demonstrate. Air defenses, which are designed to shoot down incoming PGMs, are analogous to active cyber defenses in which the defender uses its own virus (sometimes known as a white worm) to disable an attacker's. Another countermeasure in kinetic warfare is interfering with the satellite navigation signals, such as those provided by the US Global Positioning System, that many modern PGMs use. Spoofing, for example, involves transmitting fake navigation signals, which are designed to mislead a weapon about its location. Conceptually, spoofing is similar to sinkholing, a form of active cyber defense that involves redirecting data being transmitted by a virus to a computer controlled by the victim of an attack.

An entirely different approach to defending against PGMs (or, indeed, any other form of kinetic attack) is to raise the political costs of a strike. For example, both states and terrorist organizations have used civilians as human shields by hiding weapons in schools, hospitals, and mosques.[22] More prosaically, in every state, many elements of war-supporting infrastructure—including power stations, electricity grids, and oil refineries—are dual use in that they serve both civilian and military purposes. Even if attacking such facilities is legally permissible, it can still be politically costly.

In the cyber world, civilian and military networks also are often one and the same. For example, an overwhelming majority of US military communications data is believed to pass through public networks that also handle civilian data.[23] Going further, organizations that have civilian functions can also conduct offensive cyber operations. For example, China's National Computer Network Emergency Response Technical Team—a body under the Ministry of Industry and Information Technology that is nominally responsible for defending China's civilian networks from attack—may have been involved in offensive cyber operations.[24] This intermingling raises the potential political cost of cyber operations against military targets through the risk of simultaneously implicating civilian assets. The existence of such intermingling inevitably raises the question of whether it is part of a deliberate strategy designed to defend military assets in cyberspace.

The Importance of Effective Battle Damage Assessment

Battle damage assessment is a second enabling capability that is needed to exploit precision to its full extent. Knowledge that a kinetic strike has been successful can avoid wasting resources on unnecessary repeated strikes against the same target. Immediate feedback also enables the attacker to capitalize quickly on the success. For example, if timely confirmation is available that an air defense battery protecting an underground bunker has been destroyed or disabled, mission commanders can exploit the success by authorizing aircraft to attack the bunker before the adversary can take countermeasures (such as evacuation). Conversely, confirmation that the strike against the air defense system was unsuccessful can be used to authorize another attempt to destroy it. The costs of ineffective (or entirely absent) BDA in this scenario could be quite high. If the strike against the air defense system is incorrectly believed to have been successful, the lives of the pilots sent to attack the bunker will be at risk. If the strike was successful but its outcome cannot be confirmed, mission commanders may waste resources on further strikes as well as an opportunity to destroy the bunker.

As a general rule, the more discriminating a strike is, the more difficult BDA becomes. The particular challenges of BDA for PGMs became apparent in the 1991 Gulf War. To give an example, overhead imagery proved relatively ineffective at assessing the effects of attacks on hardened structures. When these attacks were successful, they generally caused extensive internal damage but very little external damage; often the only visible effect of the attack was a hole made by the incoming bomb.[25] Image analysts thus tended to seriously underestimate the effectiveness of strikes against such targets. Thirteen years later, a 2004 report by the US General Accounting Office on the wars in Afghanistan and Iraq highlighted the continued "inability of damage assessment resources to keep up with the pace of modern battlefield operations."[26] The results included the "inefficient use of forces and weapons" and ground advances that were slowed unnecessarily.[27]

In extreme cases, the lack of effective BDA can have truly major consequences. In early 2011 after the US intelligence community acquired evidence of Osama bin Laden's whereabouts, senior American officials debated whether and how to attempt to kill him.[28] Some of President Barack Obama's key advisers reportedly recommended using an aircraft-launched standoff PGM. One of the main reasons—if not the main reason—why Obama rejected this course of action was apparently its lack of any reliable way to verify the strike's success. It could, therefore, have been very difficult to justify the infringement of Pakistani sovereignty, and the United States might have wasted considerable resources in continuing efforts to find bin Laden if he escaped. Obama's decision to use special forces solved the BDA problem but created other extremely serious risks. For example, if Pakistani troops had captured the Americans, the consequences for relations between Washington and Islamabad (not to mention Obama's presidency) would have been much more serious than if a standoff munition had been used.

From a tactical perspective, BDA after a cyber attack is important for many of the same reasons as after a kinetic attack. In fact, such assessments may be even more important because cyber attacks can often produce temporary or reversible effects. Therefore, an attacker may need to discover not just whether the attack achieved the desired effect initially but also whether the target IT system is still compromised and its attack undetected.

The strategic importance of cyber BDA is likely to depend on the particular attack scenario. Because the use of cyber weapons is generally more deniable than the use of kinetic weapons and because cyber attacks may sometimes even go undetected (especially if unsuccessful), states may be less concerned about the need to provide ex post facto justifications for a strike, rendering BDA less important for cyber operations than for kinetic ones. Had some (extremely) hypothetical way to kill bin Laden with a cyber weapon been available, for example, it is conceivable that Obama might have opted for it even without a reliable means of conducting BDA. Using a cyber weapon, however, carries the risk that it might spread and infect third-party, or perhaps even friendly, IT systems. BDA would be extremely important to enable rapid action to mitigate the consequences.

Cyber BDA has been discussed very little in the open literature, so any discussion is necessarily fairly speculative.[29] Nonetheless, governments must have confronted this question. Israel, for example, is reported to have disabled Syrian air defenses with a cyber weapon, in combination with other tools, before its aircraft destroyed Damascus's clandestine plutonium-production reactor in 2007.[30] Given that the human and diplomatic costs of having its aircraft shot down would have been high, Israel presumably had some means of verifying that it had indeed disabled Syria's air defenses.

Network exploitations are presumably the principal tool for cyber BDA. (If a cyber attack has physical effects, other techniques for conducting BDA may be possible. Israel, for example, may have been able to monitor the electromagnetic emissions from Syria's radars.) Indeed, one reason why cyber BDA may be less challenging than physical BDA is that a cyber weapon can potentially be programmed either to conduct an assessment of its own effects or to expropriate information on which such an assessment can be based. By contrast, adding sensors and transmitters for BDA onto a kinetic warhead is extremely difficult, if not impossible.

On balance, however, there are good reasons to expect that cyber BDA is likely to be more challenging than physical BDA, especially for highly precise attacks. (BDA for indiscriminate cyber attacks—against critical infrastructure, say—presents far fewer challenges.) For example, a cyber attack that is designed to prevent an adversary from doing something, such as launching a missile, could present BDA challenges since the attacker might not know whether the cyber weapon had worked until the adversary tried to launch it. More generally, because the effects of many cyber attacks are temporary or reversible, effective BDA cannot rely on a "snapshot" of the target system at a certain moment; instead, continuous monitoring is required. Even if such monitoring is possible, cyber defenses may prevent the information from being sent to the attacker in a timely way. For example, if a cyber weapon is transported across an air gap in a physical storage device, information relevant to BDA could potentially be transported in the same way in the opposite direction; but such a process could be slow and perhaps too slow to be militarily useful. Finally, if using a cyber weapon reveals its own existence, the owner of the targeted IT system can take steps to secure its network and make it less visible, potentially defeating any exploitation being used for BDA. More ambitiously, the owner might even try to fool the attacker by allowing it to exfiltrate deliberately misleading information about the effectiveness of the attack.[31]

Overall, cyber BDA appears to be both important and difficult. Moreover, efforts to defeat BDA perhaps could become a significant feature of cyber warfare. To this author's knowledge, defeating BDA has not been a major focus of states' attempts to undermine advances in PGMs, but efforts to defeat ISR capabilities could have that effect. By contrast, it seems plausible that states could invest significant resources in trying to defeat cyber BDA by developing rapid response capabilities. Indeed, the US military already has in place "Cyber Protection Forces . . . [to] defend priority [Department of Defense] networks and systems," although whether these forces are tasked with attempting to foil adversary BDA attempts is unknown.[32]

Could Cyber Warfare Be Strategic?

Wars are fought for a political purpose. From almost as soon as aircraft were developed, proponents of airpower argued, or hoped, that it would prove to be strategic—that is, have the capability of effecting political objectives by itself. Before the advent of precision-guided weapons, decades of practical experience largely discredited these advocates. Large-scale conventional bombing—including during World War II, the Korean War, and the Vietnam War—may have been called strategic, but this description can be applied accurately only to its scale, not to its effects.[33] To be sure, dumb bombs have been useful military tools on occasion, but with the probable exception of the atomic bombs dropped on Japan in 1945, they never proved decisive.

The development of PGMs has revived belief in the strategic value of standoff attacks—at least if one goes by the actions of technologically advanced states. The United States and its allies have largely relied on air-delivered PGMs and ship- and submarine-launched cruise missiles as their sole or primary military tools in multiple wars: Yugoslavia in 1999, Libya in 2011, and the conflict against the so-called Islamic State that has been waged in Iraq and Syria since 2014. Additionally some senior US military officers expressed hope, both publicly and privately, ahead of the 1991 Gulf War, that the air campaign would force Iraq to withdraw from Kuwait.[34] The tendency to want to count on standoff strikes is not exclusively an American one. Israel's 2006 war in Lebanon and Saudi Arabia's ongoing involvement in the civil war in Yemen also both started as standoff operations, with ground forces deployed only after PGMs proved ineffective at achieving political objectives.

Standoff operations may be extremely attractive to decision makers, but as these examples demonstrate, they have rarely been effective. The bombing of Yugoslavia in 1999 is the one indisputable success, although it was a close-run thing. Seventy-eight days of bombing were required—much more than originally anticipated—by which time the Coalition was close to collapse. Understanding the reasons why standoff strikes with PGMs have failed to achieve their goals casts light on the question of whether cyber weapons could prove to be strategic.[35]

There are two ideal-type strategies by which the employment of PGMs or cyber weapons could effect political change.[36] A *compellence* strategy seeks to inflict pain and demonstrate the willingness to inflict more with the aim of convincing an adversary to concede. A *denial* strategy, by contrast, seeks to weaken the military forces that an enemy is using to prosecute a conflict (and, perhaps, the enemy regime's grip on power). In the real world, these strategies can become indistinct. For example, attacks against a state's military-industrial sector can be justified as denial but may also have, intentionally or otherwise, a punitive effect on civilians. Such attacks are exemplified by both Allied and Axis bombing campaigns in World War II and by much more targeted strikes, such as those against Yugoslavia's electricity and water system in 1999.[37] Conversely, a denial strategy involving strikes against exclusively military targets would administer significant punishment if the adversary's leadership values its own grip on power and its military forces more than it does its citizens' lives and well-being.[38]

Almost by definition, a denial strategy cannot precipitate political change if only standoff weapons—whether kinetic or non-kinetic—are employed. Even if standoff strikes succeed in degrading an adversary's military capabilities, deployed forces are still required to capitalize on this weakness. In 2001, at the start of the Afghanistan war, for example, US airpower played a significant role in weakening Taliban forces, but an armed opposition with broadly equivalent fighting skills to the Taliban was still needed to take and hold territory in physical battles.[39] This opposition force took the form of the Northern Alliance, assisted by US special operations forces. Conversely, Saudi-led airstrikes against Yemen, which began in March 2015, failed to restore President Abdrabbuh Mansour Hadi to power after he had been deposed in a Houthi-led rebellion in large part because he lacked a ground force to take advantage of the strikes. Riyadh apparently hoped that the strikes would spark an anti-Houthi tribal uprising, but it did not occur.[40] As a result, foreign-trained Yemeni fighters were inserted into Yemen in May 2015 and followed by forces from Saudi Arabia and the United Arab Emirates in progressively larger numbers.[41]

Similarly, even if cyber attacks prove highly effective at disrupting an enemy's military operations, physical force will almost certainly be required to exploit this disruption. To be sure, the scenarios in which cyber attacks might prove useful could be very different from the Afghan or Yemeni scenarios since potential adversaries with cyber vulnerabilities range from non-state actors to sophisticated nation-states. But in all cases, success would surely demand a physical force in addition to a cyber force. In fact, against a sophisticated state, such as Russia or China, very considerable physical force might be needed as the state's military would probably remain formidable even after its networks had been compromised—and not least because, in such a conflict, US networks would probably be compromised too.[42]

A second issue is whether cyber weapons could be used to punish an adversary until it submitted. Much of the existing debate on this point revolves around essentially technical questions.[43] How plausible are cyber attacks against critical infrastructure? If such attacks did take place, would they cause large-scale death and long-lasting damage, or would their effects be less costly and more temporary? An even more fundamental question needs to be addressed: Even if cyber attacks against critical infrastructure were relatively easy and even if such attacks caused massive and long-lasting damage, would they actually be effective at compellence?

The history of punitive kinetic attacks demonstrates that, under some circumstances, states (and non-state actors) can withstand astonishing levels of punishment without conceding. To be sure, whether highly damaging cyber attacks were effective at compellence might well depend on what was at stake and the commitment of society at large to the cause. As the bombing of Yugoslavia in 1999 demonstrates, standoff operations can sometimes be effective in forcing one state to bend to another's will. But as the conventional bombing of British, German, and Japanese cities during World War II also illustrates, much greater levels of death and destruction can prove insufficient. Given that cyber weapons are unlikely to inflict costs on anything approaching that scale—even if the direst predictions about their destructive potential are realized—it should not be assumed that they would be effective tools for compellence.

Moreover, compellence may be even more difficult with cyber weapons than with kinetic weapons for at least one reason: compellence does not rely on inflicting pain per se but on the threat to keep doing so until an adversary concedes.[44] Meting out some punishment may well be necessary to make such a threat credible, but inflicting even high levels of pain may not establish credibility if the victim believes that the attacker is unwilling or unable to continue. This theoretical problem could become a real complication in a campaign of cyber compellence since, after the first wave of attacks, the victim might be able to take steps that would make further attacks much more difficult. Most obviously, the victim could analyze the virus (or viruses) that perpetrated the attacks and the means by which its IT systems were penetrated and use this information to patch vulnerabilities. Next, it could implement enhanced cybersecurity measures to reduce generic vulnerabilities, and it could try to "hack back" against the perpetrator to disrupt further attacks. Such steps would reduce the likelihood of compellence being successful. Again, however, there could be no guarantees. The time required to analyze the cyber weapon could be too long for the results to be useful in preventing further attacks.[45] Even if the analysis could be completed quickly, its utility might be limited if the attacker had developed multiple cyber weapons that all worked in different ways. Nonetheless, the basic point remains: compared to kinetic compellence, cyber compellence faces additional challenges.

To be sure, steps to enhance the cybersecurity of critical infrastructure are highly worthwhile. Although the repeated unsuccessful attempts at compellence with kinetic weapons suggest that cyber compellence might also prove unsuccessful, it still might be attempted. Meanwhile, some actors, including terrorists, may try to attack critical infrastructure for reasons other than compellence. Nonetheless, understanding the challenges of cyber compellence is useful in constructing more effective cyber defenses. Specifically, rapid response capabilities that enable a state to analyze cyber attacks on critical infrastructure quickly and use that information to prevent further attacks would be particularly useful in defeating attempts at compellence. Indeed, the US Department of Defense has recently stood up "National Mission Forces . . . [to] defend the United States and its interests against cyberattacks of significant consequence."[46] While the exact task of these forces is not publicly known, simply their existence might contribute to deterring attempts at compellence.

* * *

Focusing on the analogy between cyber weapons and PGMs risks giving the incorrect impression that the former are simply a new kind of the latter. They are not; further, they have many important differences, both obvious and subtle. Cyber weapons can often reach their targets effectively and instantaneously, though they can also be designed to have a delayed effect. Kinetic weapons generally travel much more slowly than cyber weapons, but if and when the former reach their targets, they usually have an almost instantaneous effect. PGMs are also limited by range, a concept without much meaning in cyberspace. Some cyber weapons can create reversible effects, whereas the effects of kinetic weapons are almost always irreversible.

More subtly, cyber vulnerabilities can usually be addressed relatively quickly. Thus, it is unlikely that a cyber weapon can be used repeatedly over the course of a multiday conflict without becoming obsolete. In fact, a cyber weapon might be effective only once. As a result, even if a state has stockpiled many different cyber weapons, it likely will face strong pressures to be highly selective in their employment. By contrast, while using a kinetic weapon certainly can provide an adversary with information that is useful in developing countermeasures, exploiting such information generally takes much longer than in the case of cyber weapons (the development of a new air defense system, for example, typically takes years). Therefore, advanced states can and do stockpile PGMs of a given type in large quantities and are increasingly using such weapons by default instead of dumb bombs. Indeed, as conflicts proceed, states tend to use ostensibly precise weapons in increasingly less selective ways, vitiating the putative special purpose of these weapons and depleting their stocks.

Another potential false impression is that based on the experience of PGMs, cyber weapons are unlikely to have significant implications for warfare. While it is still far too early to assess with any confidence exactly how military operations in cyberspace will change armed conflict, that such changes will be far-reaching seems entirely possible. Indeed, for all the limitations associated with PGMs, plenty of evidence shows that their development does represent a revolution in military affairs. These weapons are not usually able to achieve war aims by themselves, but they have altered leaders' calculations about the use of force and have thus altered national strategies. Moreover, because of the challenges associated with the effective employment of PGMs, the precision revolution is still incomplete. As states further develop ISR and BDA capabilities (and overcome other barriers), PGMs can be expected to become more potent at the tactical level and perhaps even at the strategic level too.

Similarly, the advent of cyber warfare will probably further lower the threshold for the use of force. Senior officials—at least in the United States—have said as much. In 2014, for example, Eric Rosenbach, then an assistant secretary of defense, stated, "The place where I think [cyber operations] will be most helpful to senior policymakers is what I call in 'the space between.' What is the space between? . . . You have diplomacy, economic sanctions . . . and then you have military action. In between there's this space, right? In cyber, there are a lot of things that you can do in that space between that can help us accomplish the national interest."[47]

Yet the analogy with PGMs suggests that the ability of states to employ cyber weapons effectively is likely to lag their desire to use them. In fact, it may take decades not only for states to understand the limitations of cyber weapons and whether and how these limitations can be overcome but also for the full implications of cyber warfare to become apparent.

Notes

1. For example, Andrew F. Krepinevich, "Cavalry to Computer: The Pattern of Military Revolutions," *The National Interest* 37 (Fall 1994): 30–42.

2. For example, Joseph S. Nye Jr., "Nuclear Lessons for Cyber Security?," *Strategic Studies Quarterly* 5, no. 4 (Winter 2011): 18, http://www.airuniversity.af.mil/Portals/10/SSQ/documents/Volume-05_Issue-4/Nye.pdf; and Kenneth Geers, *Strategic Cyber Security* (Tallinn, Estonia: NATO Cooperative Cyber Defence Centre of Excellence, June 2011), 112, https://ccdcoe.org/publications/books/Strategic_Cyber_Security_K_Geers.PDF.

3. This comparison has been discussed in, for example, Peter Dombrowski and Chris C. Demchak, "Cyber War, Cybered Conflict, and the Maritime Domain," *Naval War College Review* 67, no. 2 (Spring 2014): 85–87, https://www.usnwc.edu/getattachment/762be9d8-8bd1-4aaf-8e2f-c0d9574afec8/Cyber-War,-Cybered-Conflict,-and-the-Maritime-Doma.aspx; and Andrew F. Krepinevich, *Cyber Warfare: A "Nuclear Option"?* (Washington, DC: Center for Strategic and Budgetary Assessments, 2012), 7–12, http://csbaonline.org/uploads/documents/CSBA_e-reader_CyberWarfare.pdf.

4. For the purposes of this chapter, a *cyber weapon* is defined as a computer program designed to compromise the integrity or availability of data in an enemy's IT system for military purposes, and a *cyber attack* is defined as the use of a cyber weapon for offensive purposes. A cyber weapon may be used by itself or in concert with other weapons (kinetic or otherwise). Its effects may be felt purely in cyberspace or in physical space too. Cyber exploitation that compromises only the confidentiality of data is not considered a form of cyber attack.

5. Tim Maurer, "The Case for Cyberwarfare," *Foreign Policy*, October 19, 2011, http://foreignpolicy.com/2011/10/19/the-case-for-cyberwarfare/.

6. In technical jargon, the target location error should not be significantly larger than the weapon's circular error probable.

7. George Tenet, "DCI Statement on the Belgrade Chinese Embassy Bombing," testimony to the Permanent Select Committee on Intelligence of the US House of Representatives, July 22, 1999, https://www.cia.gov/news-information/speeches-testimony/1999/dci_speech_072299.html.

8. Barry D. Watts and Thomas A. Keaney, "Effects and Effectiveness," in *Gulf War Air Power Survey*, vol. 2 (Washington, DC: US Government Printing Office, 1993), pt. 2, 331–32, http://www.dtic.mil/dtic/tr/fulltext/u2/a279742.pdf.

9. Ibid., 340.

10. Uzi Rubin, *The Rocket Campaign against Israel during the 2006 Lebanon War*, Mideast Security and Policy Studies 71 (Ramat Gan: The Begin-Sadat Center for Strategic Studies, Bar-Ilan University, June 2007), 19–21, https://besacenter.org/mideast-security-and-policy-studies/the-rocket-campaign-against-israel-during-the-2006-lebanon-war-2-2/.

11. FireEye and Singtel, *Southeast Asia: An Evolving Cyber Threat Landscape*, FireEye Threat Intelligence (Milpitas, CA: FireEye, March 2015), 13, https://www.fireeye.com/content/dam/fireeye-www/current-threats/pdfs/rpt-southeast-asia-threat-landscape.pdf. In one attack, for example, one state's air force was targeted with "spear-phishing emails that referenced the country's military and regional maritime disputes . . . [and that] were designed to appear to originate from email accounts associated with other elements of the military." The report implies—but does not state explicitly—that China was responsible for the attacks.

12. Kim Zetter, *Countdown to Zero Day: Stuxnet and the Launch of the World's First Digital Weapon* (New York: Crown, 2014), 337–41; and Nicolas Falliere, Liam O. Murchu, and Eric Chien, *W32.Stuxnet Dossier*, Version 1.4 (Cupertino, CA: Symantec Security Response, February 2011), especially 7–11, https://www.symantec.com/content/en/us/enterprise/media/security_response/whitepapers/w32_stuxnet_dossier.pdf.

13. David E. Sanger, "Obama Order Sped Up Wave of Cyberattacks against Iran," *New York Times*, June 1, 2012, http://www.nytimes.com/2012/06/01/world/middleeast/obama-ordered-wave-of-cyberattacks-against-iran.html.

14. William A. Owens, Kenneth W. Dam, and Herbert S. Lin, eds., *Technology, Policy, Law, and Ethics Regarding U.S. Acquisition and Use of Cyber Attack Capabilities* (Washington, DC: National Academies Press, 2009), 118; and Krepinevich, *Cyber Warfare*, 43–44.

15. Ellen Nakashima, "U.S. Eyes Preemptive Cyber-Defense Strategy," *Washington Post*, August 29, 2010, http://www.washingtonpost.com/wp-dyn/content/article/2010/08/28/AR2010082803312.html.

16. One worrying possibility is that nations with fewer resources will focus on simpler, less discriminating cyber weapons that contain fewer safeguards against their spread. Not only are such weapons likely to cause much more collateral damage than more sophisticated cyber weapons would but also, unlike with PGMs, such damage might be felt far from the physical location of the target.

17. Dudu Mimran, "The Emergence of Polymorphic Cyber Defense," dudumimran.com (blog), February 10, 2015, http://www.dudumimran.com/2015/02/the-emergence-of-polymorphic-cyber-defense.html.

18. Morphisec, "What We Do," http://www.morphisec.com/what-we-do/. Interestingly, this technology is designed to "fit around" existing Windows-based operating systems.

19. "IAI Harpy," *Jane's Unmanned Aerial Vehicles and Targets*, September 30, 2015, www.ihs.com.

20. Existing cyber weapons may be able to execute preplanned attacks autonomously, but that is a much less stressing task.

21. Defense Advanced Research Projects Agency (DARPA), "Seven Teams Hack Their Way to the 2016 DARPA Cyber Grand Challenge Final Competition," July 8, 2015, http://www.darpa.mil/news-events/2015-07-08.

22. For example, Terrence McCoy, "Why Hamas Stores Its Weapons inside Hospitals, Mosques and Schools," *Washington Post*, July 31, 2014, https://www.washingtonpost.com/news/morning-mix/wp/2014/07/31/why-hamas-stores-its-weapons-inside-hospitals-mosques-and-schools/.

23. Michael Gervais, "Cyber Attacks and the Law of War," *Journal of Law & Cyber Warfare* 1, no. 1 (Winter 2012): 78–79.

24. Bill Marczak et al., "China's Great Canon," Research Brief (The Citizen Lab and Munk School of Global Affairs, University of Toronto, April 2015), 11, https://citizenlab .org/wp-content/uploads/2009/10/ChinasGreatCannon.pdf.

25. Watts and Keaney, "Effects and Effectiveness," 30–47.

26. US General Accounting Office, *Military Operations: Recent Campaigns Benefited from Improved Communications and Technology, but Barriers to Continued Progress Remain*, GAO-54-547 (Washington, DC: General Accounting Office, June 2004), 24, http://www.gao .gov/new.items/d04547.pdf.

27. Ibid, 23–24.

28. Mark Bowden, "The Hunt for 'Geronimo,'" *Vanity Fair*, October 12, 2012, http:// www.vanityfair.com/news/politics/2012/11/inside-osama-bin-laden-assassination-plot.

29. For rare examples, see Martin C. Libicki, *Conquest in Cyberspace: National Security and Information Warfare* (Cambridge: Cambridge University Press, 2007), 87–90; and Maj. Richard A. Martino, "Leveraging Traditional Battle Damage Assessment Procedures to Measure Effects from a Computer Network Attack" (graduate research project, Air Force Institute of Technology, Air University, June 2011), http://www.dtic.mil/cgi-bin/GetTR Doc?Location=U2&doc=GetTRDoc.pdf&AD=ADA544644. Much of the available literature on BDA focuses on defensive BDA, or a victim's assessment of the effects of a cyber attack on its own networks. This task is easier than an offensive BDA since the attacker has no guarantee of being able to maintain access to the victim's networks.

30. David A. Fulghum, Robert Wall, and Amy Butler, "Cyber-Combat's First Shot: Israel Shows Electronic Prowess: Attack on Syria Shows Israel Is Master of the High-Tech Battle," *Aviation Week & Space Technology* 167, no. 21 (November 26, 2007): 28–31.

31. Libicki, *Conquest in Cyberspace*, 88–90.

32. US Department of Defense, "The Department of Defense Cyber Strategy" (Washington, DC: Department of Defense, April 2015), 6, http://www.defense.gov/Portals/1 /features/2015/0415_cyber-strategy/Final_2015_DoD_CYBER_STRATEGY_for_web.pdf.

33. See, for example, Richard Overy, *The Bombers and the Bombed: Allied Air War over Europe, 1940-1945* (New York: Viking, 2014).

34. Watts and Keaney, "Effects and Effectiveness," 15, 341, 378.

35. For the explicit or, through analogy with nuclear weapons, implicit case that cyber weapons are strategic, see, for example, Geers, *Strategic Cyber Security*, 15; and Mike McConnell, "Mike McConnell on How to Win the Cyber-War We're Losing," *Washington Post*, February 28, 2010, http://www.washingtonpost.com/wp-dyn/content/article /2010/02/25/AR2010022502493.html. For similar claims in the Russian and Chinese military literature, see Krepinevich, *Cyber Warfare*, 3–4.

36. For a classic theoretical discussion, see Robert A. Pape, *Bombing to Win: Air Power and Coercion in War* (Ithaca: Cornell University Press, 1996), ch. 2.

37. Philip Bennett and Steve Coll, "NATO Warplanes Jolt Yugoslav Power Grid," *Washington Post*, May 25, 1999, https://www.washingtonpost.com/wp-srv/inatl/longterm /balkans/stories/belgrade052599.htm.

38. Within the nuclear deterrence literature, the term "aspects of state power" has been used to describe what dictatorial regimes are hypothesized to value. Michael Quinlan, *Thinking about Nuclear Weapons: Principles, Problems, Prospects* (Oxford: Oxford University Press, 2009), 126.

39. Stephen D. Biddle, "Allies, Airpower, and Modern Warfare: The Afghan Model in Afghanistan and Iraq," *International Security* 30, no. 3 (Winter 2005/2006): 161–76.

40. David B. Ottaway, "Saudi Arabia's Yemen War Unravels," *The National Interest*, May 11, 2015, http://nationalinterest.org/feature/saudi-arabias-yemen-war-unravels-12853.

41. Michael Knights and Alexandre Mello, "The Saudi-UAE War Effort in Yemen (Part 1): Operation Golden Arrow in Aden," Policywatch 2464 (Washington, DC: Washington Institute for Near East Policy, August 10, 2015), http://www.washingtoninstitute.org/policy-analysis/view/the-saudi-uae-war-effort-in-yemen-part-1-operation-golden-arrow-in-aden.

42. The question of how militaries would fare without their IT systems is discussed in Martin C. Libicki, "Why Cyber War Will Not and Should Not Have Its Grand Strategist," *Strategic Studies Quarterly* 8, no. 1 (Spring 2014): 29–30, http://www.dtic.mil/get-tr-doc/pdf?AD=ADA602105.

43. For various different perspectives in this debate, see Jon R. Lindsay, "Stuxnet and the Limits of Cyber Warfare," *Security Studies* 22, no. 3 (July 2013): esp. 385–97, 402–4; Krepinevich, *Cyber Warfare*, 39–65; and Richard A. Clarke and Robert K. Knake, *Cyber War: The Next National Security Threat and What to Do about It* (New York: HarperCollins, 2010), 64–68, 96–101.

44. Thomas C. Schelling, *Arms and Influence* (New Haven, CT: Yale University Press, 1966), 70.

45. The attacker could also try to issue a compellent threat before using cyber weapons. However, making a credible threat not only is difficult but also would alert the victim to the (possible) presence of a cyber weapon in its IT systems. Erik Gartzke, "The Myth of Cyberwar: Bringing War in Cyberspace Back Down to Earth," *International Security* 38, no. 2 (Fall 2013): 59.

46. US Department of Defense, *Department of Defense Cyber Strategy*, 6.

47. Quoted in Tim Maurer, "The Future of War: Cyber Is Expanding the Clausewitzian Spectrum of Conflict," *Foreign Policy*, November 13, 2014, http://foreignpolicy.com/2014/11/13/the-future-of-war-cyber-is-expanding-the-clausewitzian-spectrum-of-conflict/.

4 Cyber, Drones, and Secrecy

DAVID E. SANGER

The week before Barack Obama first took the oath of office as president of the United States in January 2009, he entered the Oval Office for one of those great traditions of American democracy, the moment when the president-elect meets with the sitting president for a candid conversation about the global realities he is about to face.

There was very little that Obama and George W. Bush liked about one another. Obama had been elected on a platform of extracting America from what he once called "dumb wars" of occupation. Bush, in Obama's thinking, had two traits that did not go well together—an absence of intellectual curiosity, which led him to take questionable intelligence that crossed his desk at face value, and an over-tendency to reach for military force when it came to exerting American power. To Obama the 2008 election had been a referendum on the wars in Iraq and Afghanistan. To Bush, Obama's election was symbolic of a nation that still could not face the hard lessons of the September 11, 2001, attacks; particularly, he felt America had to think differently about conducting an aggressive, anticipatory, and sometimes preemptive defense.

But in their meeting that day, Bush told his successor that, politics aside, there were two programs he would be foolish to abandon, because they could both protect the nation and potentially save his presidency. The drone program was, by that day in 2009, the least covert program in the arsenal of counterterrorism measures of the Central Intelligence Agency (CIA). The agency used armed, remotely piloted vehicles to wipe out small clusters of militants as they plotted against US forces in Pakistan and Afghanistan. The use of drones, while still limited, surged during the Bush administration. When the 9/11 attacks happened, drones had been entirely a surveillance platform, and their utility had been largely dismissed by the US Air Force, which never viewed unmanned aerial vehicles (UAVs) as real airplanes since they were not flown by pilots who put their lives at risk. Necessity changed this way of thinking. The Predator and ultimately its much larger cousin, the Reaper, were equipped with missiles that could wipe out a living room full of suspected terrorists and usually keep the house next door untouched. Use of the Predator was limited in Bush's early days in office, with the relatively small fleet divided between the Iraqi and the Afghan-Pakistani theaters. But by the time of Obama's inauguration, the planes were rolling off the production line.

The second program the two men discussed in that meeting was, in contrast, perhaps the most covert program in the US government and known only to a very tight group of aides. It was code-named Operation Olympic Games. Obama knew a bit about it from his security briefings as a candidate, but he had not been immersed in the details. He was about to be.[1]

Olympic Games was the first truly sophisticated offensive cyber operation in American history. Like armed drones, cyber weapons had been born of necessity—in this case, the need to slow the development of Iran's nuclear program. It had been the brainchild of a fledgling offensive cyber operation built within US Strategic Command, a unit better known for tending to America's strategic nuclear weapons, and of cyber operators at the National Security Agency. Later, the CIA would play a larger and larger role. But unlike with drones, the success of the Olympic Games operation was far more uncertain.

The first tests were promising. A "worm" was inserted into the computer controllers that commanded the operation of centrifuges inside the Natanz nuclear enrichment center. By 2009 the worm was already succeeding at speeding up and slowing down the supersonic machines, sending them spinning out of control. Best of all, the Iranians had not figured it out. Unlike the targets of drone operations, they did not know they were under attack, though some harbored suspicions. But the effects so far had been limited; without a doubt, they were less dramatic than a drone strike.

In his first year in office, Olympic Games became the vehicle for Obama's primer in America's nascent offensive cyber capabilities. From the same newly renovated Situation Room where he reviewed the details of the drone program, the new president examined, time and time again, a giant schematic of the Natanz nuclear enrichment plant spread out before him. A range of officials—intelligence officers, generals, lawyers—circulated in and out and explained how the most sophisticated new weapon in the American cyber arsenal, later dubbed Stuxnet, was being aimed at critical clusters of centrifuges in Iran's nuclear program. The operation had been developed in concert with Israel's Unit 8200; indeed, when details leaked in 2012, Israeli officials, while not publicly acknowledging the effort, privately took credit for its success. In fact, some Israelis contended that they were more responsible for the operation's development than were officials in Washington.[2]

Whatever the origin, Olympic Games was seen in Washington as an alternative to bombing Iran's nuclear facilities, and to the Americans it was a way of focusing Israel's energies on a project that promised to slow Iran's progress with a limited risk of triggering yet another Middle East war. And yet, as in the drone program, there was a deep sense of the experimental nature of the enterprise. Cyber weapons had been used before but largely in computer-on-computer attacks. No one could remember a case where an American president oversaw a cyber attack on physical infrastructure, hoping that it could accomplish what previously would have required bombs or saboteurs. As with the drone program, every debate about how and when to deploy America's cyber capabilities, and under what rules of engagement, seems to carve new territory.

Today as President Trump has taken over far more mature drone and cyber programs, clearly the two weapons raise similar moral and legal issues that future presidents will have to grapple with. Indeed, Pentagon officials openly wonder whether the next major global conflict might open in cyberspace and be prosecuted by a range of new, autonomous weapons—not only aerial unmanned vehicles but also undersea ones. And yet these drone and cyber weapons, nurtured by the same policymakers and sometimes used in the same conflicts, have taken very different paths—as have the questions surrounding their use.

The drone became President Obama's weapon of choice, as an alternative to sending troops into areas of the world that Obama feared would become quagmires. He used drones in Pakistan against al Qaeda, in the Horn of Africa against al-Shabaab militants, in Yemen against a variety of extremist groups, and in Iraq and Syria against the Islamic State. In most of those cases, the CIA conducted the attacks under Title 50 authority, which outlines the authorities for "covert action." Thus, the United States could not acknowledge the strikes, and any rules of engagement would also be kept secret. Obama, in a candid discussion with law students at the University of Chicago in April 2016, conceded that this did a "disservice" to the creation of legal and ethical standards for the strikes, something he said he was trying to repair.[3]

In those same comments, Obama acknowledged that in his first years in office he was concerned that the drone program had a weak legal basis. He had never suggested it at the time as the administration alternated between ducking questions about the program and simply defending its use.

"I think it's fair to say that in the first couple of years of my presidency, the architecture—legal architecture, administrative architecture, command structures—around how these [strikes] were utilized was underdeveloped relative to how fast the technology was moving," he said. He told the students that he pushed his staff to come up with a complex set of rules and legal strictures to make sure each strike comported to the rules of law—that is, that they amounted to a proportionate use of force, that civilian casualties were limited, and that the United States was not creating as many or more adversaries as it was killing.[4] As the program sped ahead and emerged from the shadows, Obama made an effort, one that only partially succeeded, to move more of the program out of secret intelligence channels and more into the hands of the US Air Force, where strikes and their adherence to law could be more openly debated. Near the end of Obama's presidency, National Security Adviser Susan Rice, in a speech at the US Air Force Academy in Colorado Springs, thanked the academy for turning out a new generation of combatants with an understanding of the rules that governed this new weapon and with an ability to handle the many stresses of using it.[5]

It is not clear that Obama and Rice could have given the same set of speeches about cyber weapons, which they discussed far less frequently. Indeed, though cyber weapons can usually be employed in a far less lethal manner than guided missiles shot from a drone, they have been deployed less frequently and under rules of engagement that are far less clear, at least to the public. Indeed, the rules for cyber weapons have seemed more difficult to develop, for the effects of a cyber strike are not as predictable as those for drone strikes.

These differences have given rise to one of the many oddities in the way governments and policymakers think about employing drones and employing cyber weapons. Since 2004, armed drones have been aimed at killing people, mostly suspected terrorists. Cyber weapons, in contrast, have been focused on an adversary's weapons systems, facilities, and equipment that can be repaired. One would think that given the difference in lethality, there would be more hesitance to use a drone than a cyber weapon. Oddly, exactly the opposite has been the case.

Why is that? Perhaps drones seem to be an extension of any other kinetic weapon, although they unleash more precisely targeted bombs and thus may seem a more humane way to execute a military or covert operation. But the answer may also have to do with the unpredictability of cyber attacks. The concerns about what could go wrong—the possibility that code could escape and wreak havoc on a broad swath of civilians while perhaps only temporarily disabling its intended target—almost paralyze cyber operators. The president had to intervene in the fall of 2015 to get the US Cyber Command to turn its digital guns on the country's number one terrorist enemy at the time, the Islamic State.

The Islamic State example is a telling one. Mr. Obama and his staff talked about those attacks because they took place in an environment that seemed closer to traditional war than to covert action. And in a war, the law of armed conflict frames the choices. Decision makers may be more cautious and more restrained than required under the law, but the rules of war create limits. US Cyber Command constantly holds meetings to assess whether the use of cyber weapons complies with the principles of necessity, discrimination, and proportionality—that is, the rules that would apply in the case of an armed attack.

But in peacetime, or in times of confrontation that have not yet turned to open conflict, calculating what types and levels of coercion are legitimate is far more difficult. This creates a political and strategic challenge as much as a legal and tactical one. Conducting attacks with drones or malware that seem shocking, callous, unfair, misdirected, or damaging to "bystanders" can easily sow backlash at home or inside friendly states. It can spur terrorist recruitment, and it can encourage retribution. The choices framed in this chapter assume that most cyber attacks, like most drone attacks, will take place in the gray area between declared state-on-state war and peacetime operations. In short, the decisions President Donald Trump will have to make will likely be similar to the vexing choices Mr. Obama faced on whether to use cyber weapons as an alternative to more traditional forms of low-level warfare, even while recognizing that, sooner or later, that may escalate the global use of cyber weapons. Only at the end of the Obama administration was it possible to start comparing the issues, and lessons, raised by using drones and cyber weapons. Much more analysis will likely be possible years from now, as details of the two programs are declassified or leaked, but now one can reach some tentative conclusions, rooted in the early histories of both programs.

The Light-Footprint Strategy

From the early years of the Obama administration, these two weapons—drones and cyber—became keystones of what is known as the "light-footprint" strategy. (The third element was the increased reliance on special operations forces.) Both drones and cyber weapons allowed for remote-control attacks, which are politically attractive in a country tired of casualties from two grinding ground wars. Both weapons were stealthy. Both were cheap, at least by the standards of a Pentagon burning through more than $600 billion a year. That combination soon made them as irresistible to Obama as they had been to his predecessor. The low cost meant that the president could experiment with them without needing a big appropriation from Congress. Moreover, they were swathed in secrecy, meaning, among other things, that it was still possible to hide mistakes.

They were also, as Obama would soon learn, capable of redefining a president's view of how to exercise military power. Obama reserved to himself the right to authorize the use of both weapons. By 2012 the president was personally overseeing the "kill list," the list of potential human targets for the drones. He was also overseeing the first use of cyber weapons and worrying about the precedents that he was setting for a new age of cyber conflict, one in which America would be vulnerable.

Yet how these two programs were viewed and employed would diverge during the Obama presidency. Drones became caught up in a political firestorm. At home it became clear that the "clean" kills of these precise weapons often involved considerable collateral damage, a fact the Obama administration went to some lengths to downplay. The remote-control nature of the killings raised moral and ethical questions, some voiced by former UAV pilots. And America's monopoly on the technology—and thus its ability to set some norms about their use—faded as other countries raced to match the US drone capability. Israel and China both have advanced programs; most others are still catching up.

Cyber attacks, because they are so stealthy and their effects often so hard to see, were treated more as a state secret—and were less politically charged. But the barrier to entry for other countries was low, and as Obama entered the twilight of his presidency, many sophisticated players existed. Olympic Games proved to be the start of a new era of state-sponsored cyber attacks. Seven years after Obama's meeting with President Bush, Russia, China, North Korea, and Iran had all launched state-sponsored attacks and had become far more skilled at cyber exploits and espionage.

The policy of secrecy surrounding the drone and cyber programs diverged as well. The administration wisely gave up its futile effort to avoid talking about drones; since each attack was reported in the media, the technology was not deniable. In the White House press room, where officials had been instructed by lawyers not to discuss drones at all, or at least not those operated by the CIA, the fiction that the weapon did not exist gradually faded. By 2013 the president himself had delivered a detailed speech about the legal basis for the drone program and had joked about the weapon on a late-night comedy show and at the 2010 White House Correspondents Dinner. His administration worked to develop what one aide called a regular order of procedures and considerations about using the unmanned, remotely piloted killing machines—and of special rules for when they were used, in rare cases, against an American citizen. The president, in short, was trying to get out ahead of the public debate, to define a set of rules for a new era of conflict, and to normalize use of the weapon for his successor.

But he could not—or at least did not—do the same for cyber. Other than acknowledging, almost in passing, that the United States makes use of offensive cyber weapons, he refused to discuss them in public, save for acknowledging in 2016 that cyber weapons were being employed against the Islamic State. (His last deputy secretary of defense, Robert Work, even went to some rhetorical lengths when he told reporters that the United States was "dropping cyberbombs" on the extremist group, an effort to analogize the imagery of kinetic activity to what appeared to be fairly standard efforts to disrupt the command-and-control mechanisms of the jihadist group.)[6]

When cyber capabilities were discussed, it was in the vaguest terms. The whole technology was still considered to be classified. To this day, Olympic Games has never been officially acknowledged, nor have the lessons of its successes and failures been publicly debated. The result is a stilted conversation that impedes debate about how, when, and under what authorities the United States could use cyber weapons. Worse yet, there is little public debate of how to develop a doctrine for both deterrence and offensive uses. The admiral who commands both the National Security Agency and US Cyber Command—the heart of the offensive cyber weapons effort—conceded in many public forums that the United States was still struggling to develop a theory of cyber deterrence that is analogous to the deterrence theory that surrounds nuclear weapons. In short, the United States found itself, at the time of this writing, still unable to do what it managed to accomplish in the nuclear age: keep the details of the weapon itself classified but hold a vibrant, unclassified debate about its use.

To understand the commonalities and the differences in American debates over how to use drones and cyber weapons, it is first necessary to consider how the weapons differ.

A Short History of Drones and the Myth of Perfect Aim

It did not take long for Obama to embrace his predecessor's advice to retain and accelerate the two secret programs.

For a new president seeking a way out of Iraq and debating a brief surge in Afghanistan, drones were particularly attractive. The more Obama learned about the technology, the more he turned to it. The first term of his administration saw about three hundred drone strikes in Pakistan alone, roughly a sixfold increase from the number during the entire Bush administration.

The increase was striking. But it did not become an issue in President Obama's re-election bid in 2012, and that alone is telling. Obama's liberal base preferred to ignore that the man they elected to change the course of American national security had embraced a vivid symbol of the Bush era. The right wing, which wanted to portray Obama as a "community organizer," did not want to cite evidence that he was using the drone as a killing machine more frequently than his predecessor had. Since it fit no one's preferred electoral narrative, it was rarely an issue.

My *New York Times* colleague Charlie Savage put the appeal of the UAVs quite well in *Power Wars*, his study of the legal struggles of the Obama administration to create a legal framework for these new weapons:

> Under Obama, remote-controlled aircraft were becoming the weapon of choice for strikes away from the traditional battlefields. In part this is because he had far more of them to deploy than Bush had had—the technology was brand new and it had taken time to ramp up production. But Obama was also enraptured by their potential for risk reduction. Conventional air power strikes put American pilots—and sometimes Special Operations spotters on the ground—at risk. By contrast, if a drone crashed or was shot down, its pilot still went home for dinner. They also enabled operators to watch a target for a long period before unleashing a missile, which held out the promise of greater precision and fewer civilian deaths.[7]

That set of qualities in the drone—its clandestine operation (with the attendant plausible deniability and limited oversight), casualty-free nature (for the attacker), persistence over the target, and apparently dead-true aim—was both attractive and deceiving. Persistence allowed a pilot with a joystick somewhere in the Nevada desert to watch a target for days, but it also offered false confidence that the trigger would be pulled at just the right time. Often it was not. American efforts to dispute the existence of collateral damage soon collapsed under further study.

The trouble became particularly acute after the Bush administration authorized "signature strikes" in 2008. No longer would drone operators have to hunt an identified individual terrorist or militant. Instead, a motorcade that appeared to be carrying a group of Taliban or al Qaeda militants became a legitimate target, even if there was no great certainty about who was inside the cars.[8] So would a gathering in which a terror suspect met with colleagues, even if the latter were unknown to the drone operator. When my colleague Eric Schmitt and I first reported on this change in policy, in which "patterns of life movement" made for new target sets, our story explained why the pace of strikes rose so quickly during the last months of the Bush administration and at the beginning of the Obama administration: the weapon had moved from one used for "targeted killing" to one used for protecting the US military.

But the wider use of signature strikes created problems similar to the attribution problem that plagues the investigation of cyber attacks. By definition, the intelligence surrounding a signature strike is imprecise. That motorcade could be full of al Qaeda terrorists. Or it could be full of teenagers or guests for a wedding. In the cyber realm, it is difficult to imagine that the president would allow a strike against, say, a Chinese computer server simply because an attack on the Pentagon looked typical of Chinese behavior. It could, in fact, be an elaborate ploy. Yet in the drone world, the United States was essentially doing exactly that—striking cars that moved in ways "typical" of al Qaeda. No wonder Obama expressed misgivings about the rules of engagement.

Signature strikes quickly raised questions about the standards of intelligence used to authorize an attack. Precisely because the identities of the targeted individuals were unknown, the success of the missions became far harder to measure and often far more dubious. Collateral damage soared. By one outside estimate, 482 drone strikes in Pakistan, Afghanistan, and Yemen resulted in around 289 civilian casualties.[9] And even after a strike the CIA oftentimes had no idea who some of the occupants in the motorcade or a meeting actually were.

This posed both a statistical and a political problem: suddenly the precision killing tool looked less precise. The solution was a piece of statistical innovation. Soon all males older than grade-school age who were killed in a strike were counted as "presumed combatants," even if their identities were unknown.[10] As intended, this vastly lowered the collateral damage statistics. In congressional oversight hearings and in speeches, the UAV was once again described as an instrument of remarkable wartime precision.[11]

Of course, in places like Pakistan the level of resentment against drones soared. Pakistani media often exaggerated the collateral damage and failed to report the killing of true militants and terrorists. At other times, misjudgment by American operatives led to tragedy. Local tribesman told stories of an attack on a *jirga* (assembly) at which some Taliban members came to act as mediators: when the drone-launched missile struck, the Taliban died, and so did many of the tribal elders. Over time the drone became the symbol of an aggressive, heartless America, its image sketched out in markets in Peshawar. Its constant overhead buzz, with its suggestion of imminent death and wreckage, became associated with a distant power that did not know how to control the weapon in its hands.

One such strike resulted in a considerable tightening of the targeting roles. After a minor rebellion in the Obama war cabinet, led by Secretary of State Hillary Clinton and Chairman of the Joint Chiefs Adm. Michael Mullen, the administration concluded that the signature strike standard was simply too loose.[12] American ambassadors were given far more authority to sign off on—or halt—a strike in "their" territory, an imposition of State Department control that quickly resulted in serious conflicts with the CIA and the Pentagon. The latter agencies were not accustomed, or very happy, with the idea of career foreign service officers limiting their ability to pull the trigger.

Over time, Obama himself became deeply involved in debating what was a legitimate target and what was not, a subject that appealed to his legal training but also left him feeling deeply uneasy. My colleagues Scott Shane and Jo Becker captured this in a story about the president's direct role in approving the kill list, or those targeted for assassination from the air.

> Mr. Obama is the liberal law professor who campaigned against the Iraq war and torture, and then insisted on approving every new name on an expanding "kill list," poring over terrorist suspects' biographies on what one official calls the macabre "baseball cards" of an unconventional war. When a rare opportunity for a drone strike at a top terrorist arises—but his family is with him—it is the president who has reserved to himself the final moral calculation.
>
> "He is determined that he will make these decisions about how far and wide these operations will go," said Thomas E. Donilon, his national security adviser. "His view is that he's responsible for the position of the United States in the world." He added, "He's determined to keep the tether pretty short."[13]

In an interview late in his presidency, Obama reflected on how American leaders found themselves making these life-and-death decisions from conference rooms in Washington. It started, he said, with the best of intentions after 9/11. Referring to both the military and the CIA, he noted that "they started just going because the goal was let's get al Qaeda, let's get these leaders. There's a training camp here. There's a high-value target there. Let's move."

Obama then went on to offer the most detailed explanation any American president has given of the kind of structure he tried to impose around the rules of engagement for using of drones:

> Given the remoteness of these weapons and their lethality, we've got to come up with a structure that governs how we're approaching it. And that's what we've done. So I've put forward what's called a presidential directive. It's basically a set of administrative guidelines whereby these weapons are being used.
>
> Now, we actually did put forward a non-classified version of what those directives look like. And it says that you can't use these weapons unless you have near certainty that there will not be civilian casualties; that you have near certainty that the targets you are hitting are, in fact, terrorist organizations that are intending to do imminent harm to the United States. And you've got all the agencies who are involved in that process, they have to get together and approve that. And it goes to the highest, most senior levels of our government in order for us to make those decisions.
>
> And what I've also said that we need to start creating a process whereby this—whereby public accountability is introduced so that you or citizens or

members of Congress outside of the Intelligence Committee can look at the facts and see whether or not we're abiding by what we say are these norms.

And we're actually—there's a lot of legal aspects to this because part of the problem here is, is that this drone program initially came through the intelligence side under classified programs, as opposed to the military. Part of what I've also said is I don't want our intelligence agencies being a paramilitary organization. That's not their function. As much as possible this should be done through our Defense Department so that we can report, here's what we did, here's why we did it, here's our assessment of what happened.

And so slowly we are pushing it in that direction. My hope is, is that by the time I leave office there is not only an internal structure in place that governs these standards that we've set, but there is also an institutionalized process whereby the actions that the U.S. government takes through drone technology are consistently reported on, on an annualized basis so that people can look.

And the reason this is really important to me—and this was implied in your question—is there is a lot of misinformation about this. There is no doubt—and I said this in an interview I think recently—there is no doubt that some innocent people have been killed by drone strikes. It is not true that it has been this sort of willy-nilly, let's bomb a village. That is not how folks have operated. And what I can say with great certainty is that the rate of civilian casualties in any drone operation are far lower than the rate of civilian casualties that occur in conventional war.[14]

Does Obama's Drone Standard Work for Cyber Strikes?

Obama's explication of how he tried to develop rules for the use of drones is worth considering as parallel questions are addressed on the use of cyber weapons.

The initial parallels are striking. In the debate over Olympic Games, Obama showed a similar concern over making sure "you can't use these weapons unless you have near certainty that there will not be civilian casualties." He wanted to ensure that local Iranian hospitals, for example, were not taken out along with the Natanz centrifuges. He succeeded in avoiding such a scenario in that case, but in many other cases where cyber will be most useful—for example, taking out a power grid or a cell phone network—the ultimate effects are unpredictable and could well result in the loss of civilian lives.

In cyber attacks, the decision-making also "goes to the highest, most senior levels of government." Under existing directives, only the president can authorize offensive cyber strikes in both peacetime and wartime. Of course, reality is more complicated. Is a preventive or preemptive effort to block an attack on the United States—say, by taking out a server in China or North Korea—an "offensive" strike? Or is it simply "active defense," even if it looks to the target country like an act of offense?

But one of Obama's standards—that the "targets you are hitting are, in fact, terrorist organizations that are intending to do imminent harm to the United States"—does not fit at all. That criterion clearly did not fit the Iran case. Its government might be a state sponsor of terror, but it was not a terrorist organization. Its drive to produce a nuclear weapon was a threat to the United States but not an imminent one, for it was months, if not years, away. (The cyber attacks on the Islamic State, of course, were far more in line with the Obama test for drones.)

In fact, the drone standard laid out by Obama may be of only limited use when applied to cyber conflict. And it is worth remembering that cyber capabilities have some unique advantages over drones:

- Cyber strikes can be dialed up and dialed back; thus, the level of damage is, theoretically, much easier to control.
- Most cyber strikes do not necessarily lead to fatalities.
- If the strike is well hidden, it is entirely possible that the same target can be hit repeatedly, giving the president the option of starting with a small attack and increasing the level as needed.
- Not only is a cyber strike more plausibly deniable than a drone strike, it is possible to make it look as if someone other than the true attacker is responsible. For that reason, the levels of transparency that President Obama ultimately sought to achieve for US drone operations would likely be resisted by both the Pentagon and the intelligence community.

This last difference is worth a bit more consideration. The availability of outright secrecy surrounding cyber strikes—or, at a minimum, plausible deniability—may create an incentive particularly for intelligence agencies to use the weapon more aggressively. So far there is no way to measure that accurately from the outside, as no records of how many cyber strikes take place (and definitions of a "strike" would likely vary) and no discussion of what percentage of proposed strikes are actually executed are made public. Some anecdotal evidence suggests a reluctance to conduct cyber strikes because the code could act in unpredictable ways and perhaps reveal the identity of the country that launched it. It is a legitimate fear, for that is exactly what happened in the case of Olympic Games.

In short, one significant disadvantage of using cyber weapons is that the results can be significantly more unpredictable than many advocates of the technology admit. The immediate impact is difficult to assess because so much depends on the inner workings of the target and the quality of the intelligence about that target. Further, its long-term consequences are almost impossible to anticipate. Code mutates. It is cut up into smaller pieces and used for other purposes. The list of what can go wrong—or at least of what is unexpected—is a long one.

Moreover, the experience with drones has made some in Washington worry about the international reaction if the United States is linked to an attack on a state. Drones fostered the image of the United States as an uncaring superpower using its technological edge to rule the world; cyber attacks could make it look callous should they affect a power grid or a hospital. (American officials found it easier to boast about cyber attacks on the Islamic State for a few reasons: the target group is widely reviled, and because it has no real infrastructure of its own, apart from stolen oil rigs, an attack carried little risk of worsening the lives of ordinary Iraqis and Syrians.)

It is not clear whether the ability to cloak the identity of the cyber attacker encourages nations to conduct attacks or whether the fear of revelation acts as a brake. The latter appears to have held sway, according to anecdotal evidence, in some cases when the US Cyber Command drew up plans for possible attacks. Doubtless the calculus is different in every case. As one senior American official told the author, "If we are aiming at a state, with all the questions of violating sovereignty, the standards for assuring you can act covertly are much higher."

Geography Still Matters

In the run-up to the Iraq War in 2002, George W. Bush gave a set-piece speech in the grand hall of the Cincinnati train station, warning of the evils Saddam Hussein could impose on the United States.[15] Many shaded truths in the speech were worthy of question, especially in retrospect. But even at the time, one section appeared particularly fanciful. Bush raised the specter that the Iraqi leader would float a ship off the coast of the United States and launch an attack with a secret fleet of UAVs armed with chemical or biological weapons that could be dropped on American cities. The idea never quite garnered the ridicule it deserved, for the ship would have been a sitting duck for the US Coast Guard or the navy, and biological and chemical weapons are hard to disperse effectively. But the imagery of a flotilla of drones illustrates the first significant difference between drones and cyber: with drones, geography still matters, as it limits the weapon's effectiveness.

Fourteen years after that speech was delivered, it is still hard to imagine how a foreign power could launch such an attack. A base near the United States would be necessary, and, of course, hiding such an air base would be difficult. Thus, the Bush administration needed to conjure up a ship-based solution to the problem.

Cyber capabilities, in contrast, defy all geographic limits. Had he possessed the coding talent, Saddam Hussein would have been able to launch cyber attacks from one of his garish palaces. (He never seemed interested, but perhaps he would have been if he had survived another decade.) Moreover, cyber attacks are much easier to hide than a fleet of UAVs. Attacks can emanate from a People's Liberation Army tower in Shanghai, from a cell of North Koreans operating in Thailand, from a bot reproducing itself on Grandma Smith's home computer in Des Moines, or from a university in the American South. In fact, each of those places has been the source of attacks seen in the United States since 2012.

Certainly, the ability to detect such cyber threats—addressing the "attribution problem"—has come a long way in recent years. Walking into the Department of Homeland Security's giant, space-age-looking command post in suburban Virginia, it looks like a flight control center, with screens monitoring levels of Internet activity and malware across the nation. In the case of the North Korean attack on Sony Pictures Entertainment in late 2014, it took US intelligence agencies only a few days to conclude with high certainty who the attackers were. (That process was aided by the fact that the United States had pierced North Korea's networks, for other reasons, four years before and could detect post facto evidence of the attack in its systems.) Figuring out the source of attacks on banks that use the SWIFT banking system took longer although some of the code appeared to be very similar to what was used against Sony. In that case, the US government was more cautious. It was all a reminder that in the cyber realm, attribution remains as much art as science.

For now cyber can create a much larger swath of destruction than can drones, though the effects of a cyber attack are more reversible, or at least fixable, than those of a drone strike. That may not remain true for long. Eventually drones may be capable of being armed with weapons of mass destruction of some form, but the very thing that makes drones stealthy—their ability to fly "under the radar"—also currently makes them too small to carry much punch. So today they pose a direct threat to small clusters of terrorists and to commercial aviation. Their ability to create mass casualties is limited. In short, drones are a containable problem—or at least a manageable one. For a foreign state, they are not a weapon of choice: their range is too limited, and over time the chances of discovery are high. For a terrorist seeking to do high-publicity damage, it is cheaper and easier to mount an automatic weapons attack.

Cyber capabilities, in many ways, pose the opposite problem. The oceans and wide spaces that give America protection from an overwhelming attack of UAVs offer no such protection in the cyber realm. From a keyboard in Moscow or Shanghai, Pyongyang or Tehran, the world is borderless, and information travels nearly instantaneously. No local crews are needed to maintain the weapon or refuel it. If adjustments need to be made to the weapon, the work can be done from half a world away. As one cyber warrior put it, "With cyber you don't have a Djibouti problem"—a reference to the problem of negotiating, then maintaining, a launch base in a faraway nation. All cyber war can be distant, yet its effects can be local.

Even if an attacker is identified with high certainty, however, the barriers to taking preemptive defensive action remain extremely high and even far higher, at least for now, than if the United States saw a physical attack massing on the border. Consider these scenarios: If North Korea were to launch a missile in the general direction of the West Coast of the United States, there is little question that the United States would try to shoot it down. (In fact, Defense Secretary Donald Rumsfeld ordered Pacific Command to be ready to do just that when the North Koreans were threatening a launch in 2006.[16]) If Mexico sent an armed drone over the border, undoubtedly the United States would again try to shoot it down. But what if the National Security Agency saw a Sony-like attack massing in a North Korean server or in the computers of a Mexican drug lord? It is far from clear what the US response would be. The Defense Department doctrine on cyber activity states that in certain cases an attack would merit a national response, but the threshold is understandably vague. So is the nature of the response. Would the United States simply block the attack—something companies and virus protection software do every minute of every day? Or would it seek to take out the source, which could be a server, a den of hackers, or a military cyber unit?

The legal debate in the United States about what constitutes a threat worthy of preemption goes back to the missives exchanged by Daniel Webster and his British counterpart after a half-baked scheme to attack Canada resulted in Britain's burning the steamboat *Caroline* in 1837.[17] But the hesitance to, say, fry a server in Shanghai where malware is being assembled to attack an American company raises a new level of complexity. While the armed drone coming over the border is an obvious threat, the code flowing across a fiber-optic cable beneath the sea is not so obvious. The Chinese attacker would doubtless say that the code he sent around the world was benign, and it may, at first glance, appear that way. Proving the code was Chinese, sanctioned by the Chinese government, contained malware, and was intended to do great harm (rather than merely facilitate espionage) would be enormously difficult and likely pit one group of experts against another.

Indeed, it already has. The opening attack on Sony Pictures was an implant that surveyed the company's computer systems, probing its vulnerabilities. For months it acted more like an unarmed drone, interested only in espionage, than an armed one. But once in place and after it had surveyed Sony's weaknesses, the malware turned to another purpose—an attack. In short, it morphed from benign to deadly at the flip of a switch.

No one saw that coming, least of all Sony's leadership. But one senior US intelligence officer told me that "even if we had known in advance what the code could do, it's not clear we would have struck back." It is unclear whether he thought President Obama acted cautiously because of the absence of a smoking gun to prove North Korea culpable or because the administration debated and doubted how aggressively the US government should retaliate for an attack on what was a private entity rather than a public one.

Moreover, when the United States publicly declared North Korea was the culprit, many experts responded it was wrong.[18] The attack wasn't from the North, they said, but from a group of hackers pretending to be under the command of North Koreans. Others said Sony was the victim of hacktivists or teenagers.

It is not surprising that the Pentagon is spending so much time thinking about asymmetric responses to cyber attacks; this retaliation may be of an entirely different nature than counterattacking in the cyber realm. In the case of China, US retaliation came in the form of indicting officers from Unit 61398 of the People's Liberation Army. While it made Justice Department officials feel good, whether it made much difference in Chinese thinking about future attacks is debatable. Some believe that the public shaming chastened President Xi Jinping and other Chinese officials and forced them into an agreement with the United States. Others think that agreement was largely for show.

In the North Korea–Sony Pictures case, modest sanctions were imposed; it is unclear whether covert action was taken to disable North Korea's limited number of Internet protocol addresses. In other cyber attacks—the Iranian attacks against Saudi Aramco (2012), the Sands Casino and American banks (2014), and even the New York dam (2013)—there has been no noticeable response or at least no public one.[19]

Then another debate faced the Obama administration in developing cyber weaponry: what to do about collateral damage?

In approving the Olympic Games attacks, President Obama asked detailed questions about whether the computer worm could hit a hospital or a school, or whether it could end up taking out the local power grid. He was assured it could not, and there is no evidence of collateral damage resulting from the strikes themselves.

But another unanticipated form of that attack's damage came later. When the Stuxnet worm leaked out, the whole world could see the code that created it. Moreover, it could see the "modules" of the code itself. Many of these modules have served as the building blocks of other weapons, which other people have used in other attacks. One security researcher recently told the author that five years later, "elements of the Stuxnet worm are still being used in other malware." The weapon itself wasn't reproduced. But its parts were.

In short, drones and cyber weapons pose different kinds of collateral damage challenges. In the case of drones, there is no doubt where the strike will occur; the only question is whether civilians in the area may also be killed. In the case of cyber attacks, the question is whether the weapon can be contained days, weeks, or months after its launch. It does not expire after the first contact with the target, and it may take many forms, and reach many places, that its designers never intended. As the Stuxnet attacks showed, the United States and Israel went to some lengths to prevent collateral damage by having the code expire after a set date. Such efforts can mitigate the damage. But even today, elements of that attack live on, repurposed for different kinds of attacks.

The Secrecy Conundrum

One additional conundrum arises when comparing the drone program to the cyber program—namely, secrecy.

As noted earlier, whatever secrecy surrounded the drone program eroded quickly, and by 2013 the president spoke openly about the limits he wanted to place on the weapon's use. Whether one agrees with those limits—or with the concept of moving more drones to the Pentagon so that missions must be accounted for publicly rather than be hidden by the CIA—the doctrine can now be openly debated.

Not so with cyber capabilities, at least not yet. The Pentagon has published some policy and standards, but all the hard questions—including the conditions under which to use the weapons, as well as the kinds of targets that are considered legitimate—have been avoided in public.

Clearly, a centrifuge facility is on the list of legitimate targets, and so would be a launch facility. But what about a central bank, given the enormous vulnerability of the United States to financial disruption? And if a central back could be targeted, what purpose would be deemed permissible: to gain private information that could be publicized, to deny service for some time, or to corrupt the integrity of financial data? Each type of attack would carry different implications for the object of the attack and for the international standard that it could set. Would an enemy nation's utility grid be an acceptable target, given the vulnerability of our own grid to counterattack? This question does not arise as often in the context of UAVs, which could take out a power station but not an entire electrical network.

And who gets to retaliate? While many Americans now own drones, they are almost all—thankfully—unarmed. Developing malware has fewer barriers. Does it make sense to keep the current US legal ban on "hacking back"? If so, does this prohibition harm the nation's ability to build robust cyber defenses and create a deterrent? Could other forms of active defense, such as inserting beacons to trace the location of stolen data, be developed that would be less risky?

And when would the government intervene to use its own power? If there was a drone attack on New York City, undoubtedly the Air National Guard would step in to protect the air space. But cyberspace is different. Former defense secretary Ashton Carter said he could imagine a federal response in only about 2 percent of all cyber attacks; given the huge number of cyber attacks that take place daily on government and civilian targets, that percentage seems high. So far, only a handful of publicly known cases have merited any significant federal response.

Sparking the Debate

There is no reason these questions cannot be debated by the American people. From 1945 through the Cold War, almost everything about nuclear weapons was classified top secret or above. Yet the country managed to debate nuclear doctrine in the open, determining a set of rules over when the United States would use them. That debate ended in a very different place than it began. The world gained confidence in the US government's ability to control nuclear weapons because of that debate. And the United States set standards that others now follow.

The public is now having a similar debate about drones, though a vigorous press had to help drag the government to that point. Yet in the cyber arena, similar discussions have only just begun. US Cyber Command has started to discuss publicly, in the most halting way, the issues confronting the use of cyber weapons. The fear of revealing the size and scope of the US investment in cyber capabilities, however, has frozen many of the most important discussions. Sooner or later that debate will be opened up just as it was in the nuclear era and just as it has, more recently, on the issue of drones. Such debates have proved critical in defining the use of state-of-the-art technologies that can be turned into new kinds of weaponry. Such discussions will be vital in the cyber realm.

Notes

David E. Sanger, a national security correspondent for the *New York Times*, is an adjunct lecturer at the Kennedy School of Government at Harvard University and a senior fellow at the school's Belfer Center for Science and International Affairs. The views expressed in this chapter are his own.

1. The author has written extensively about the involvement of Presidents Bush and Obama in the Olympic Games program. See David E. Sanger, *Confront and Conceal: Obama's Secret Wars and Surprising Use of American Power* (New York: Crown, 2012), prologue, ch. 8; and David E. Sanger, "Obama Order Sped Up Wave of Cyberattacks against Iran," *New York Times*, June 1, 2012, 1.

2. See David E. Sanger, "A Spymaster Who Saw Cyberattacks as Israel's Best Weapon against Iran," *New York Times*, March 22, 2016. The late Meir Dagan, who was the director of the Mossad and a key player in the Israeli side of Olympic Games, chastised the author for not giving Israel enough credit for its role in the operation.

3. See transcript of Obama's talk. Office of Press Secretary, "Remarks by the President in a Conversation on the Supreme Court Nomination," University of Chicago Law School, April 8, 2016, https://www.whitehouse.gov/the-press-office/2016/04/08/remarks-president-conversation-supreme-court-nomination.

4. Ibid.

5. Susan E. Rice, "The Global Campaign against ISIL—Partnerships, Progress, and the Path Ahead," remarks as prepared for delivery at the US Air Force Academy, Colorado Springs, April 14, 2016, https://www.whitehouse.gov/the-press-office/2016/04/14/remarks-national-security-advisor-susan-e-rice-us-air-force-academy.

6. Both Mr. Obama's and Mr. Work's comments were widely reported in April 2016. For example, see David E. Sanger, "U.S. Cyberattacks Target ISIS in a New Line of Combat," *New York Times*, April 24, 2016, http://www.nytimes.com/2016/04/25/us/politics/us-directs-cyberweapons-at-isis-for-first-time.html?_r=0.

7. Charlie Savage, *Power Wars: Inside Obama's Post-9/11 Presidency* (New York: Little, Brown, 2015).

8. Eric Schmitt and David E. Sanger, "Pakistan Shift Could Curtail Drone Strikes," *New York Times*, February 22, 2008.

9. Micah Zenko and Amelia Mae Wolf, "Drones Kill More Civilians Than Pilots Do," *Foreign Policy*, April 25, 2016.

10. Jo Becker and Scott Shane, "Secret 'Kill List' Proves a Test of Obama's Principles and Will," *New York Times*, May 29, 2012, http://www.nytimes.com/2012/05/29/world/obamas-leadership-in-war-on-al-qaeda.html?pagewanted=9&_r=1&hp&adxnnlx=1338289213-gFazCDrgzwY2RtQCER9fGQ&pagewanted=all.

11. See Steve Coll, "The Unblinking Stare," *New Yorker*, November 24, 2014, http://www.newyorker.com/magazine/2014/11/24/unblinking-stare.

12. Adam Entous, Siobhan Gorman, and Julian Barnes, "U.S. Tightens Drone Rules," *Wall Street Journal*, November 4, 2011.

13. Ibid.

14. Office of Press Secretary, "Remarks by the President."

15. David E. Sanger, "Bush Sees 'Urgent Duty' to Pre-empt Attack by Iraq," *New York Times*, October 8, 2002.

16. For further discussion of the US consideration of shooting down a Taepodong missile over the Pacific, see my account in *The Inheritance: The World Obama Confronts and the Challenges to American Power* (New York: Three Rivers Press, 2010), 322–23.

17. I discussed the incident, and the difference between preemption and preventive war, in David E. Sanger, "Beating Them to the Prewar," *New York Times*, September 28, 2002.

18. David E. Sanger, Michael S. Schmidt, and Nicole Perlroth, "Obama Vows a Response to Cyberattack on Sony," *New York Times*, December 19, 2014.

19. The Bowman Avenue Dam hack, allegedly carried out by a group of Iranian hackers, took place in Rye Brook, New York, in March 2013. Authorities alleged that seven Iranian hackers penetrated the computer-guided controls of the dam, which was under repair and offline at the time. The Manhattan US Attorney indicted the hackers in March 2016 for both the dam hack and a series of cyber attacks on major US financial institutions. For more information, please see Joseph Berger, "A Dam, Small and Unsung, Is Caught Up in an Iranian Hacking Case," *New York Times*, March 25, 2016, http://www.nytimes.com/2016/03/26/nyregion/rye-brook-dam-caught-in-computer-hacking-case.html.

PART II

What Might Cyber Wars Be Like?

5 Cyber War and Information War à la Russe

STEPHEN BLANK

Originally this chapter was to explore an analogy between cyber warfare and Russia's traditional conception and practice of information warfare (IW). However, an examination of Russian strategy, argumentation, and practice in Estonia, Georgia, and Ukraine between 2006 and 2016 demonstrates that the relationship between cyber warfare and IW is not analogous but rather something more. Russia has integrated cyber and information warfare organically into its planning and capabilities to project power. As the US director of National Intelligence Gen. James Clapper testified in 2015, before the cyber attack on the Ukrainian electricity sector, Russia was establishing a cyber command to conduct

> offensive cyber activities, including propaganda operations and inserting malware into enemy command and control systems. Russia's armed forces are also establishing a specialized branch for computer network operations.
>
> Computer security studies assert that unspecified Russian cyber actors are developing means to access industrial control systems (ICS) remotely. These systems manage critical infrastructures such as electric power grids, urban mass transit systems, air traffic control, and oil and gas distribution networks. These unspecified Russian actors have successfully compromised the product supply chains of three ICS vendors so that customers download exploitative malware directly from the vendors' websites along with routine software updates, according to private sector cybersecurity experts.[1]

Clearly, Russian national security agencies are preparing for contingencies in the cyber domain as much as their counterparts in the United States, China, Israel, the United Kingdom, France, and other states are. What may be distinctive in Russia, as the examples presented in this chapter suggest, is the conception of cyber attacks as an organic element of a long-standing approach to political warfare and information operations (IO).

In Russian discussions and practice, distinguishing cyber war from IO is virtually impossible.[2] For Moscow they both come under the heading of attributes of information confrontation (Informatsionoye protivoborstvo [IP]), or IW, and are to be fully integrated in any campaign with military operations.[3] Beginning with Chechnya in 1999–2000 and through the conflicts in Estonia, Georgia, and Ukraine, Moscow has systematically employed its concepts of IW.[4] As is discussed in the following sections, the December 2015 malware assault that shut down several Ukrainian electricity transmission facilities was the most vivid example.

Russian conduct of both IW and cyber war builds on earlier foundations of what George Kennan called political warfare.

> Political warfare is the logical application of Clausewitz's doctrine in time of peace. In broadest definition, political warfare is the employment of all the means at a nation's command, short of war, to achieve its national objectives. Such operations are both overt and covert. They range from such overt actions as political alliances, economic measures (as ERP [European Recovery Plan]—the Marshall Plan), and "white" propaganda to such covert operations as clandestine support of "friendly" foreign elements, "black" psychological warfare and even encouragement of underground resistance in hostile states.[5]

Tactics and strategies developed and employed during the Soviet period have served as a foundation for establishing new strategies that incorporate some of the century-old Leninist repertoire and new trends like IW, as defined by Moscow, for the conduct of continuous political warfare against hostile targets. Although some Russian and foreign observers use new terms such as "nonlinear" or "new generation" warfare to describe Russia's practice and often say they merely mimic techniques used by the United States to interfere in other states, the IW that Russia conducts today follows the logic of past Soviet and Russian political warfare.

This chapter sketches those historical patterns and explores how newer forms of cyber operations fit into them, drawing on the experiences of Estonia in 2007, Georgia in 2008, and Ukraine in 2014–15.[6] For Russia, cyber operations may represent new forms of military operations, but they have grown organically out of Soviet thinking and tactics about political warfare.[7] This observation raises, among other things, the question of whether Russia regards and treats cyber capabilities differently than do other states and, if so, how these differences might be managed if not reconciled. It also may suggest that states will tend to utilize new technologies, including cyber, according to familiar strategies, cultural, and institutional predilections.

Russia's Permanent Siege Mentality

Russian national security policy begins with the perception that Russia lives in a constant state of siege that includes intelligence operations and the overall national security challenges posed by adversaries that are led by the United States. As Christopher Andrew and Vasili Mitrokhin wrote about the Soviet regime's abiding mind-set, "All authoritarian regimes, since they regard opposition as fundamentally illegitimate, tend to see their opponents engaged in sub-versive conspiracy." President Vladimir Putin and his associates, like their forebears, have frequently expressed their belief that the conspiracies directed against them are mainly foreign in origin.[8]

In Russia, there is no hard-and-fast distinction between peace and war as there is in American strategic thinking. The US military has a concept of "phase zero," or the stage antecedent to war. Rather, given its perception of permanent and protracted conflict, Russia is every day preparing for war by deploying all the instruments of state power globally to enhance its security and interests.

Observing the operations of Russian state and associated criminal actors on a day-to-day basis demonstrates that the entire Russian state participates in polit-ical warfare, IW, and actual military operations. Russian official documents on national security since 2009 have all been plans for mobilizing the entire state for conflict.[9] If one reads the 2009 document and the 2015 national security strat-egy and tracks behavior of the regime since 2009, then it becomes clear that the entire state is being put on a mobilization footing. Not only do they systemati-cally reinforce the message that Russia is under attack from both US-led IW and military threats but also the regime has allocated massive resources to spend on information operations like Russia Today and "troll factories" in Russia.[10] Defense spending and the industry it supports are portrayed as locomotives of economic growth as well as security measures.[11]

The instruments by which Russia conducts its operations are fundamentally nonmilitary and represent a Russian version of the term "whole of government." Although Moscow is clearly willing to use force as in Georgia and Ukraine, those military operations represent the culmination (or at least the intended culmi-nating point) of a strategy premised on years-long operations using coordinated nonmilitary instruments and military threats to subvert targeted governments from within. In other words, IW, which includes cyber warfare, saturation of the media, and psychological operations, is intended to achieve the results that direct force would otherwise have to accomplish. Just as some Russian commen-tators maintain that the end of the Cold War and even the US occupation of Germany and Japan after 1945 were massive information operations leading to strategic victories without firing a shot, they maintain that properly conducted IW can give Moscow much, or all, of the victory it currently seeks at much lower cost. To the extent that hostile interventions in other states cannot be certainly attributed to Russia, Moscow can avoid or complicate any reprisals by its adver-saries. And given that cyber operations do not rise to a level of violence that the North Atlantic Treaty Organization (NATO) regards as "military operations," NATO leaders are hesitant to respond.[12]

The concepts underlying these operations evolved in response to the fiscal, moral, and intellectual trauma that the Soviet and Russian military establish-ments experienced from at least 1991–2000 due to the discrediting effects of their opposition to reform and their malfeasance in the First Chechen War between 1994 and 1996. The Russian establishment saw the United States and NATO as mounting an unstoppable threat to its interests and identity as an imperial great power. NATO enlargement, the 1998–99 war in Kosovo, and Western support for the democratization of former Soviet states manifested this threat and intensified Russia's feeling of being under siege. In the Second Chechen War, from 1999 to 2007, Moscow effectively insulated the Russian information space from outside influence. Since Russian elites clearly believed that the loss of domestic public support during the First Chechen War was a major factor in Russia's defeat, they were determined to prevent that from happening again. State efforts to curtail media access and reporting in the Second Chechen War ensured that popular support for Russia's military operations would be staunch and enduring. The government waged a systematic campaign to capture Russian hearts and minds, recognizing that target as the true center of gravity. Public support was an invaluable lubricant of the armed forces, and a media campaign was mounted to mobilize that public support, to isolate the insurgents from domestic and foreign support, and to frame the war as an antiterrorist campaign.[13] Enduring public support allowed Putin to give the military a freer rein to fight a long war without any hint of public opposition. Russia's effective insulation of the theater and of the Russian media space demonstates how important control of the media and of the "narrative," or "framing," is to any war-winning strategy. This tactic enabled Russia to pursue and conduct sustained and vicious operations that included the use of thermobaric weapons, among other instruments.[14]

Building on the brutal success of the second Chechen campaign, Putin sought to rethink contemporary warfare and rebuild an effective military. If the United States is seen as the world's dominant military power with its array of sophisticated weapons platforms, the challenge is to find ways for Russia to win on different terms. It should not be surprising that the current strategy, much like that of 1921–39, identifies surrogate forms of power to compensate for deficiencies in sophisticated armaments. Thus, asymmetric war, including IW and IO, gained adherents because it increasingly seemed a safe alternative at much lower cost than direct military confrontation with NATO or the United States, given Russia's economic inferiority and military shortcomings. Russian writers also clearly believed that Russia itself was under information attack and that retaliation was obviously justified.[15] Moreover, the lack of enemy capabilities for definitive attribution and the fact that information warfare tools could be used like a thermostat, with the temperature being constantly adjusted as needed, lowered the risk to Russia.

The analogy to the Soviet period here is quite striking. As Jonathan Haslam has observed, "Special operations were used by the Soviet Union against prewar Poland. So special operations, or war by other means, were very much a feature of the 1920s, which was also a time of relative Soviet military weakness. Asymmetrical activities with covert operations were a substitute for not having the use of direct military power."[16]

Indeed, in 2007 Defense Minister Sergei Ivanov suggested,

> The development of information technology has resulted in information itself turning into a certain kind of weapon. It is a weapon that allows us to

carry out would-be military actions in practically any theater of war and most importantly, without using military power. That is why we have to take all the necessary steps to develop, improve, and, if necessary—and it already seems to be necessary—develop new multi-purpose automatic control systems, so that in the future we do not find ourselves left with nothing.[17]

The creative adaptation of earlier concepts and practices has proven useful to Moscow. Despite its strategic inferiority vis-à-vis the United States and NATO, Russia has won every war in which it has participated since 2000. Its successful use of IW in all these operations attests to its improved grasp of how information and cyber operations (which are a unified phenomenon in its thinking) contribute to victory.

Estonia

In 2007 the Estonian government moved a statue commemorating the Soviet Union's liberation of Estonia in World War II, defying Russian threats of reprisal if it did so. Immediately, Russians in Estonia demonstrated en masse. A widespread cyber attack was launched on Estonia's essential information and computer technology infrastructure: banks, telecommunications, media outlets, and name servers.[18] The offensive included denial of service, botnets, hacking, and systematic attacks on government offices, banks, and communications networks.[19] This "war" lasted from April 26, 2007, until mid-May 2007.

While these attacks were occurring, Moscow instituted sanctions on Estonia, demanded a revision of its laws concerning its Russian minorities, and called it a fascist or pro-fascist regime.[20] Moscow organized violent demonstrations in Tallinn among the Russian diaspora there. Meanwhile, the Russian youth organization Nashi (Ours) demonstrated at the Estonian Embassy in Moscow. This organization, like other such youth groups in Russia, is a creation of the Putin regime. Moscow has also employed Nashi and similar groups against other foreign embassies and domestic dissidents.[21]

The use of botnets precludes definitively identifying the source of attacks. Yet, although the cyber attack on Estonia cannot be conclusively traced to the Russian government, the available evidence is overwhelming: it was a predesigned Russian attack. Duma deputy Sergei Markov, a frequent Russian governmental spokesman, boasted in 2009 that his assistant and office were behind the attacks and that more such events would happen.[22] (President Putin has admitted that he began planning the Georgian war of 2008 in 2006. Estonia may possibly have been a cyber dress rehearsal for that war as well as a probe of Estonian defenses and NATO's response.[23])

Estonian authorities' investigation of the April–May 2007 incidents revealed that planning for the demonstrations and cyber strikes in Tallinn began in 2006, or well before any sign that the monument would be removed, which was the ostensible pretext for the Russian attack.[24] "They were planned in advance and at least somewhat coordinated, as Russian-language forums were full of the preparations and planning in the days leading up to the attacks. The Estonian government even planned to release news of the strike three days before it began but was dissuaded by the European Union (EU) because of an upcoming meeting between then EU president and German chancellor Angela Merkel and Russian president Vladimir Putin."[25] Indeed, in a 2006 article, Russian scientists forecast the exact nature of how botnets would be used to achieve denial of service in targeted computers.[26]

Estonian authorities observed that the demonstrations in Tallinn resembled earlier tactics and efforts by Soviet and Russian Federation authorities to destabilize or even unseat governments deemed insufficiently friendly or obedient—for example, the Czechoslovak and Bulgarian governments in postwar Europe.[27] Estonian authorities recorded the presence of Russian special forces in civilian clothes at the demonstrations, though it is not clear which of the many different kinds of the Russian special forces they meant.[28]

By disrupting and possibly unhinging the Estonian government and society, and by demonstrating NATO's incapacity to protect Estonia against this novel form of attack, the cyber attacks aimed to compel Estonia to consider Russian interests in its policies. In other words, it had a classically Clausewitzian objective of bending the enemy—Estonia, in this case—to Russia's will. In Estonia, as perhaps in the later case of Georgia, the attack may have reflected not only an effort to correct Estonia's behavior or influence its orientation but also a desire to punish it and deter others from following suit by making it an example of the risks to anyone who crosses Russia.[29]

Estonian authorities (and others) believe that Russia aimed to incite large enough demonstrations that they would provoke violence. Then, they argue, Moscow could have used the ensuing violence as a pretext for launching an anti-Estonian insurgency that could have justified either direct Russian support for the insurgents or even Russian military intervention, as occurred in Crimea in 2014. Though Western audiences might consider such threat assessments and scenarios far-fetched, the Estonians and other neighbors of Russia do not. The resemblances to earlier Soviet operations, the nature of the attacks, and the foreshadowing of the Crimean operation are more than suggestive. Indeed, the use of disaffected ethnic minorities and anger at the Baltic states' "lack of gratitude for independence" are long-standing Russian and Soviet tactics as are the organization of minority or other mass demonstrations.[30] Of course, it is inherent in the nature of cyber operations that they frequently cannot be definitively attributed to a particular source. It is but one aspect of their value.

A main goal of hybrid war—including its information and cyber dimensions—is to instill a feeling of constant political and economic insecurity among the target state's population. This pressure—in the form of trade wars, energy blackmail, propaganda, diplomatic deceit, and coercion to join alternative regional integration projects—has been felt by the post-Soviet states of the EU Eastern Partnership: Armenia, Azerbaijan, Belarus, Georgia, Moldova, and Ukraine. Since their independence most of the Eastern Partnership states have felt that they live in an insecure environment because of existing frozen conflicts, among other reasons. In this region, hybrid warfare aims less at the security and more at the stability of the region. In such conditions, the desire for stability is intense and can easily be manipulated. The campaign's main idea is that there is no stability without Russia.[31]

In the words of Estonian defense minister Jaak Aaviksoo, "It is true to say that the aim of these attackers was to destabilize Estonian society, creating anxiety among people that nothing is functioning, the services are not operable. This was clearly psychological terror in a way."[32] This observation confirms John Arquilla's insight: "Terror has been a part of war for a long time, and many centuries ago began to slip the bonds of the national and/or imperial 'monopolies' on its practice. Beyond this sense of its lasting presence in history, there are also abundant signs of terror's conceptual similarity with war as we have generally conceived of it for millennia."[33]

Another strategic purpose of the cyber operation against Estonia, as perceived by Estonian authorities, was to test to what degree European security institutions like the EU, NATO, and the Council of Europe would stand by the country. In this regard, Estonians say, Russia not only was surprised to find the strong if somewhat belated response by the EU and Council of Europe but was also disappointed by the lack of support by Russians living in Estonia.[34] However, NATO's response was late in coming. This also could have been instructive to Moscow and its neighbors.

Moving from Russia's strategic objectives to its operational practice, new forms of IW or of large-scale influence buying can be seen as updated analogues of Soviet ideological warfare and subvention of foreign communist parties and their media after 1921 that was intended to keep enemies "off balance." This strategy often involves the collaboration of Russia's largely state-owned energy firms, intelligence agencies, organized crime, and embassies in buying up key businesses in targeted states; in donating funds to political movements and politicians, thereby compromising them; and in general exercising a covert influence on local politics. This strategy informs Russian policy from the Baltic to the Black Sea and in the war against Ukraine.[35] This strategy goes beyond Russia's tense relations with its neighbors and encompasses the potential for waging such war farther afield against hostile governments in Europe and elsewhere or as part of an insurgency within a state.

Cyber attacks may play a role as needed in implementing such a strategy or may be self-standing operations in their own right that can be endlessly repeated and turned on or off. Indeed, the cyber attacks on Estonia occurred within the context of Russia's unyielding efforts to exploit the energy dependencies of all three Baltic states—Latvia, Lithuania, and Estonia—and used the combination of energy monies, bought and subverted politicians, intelligence penetration, and organized criminal syndicates to exert constant pressure on the Baltic, East European, and Central Asian states.[36]

In Estonia and in subsequent manifestations of IW and IO, the Russian government has cooperated with organized crime structures such as the Russian Business

Network (RBN) to launch attacks. According to researchers Eli Jellenc and Kimberly Zenz:

> RBN is a cyber crime organization that ran an Internet service provider (ISP) until 2007 and continues to be heavily involved in cyber crime such as phishing, malware distribution, malicious code, botnet command and control, DDOS [distributed denial of service] attacks, and child pornography. . . . While it is not certain that RBN is directly connected to the Russian mafia, it is highly likely. RBN is heavily involved in child pornography, which is traditionally controlled by the Russian mafia, and its official leader, who goes by the alias "Flyman," is suspected of running those operations (and of possibly being a pedophile himself). It is also known that Flyman has family connections to the government: his father or uncle was involved in politics in St. Petersburg before taking an important position at a ministry in Moscow. Another RBN member, Aleksandr Boykov, is a former lieutenant colonel in the *Federalnaya Sluzhba Bezopasnosti* (FSB, the successor agency to the KGB). While it is currently not possible to prove that RBN has worked in tandem with the FSB or other security services (collectively, the *siloviki*), it is likely that they are at least connected.
>
> When RBN officially hosted Internet services between early 2006 and November 2007, it was linked to 60 percent of all cyber crime.[37]

RBN may have suspended its operations since 2007. But cybercrime has grown significantly since 2007 and has spread across numerous ISPs. Therefore, cybersecurity experts continue to use the term "RBN" to refer to the loosely organized group of cyber criminals based in Russia, and cyber activity and crime by this group continue to remain high.[38]

The Estonian case reflects the logic of political warfare and its information warfare components that have long been part of Soviet and Russian strategy and practice. While the post-communist era brought changes, including the prominence of criminal syndicates and the use of unofficial groups such as hacktivists, the tactic of exerting coercive pressure on neighboring nations and states is not new. One year after the cyber attacks on Estonia, Georgia experienced a similar campaign.

Georgia

In Georgia Russia first attempted to combine kinetic and cyber attacks against command-and-control and weapons systems on the one hand, and information-psychological attacks against media, communications, and perceptions on the other hand. In other words, Russia organically integrated what Western sources would consider cyber attacks into a broader information and military operation. Although the results were mixed, the Russian political-military leadership has deeply studied this campaign and sought to refine for future use the tactics used in both aspects of its IW campaign against Georgia. Richard Weitz of the Hudson Institute observed,

> The techniques used by the Russian attackers suggest they had developed a detailed campaign plan against the Georgian sites well before the conflict. The attackers did not conduct any preliminary surveying or mapping of sites [which might have prematurely alerted Georgian forces], but instead immediately employed specially designed software to attack them. The graphic art used to deface one Georgia Web site was created in March 2006 but saved for use until the August 2008 campaign. The attackers also rapidly registered new domain names and established new Internet sites, further indicating they had already analyzed the target, written attack scripts, and perhaps even rehearsed the information warfare campaign in advance.[39]

Capt. Paulo Shakarian noted similarly that beyond the direct attacks on Georgian state institutions, the cyber campaign was part of a larger information battle between Russian media and the Georgian and Western media for control of the narrative. Here Russian bloggers were able to flood a CNN-Gallup poll with posts stating that Russia's cause was justified, and attempted to prevent Georgian media from telling Tbilisi's story.[40] In the early stages Russian "hacktivists" shut down the websites of Georgia's president, Ministry of Defense, Ministry of Foreign Affairs, Parliament, National Bank, the English-language online news dailies *The Messenger* and *Civil Georgia*, and the online Rustavi 2 television channel while also defacing the websites of the Ministry of Foreign Affairs and the National Bank.[41]

In the Georgia war of 2008, clearly Russian proficiency at IW had substantially improved from the Estonian operation. Russian military commanders, working with hackers, in both cases directed computers from locations throughout the world to attack Estonian and Georgian sites, thereby creating botnets.[42] Other studies underscore the sophistication of the IOs directed against Georgia. Most attacks were actually carried out by civilians with little or no direct (or certainly traceable) involvement by the Russian government or military. These cyber attackers were recruited through the Internet and social technology. As in Estonia, attackers were aided by Russian organized crime even to the point of hosting software ready for use in other cybercrime activities. The organizers of the cyber attacks seem to have had advance notice of Russian military intentions and were tipped off about the timing of Russian military operations while they were taking place. The absence of reconnaissance or mapping of sites at the onset of the operation signified that Russian intelligence had already deeply penetrated the Georgian networks. The number of attackers working against Georgia was much greater than those who had attacked Estonia even though far fewer computers were involved.[43] Jeff Carr, an investigator for Project Grey Goose, an organization of a hundred American volunteer security experts from the private and government sector, concluded that "the level of advance preparation and reconnaissance strongly suggests that Russian hackers were primed for the assault by officials within the Russian government."[44]

The first wave of cyber attacks on August 6–7, 2008, was carried out by botnets and command-and-control systems that were associated with Russian organized crime. Twenty-four to forty-eight hours later, Russian military operations commenced. Afterward, the second wave hit mainly, though not exclusively, through postings on websites, again a carryover from Estonia. These postings contained both the cyber attack tools and the lists of suggested targets to attack. Although cyber attacks were limited to denial of service and website defacements, which are relatively unsophisticated types of attacks, they were carried out in a very sophisticated manner.[45] Once Russian troops had established positions in Georgia, the attack list expanded to include many more websites of government agencies, financial institutions, business groups, educational institutions, news media, and a Georgian hacking forum to preclude any effective or organized response to the Russian presence and to induce uncertainty regarding what Moscow's forces might do. These attacks significantly degraded the Georgian government's ability to deal with the invasion by disrupting communications between it and Georgian society, by stopping many financial transactions, and by causing widespread confusion. It is also possible that spyware and malware were inserted into the Georgian systems for future criminal or military-strategic use.[46] The clear objective of the cyber strikes was to support and further the goals of the Russian military operations as they were timed to begin on a large scale within hours of the first Russian military strike. The attacks ended just after those operations did.

Subsequent reporting found that cyber attacks on Georgian websites and online discussions of upcoming military operations began weeks before the actual onset of hostilities. Such preparatory action included a "dress rehearsal" of the upcoming cyber attacks, providing further evidence of the unprecedented synchronization of cyber with all other military combat actions.[47] The comparative restraint in not attacking key infrastructural targets—including energy installations—but demonstrating the ability to do so, both in Georgia and beyond, signaled a broader strategy to deter Georgia or others from escalating the conflict.[48]

This last point is particularly important. Simultaneously displaying the capacity to destroy key infrastructural targets while withholding orders to do so makes an impression on both the directly targeted states and the interested but heretofore uninvolved observers. Those observers could, of course, ultimately become targets themselves. Restraint in exercising coercive options aims to de-escalate the conflict by simultaneously conveying moderate Russian intentions while demonstrating Russia's potential to do more harm, thereby deterring the target and third parties from retaliating.

The Georgian IW campaign highlights the returns that Moscow gained on its substantial investment in the resources needed to conduct IO and IW. Moscow struck to prevent Georgian accession to NATO and prove Russia's primacy in the former Soviet space, and it seems to have achieved both objectives. In addition, the Georgian war highlighted Russia's advancing cyber capabilities.

Crimea and Eastern Ukraine

Russia's interventions in Crimea and Eastern Ukraine, beginning in 2014, followed the patterns of the 2007 and 2008 exertions in Estonia and Georgia. The greater duration and intensity of the Ukraine conflict reflects the deeper political-economic connections between Russia and Ukraine and the greater stakes Russia perceives in repelling Western influence over Ukraine's future. Russian leaders perceived the onset of the crisis—that is, the demonstrations against Viktor Yanukovych's government—and the subsequent departure of Yanukovych as a coup conducted with, at least, the collusion of the West. As such, the situation provided stark confirmation of the Kremlin's portrayal of existential US-led hostility to Russian interests.

As in Estonia, Russian actors mounted intense IO to shape how Ukrainians, Russians, and international audiences perceived the unfolding events. These operations were conducted through all media, especially Web-based outlets. Opinion surveys and anecdotal reporting in Russia indicate the effectiveness of these efforts in shaping perceptions in Russia (if not elsewhere).[49] For example, a 2014 Levada Institute survey found that 69 percent of Russians believed that this media provided "an objective picture" of the crisis in Ukraine. A full 88 percent believed the United States and the West were "conducting an information war against Russia."[50] In June 2015 the Pew Research Center found that 50 percent of Russians blame the West for the conflict in Ukraine.[51] Russia also mounted economic pressure on Ukraine and on western European states and consumers who rely on energy inputs flowing through Ukraine. Connections between the Russian state, businesses, individual elites, and their Ukrainian counterparts also were exploited to solidify both Russia's hold on Crimea and the pro-Russian elements' hold on parts of Eastern Ukraine.

NATO's Cooperative Cyber Defence Centre of Excellence published an account of the cyber war in Ukraine through November 2015.[52] It demonstrates the occurrence of cyber strikes and IW against Ukraine's revolution in 2013–14 as well as during the war, though not at the level that had occurred earlier in Estonia and Georgia, where the denial of service and disruption attacks were largely seen as "symbolic" in nature.[53] In Ukraine investigations revealed a Russian cyber campaign known as Operation Armageddon, which reportedly began in mid-2013. According to a US cybersecurity firm, attackers used spear-phishing emails with attachments that appeared official to lure Ukrainian officials and other high-level targets. Malware then infected the victims' computers and was used to "identify Ukrainian military strategies" in order to advance Russian war objectives.[54]

In July 2014 a pro-Russia hacktivist group reportedly hacked into one of Ukraine's largest commercial banks and published stolen customer data on a Russian social media website. Earlier the bank's co-owner had offered a $10,000 bounty for the capture of Russian-backed militants in Ukraine. The circumstances and rudimentary quality of this operation left some commentators doubting a link to Russian state authority.[55] The same hacktivist group entered the Ukrainian Finance Ministry network in May 2015 and posted what it claimed were documents stolen from the network that revealed Ukraine was unable to service its external debt.[56] Seen from a different angle, these activities amount to an electronic campaign to prepare the battlefield.

If the information operations in Ukraine through late 2015 resembled those in Estonia and Georgia earlier, a new and more forceful application occurred on December 23, 2015, when sophisticated cyber attacks shut down three regional electric power distribution companies, affecting approximately 225,000 customers. As an investigative report of the US Department of Homeland Security recorded, these synchronized and coordinated attacks were conducted remotely, exploiting legitimate credentials of Ukrainian operators "via unknown means." Multiple human actors remotely hijacked the operation of breakers at more than fifty regional substations. According to the department's report, "The primary access pathway was the use of legitimate remote access pathways such as VPN [virtual private network] to access local systems. . . . The exact nature of the credential harvesting remains unknown. It is likely that the credentials were obtained well ahead of the December 23, 2015, event."[57]

The deep knowledge and advanced penetration of the Ukrainian electricity providers follow the patterns seen in Estonia and Georgia. In the Ukrainian conflict, given the deeper ongoing human connections with Russia, it is possible that human agents and collaborators were involved. In any case, the cyber attack reflected strategic thinking and operational planning, befitting Russia's articulated general approach to information warfare. Importantly, in this regard, while the preparations for the attack were begun well in advance—perhaps shortly after the deposition of the Yanukovych government in February 2014—the attack itself was unleashed one month after Ukrainian nationalists and Crimean Tartars disabled electricity transmission lines to Crimea, beginning on November 22, 2015.[58] The Crimean total power outage lasted two weeks.[59]

Thus, if the cyber attack on Ukraine was instigated directly or indirectly by Russian authorities, then it suggests a strategic logic that also was seen in Georgia, where capabilities to attack the energy infrastructure were put in place but not activated. In Georgia the Georgian state did not escalate the conflict, and Western powers did not intervene. Russian cyber operators did not then have cause to attack Georgia's energy supply system. Conversely, the attack on the energy supply to Russian-held Crimea was, in Russian eyes, an escalation that invited a somewhat symmetrical response. Ukraine's energy supply was cut off—the symmetrical part—but the method was a sophisticated cyber penetration and attack when compared to the simple toppling of transmission towers. Taken together, the Georgian and Ukrainian examples reflect a logic of deterrence and compellence by cyber means. A capability to do harm is emplaced to deter adversaries from acting against Russian interests. When the adversary is restrained, the cyber attack is not unleashed, but when the adversary attacks Russian interests, Russian actors inflict a roughly proportionate response.

Conclusion

The Russian deep state clearly has incorporated cyber strikes and information operations into information warfare, as it defines the term. IW assumes growing importance as a war-winning strategy that avoids attribution, inhibits enemy reactions, and minimizes expenses—all crucial strategic issues for Russia. These trends in IW also appear to Russia's leaders as an equal and opposite, if possibly asymmetric, reaction to what they believe is an all-encompassing political and information war being conducted against them and Russia.[60]

Russia's government thus defines IW as a strategic war-winning force in its own right and as an indispensable weapon for the intelligence preparation of the battlefield over many years. The ensuing subversion of the enemy from within, before a shot is fired, is an essential strategic operation. Other countries' military and political leaders appear to overlook these points at their own and our allies' peril. The instruments themselves may not be new, but their combination and the uses for which they are deployed strongly diverge from Western thinking and practice. Russia's strategy and operations in the information and cyber warfare domain continue to confound Western governments and audiences who have yet to devise a compelling strategy with which to meet Russia's exertions.[61]

Dismissing Russia's view as paranoid may be emotionally satisfying, but it leaves the rest of the world ill prepared for actual cyber war, which for Russia is a constant ongoing phenomenon whether direct force is being used or not. To paraphrase Leon Trotsky, we may not be sufficiently interested in differing views about IW like Russia's, but Russia's view of IW is very interested in us. Russia has already engaged its adversaries in information warfare; thus, its adversaries must understand and learn from it for their own security.

Notes

All of the foregoing was written before the final months of the 2016 US presidential election. Russian information warfare operations in the US campaign followed the logic and patterns of episodes recounted in this chapter and require a separate analysis.

1. James R. Clapper, "Statement for the Record: Worldwide Threat Assessment of the US Intelligence Community," Senate Armed Services Committee, February 26, 2015, http://fas.org/irp/congress/2015_hr/022615clapper.pdf, 2–3.

2. Kenneth Geers, ed., *Cyber War in Perspective: Russian Aggression against Ukraine* (Tallinn, Estonia: NATO Cooperative Cyber Defence Centre of Excellence, 2015); Roland Heickerö, *Emerging Cyber Threats and Russian Views on Information Warfare and Information Operations* (Stockholm: Swedish Defense Research Agency, 2013), www.foi.se; and Tim Thomas, "Russian Strategic Thought and Cyber in the Armed Forces and Society: A Viewpoint from Kansas," presentation at the Center for Strategic and International Studies, Washington, DC, January 20, 2016.

3. Thomas, "Russian Strategic Thought."

4. Geers, *Cyber War in Perspective*; and Stephen Blank, "Russian Information Warfare as Domestic Counterinsurgency," *American Foreign Policy Interests* 35, no. 1 (2013): 31–44.

5. Max Boot and Michael Doran, "Political Warfare," Policy Innovation Memorandum no. 33 (Washington, DC: Council on Foreign Relations, June 2013), http://www.cfr.org/wars-and-warfare/political-warfare/p30894.

6. Stephen Blank, "Class War on the Global Scale: The Culture of Leninist Political Conflict," in *Conflict, Culture, and History: Regional Dimensions,* ed. Stephen Blank et al. (Maxwell Air Force Base, AL: Air University Press, 1993), 1–55.

7. Ibid.; Maria Snegovaya, *Putin's Information War in Ukraine: Soviet Origins of Russia's Hybrid Warfare,* Russia Report I (Washington, DC: Institute for the Study of War, 2015); and Timothy L. Thomas, "Russia's Reflexive Control Theory and the Military," *Journal of Slavic Military Studies* 17, no. 2 (2004): 237–56.

8. See Christopher Andrew and Vasili Mitrokhin, *The Sword and the Shield: The Mitrokhin Archive and the Secret History of the KGB* (New York: Basic Books, 1999), 31. They explicitly tie this insight to the Bolshevik Party from the moment it took power until its demise in 1991.

9. Stephen Blank, "'No Need to Threaten Us, We Are Frightened of Ourselves': Russia's Blueprint for a Police State—the New Strategy," in *The Russian Military Today and Tomorrow: Essays in Memory of Mary Fitzgerald,* ed. Stephen J. Blank and Richard Weitz (Carlisle Barracks, PA: Strategic Studies Institute, US Army War College, 2010), 19–150; "Military Doctrine of the Russian Federation"(Washington, DC: Carnegie Endowment for International Peace, February 5, 2010), www.carnegieendowment.org/files/2010russia_militarydoctrine.pdf ; "Military Doctrine of the Russian Federation, February 5, 2010," Foreign Broadcast Information Service–Central Asia (FBIS-SOV), February 9, 2010; Voyennaia Doktrina Rossiiskoi Federatsii, December 26, 2014, www.kremlin.ru; *Natsional'naya Strategiya Bezopasnosti Rossii, do 2020 Goda* (Moscow: Security Council of the Russian Federation, May 12, 2009), www.scrf.gov.ru, and available in English from FBIS-SOV, May 15, 2009, in a translation from the Security Council website (henceforth NSS); and *Natsional'naya Strategiya Bezopasnosti Rossii,* December 31, 2015, www.kremlin.ru.

10. Russia Today, a Russian news agency, is widely considered as acting as the voice of the Kremlin. The channel receives funding from the Russian government, which Russian president Vladimir Putin acknowledged in a June 2013 comment: "Certainly the channel is funded by the government, so it cannot help but reflect the Russian government's official position on the events in our country and in the rest of the world one way or another." A sizable portion of its budget comes from the Russian government, including a reported $307 million in 2016. For more, please see Max Fisher, "In Case You Weren't Clear on Russia Today's Relationship to Moscow, Putin Clears It Up," *Washington Post,* June 13, 2013, https://www.washingtonpost.com/news/worldviews/wp/2013/06/13/in-case-you-werent-clear-on-russia-todays-relationship-to-moscow-putin-clears-it-up/?utm_term=.a2ca4bb6d35e; and "Russia Cuts State Spending on RT News Network," *Moscow Times,* October 11, 2015, https://themoscowtimes.com/articles/russia-cuts-state-spending-on-rt-news-network-50194.

11. "Russia Cuts State Spending," *Moscow Times*; Blank, "'No Need to Threaten Us,'" 19–150; Andrew Monaghan, "Defibrillating the Vertikal? Putin and Russian Grand Strategy" (London: Chatham House, October 7, 2014); Presentations by Kirill Rogov and Sergei Aleksashenko at the German Marshall Fund, Washington, DC, April 25, 2015; "Russian Military Production Up by Nearly 20% in 2015," Sputniknews.com, April 19, 2016; Steven Rosefielde, "Russia's Military Industrial Resurgence: Evidence and Potential," paper presented to the Conference on the Russian Military in Contemporary Perspective, Washington, DC, May 9–10, 2016; and Aleksandr Bastrykin, "Pora Postavit'

Deistvennyi Zaslon Informatsionnoi Voine," April 18, 2016, http://www.kommersant
.ru/doc/296157816.

12. Vladimir Vasilyevich Karyakin, "The Era of a New Generation of Warriors—Information and Strategic Warriors—Has Arrived," *Nezavisimaya Gazeta Online*, Moscow (in
Russian), April 22, 2011, Open Source Committee, FBIS-SOV, September 11, 2012; M. A.
Gareyev, "Russia's New Military Doctrine: Structure and Substance," *Military Thought* no.
2 (2007): 1–14; and Y. N. Baluyevsky, "Theoretical and Methodological Foundations of the
Military Doctrine of the Russian Federation (Points for a Report)," *Military Thought* no. 1
(2007): 15–23, exemplify this point.

13. Blank, "Russian Information Warfare."

14. Ibid.

15. Karyakin, "Era of a New Generation"; Gareyev, "Russia's New Military Doctrine";
and Baluyevsky, "Theoretical and Methodological Foundations."

16. Quoted in Thomas K. Grose, "Russia's Valuable Weapon: Vladimir Putin's Spies
Are Critical to His Strategy in Syria and Ukraine," *U.S. News & World Report*, October 14,
2005, http://www.usnews.com/news/the-report/articles/2015/10/14/russias-spies-are
-critical-to-putins-operations-in-syria-ukraine.

17. Sergei Ivanov, NTV, Moscow (in Russian), August 15, 2007, Open Source Center,
FBIS-SOV, August 15, 2007.

18. Joshua Davis, "Hackers Take Down the Most Wired Country in Europe," *Wired*,
August 21, 2007, 163.

19. A concise description of the attacks may be found in Rebecca Grant, *Victory in
Cyberspace* (Washington, DC: US Air Force Association), 3–9.

20. James A. Hughes, "Cyber Attacks Explained," *CSIS Commentary* (Washington, DC:
Center for Strategic and International Studies, June 15, 2007).

21. Stephen Blank, "Towards the Police State: Increasing Authoritarianism in Putin's
Russia," *Acque e Terre* no. 4 (2007).

22. William J. Dobson, *The Dictator's Learning Curve: Inside the Global Battle for Democracy*
(New York: Random House, 2012), 31.

23. "Putin Admits Moscow Planned Military Actions in Georgia in Advance," Rustavi
2, Tbilisi, August 8, 2012, http://www.rustavi2.com/news/news_text.php?id_news=462
58&pg=1&im=main.

24. Conversations with Estonian authorities in Tallinn, October 2007.

25. Gadi Evron, "Battling Botnets and Online Mobs: Estonia's Defense Efforts during
the Internet War," *Georgetown Journal of International Affairs* 9, no. 1 (Winter/Spring 2008):
122–23.

26. Igor Kotenko and Alexander Ulanov, "Agent-Based Modeling and Simulation of
Network Softbots' Competition," in *Proceedings of the Seventh Joint Conference on Knowledge-
Based Software Engineering*, ed. Enn Tyugu and Takahira Yamaguchi (Amsterdam: IOS Press,
2006), 243–52.

27. Conversations with Estonian authorities in Tallinn, October 2007.

28. Ibid.

29. Cory Welt, "Appendix 5: Russia and Its Post-Soviet Neighbors," in *Alternative
Futures for Russia to 2017*, ed. Andrew C. Kuchins (Washington, DC: Center for Strategic and
International Studies, 2007), 54–60.

30. Vladislav M. Zubok, *A Failed Empire: The Soviet Union in the Cold War from Stalin to
Gorbachev* (Durham: University of North Carolina Press, 2009); "Foreign Minister Sergey
Lavrov's Interview with Swedish Newspaper *Dagens Nyheter*, Moscow," April 28, 2016,

http://www.mid.ru/foreign_policy/news/-/asset_publisher/cKNonkJE02Bw/content/id/2258885?p_p_id=101_INSTANCE_cKNonkJE02Bw&_101_INSTANCE_cKNonkJE02Bw_languageId=en_GB; and Thomas T. Hammond, *Anatomy of Communist Takeovers* (New Haven, CT: Yale University Press, 1975).

31. Hanna Shelest, "Hybrid War & the Eastern Partnership: Waiting for a Correlation," *Turkish Policy Quarterly* 14, no. 3 (Fall 2015): 46, www.turkishpolicy.com/article/772/hybrid-war-the-eastern-partnership-waiting-for-a-correlation.

32. "Looking West—Estonian Minister of Defense Jaak Aaviksoo," *Jane's Intelligence Review,* October 2007.

33. John Arquilla, "The End of War as We Knew It? Insurgency, Counterinsurgency, and Lessons from the Forgotten History of Early Terror Networks," *Third World Quarterly* 28, no. 2 (2007): 382.

34. Conversations with Estonian authorities in Tallinn, October 2007.

35. Ibid.; Janusz Bugajski, *Cold Peace: Russia's New Imperialism* (Washington, DC: Center for Strategic and International Studies, Praeger, 2004); Richard Krickus, *Iron Troikas: The New Threat from the East* (Carlisle Barracks, PA: Strategic Studies Institute, US Army War College, 2006); and Keith C. Smith, *Russian Energy Politics in the Baltics, Poland, and the Ukraine: A New Stealth Imperialism?* (Washington, DC: Center for Strategic and International Studies, 2004).

Magdalena Rubaj and Tomasz Pompowski, "What Is the KGB Interested In?," *Fakt,* Warsaw (in Polish), October 19, 2004, in Open Source Center, FBIS-SOV, October 19, 2004; Jan Pinski and Krzystof Trebski, "The Oil Mafia Fights for Power," *Wprost,* Warsaw (in Polish), October 24, 2004, in FBIS-SOV, October 24, 2004; *Polish Radio 3,* Warsaw (in Polish), October 15, 2004, in FBIS-SOV, October 15, 2004; *PAP,* Warsaw (in Polish), December 13, 2004, in FBIS-SOV, December 13, 2004; and Open Source Center Analysis, "Lithuania: Businessman Stonys Wields Power with Russian Backing," FBIS-SOV, October 1, 2007.

36. Bugajski, *Cold Peace*; Krickus, *Iron Troikas*; and Smith, *Russian Energy Politics.* In Central Asia, when the Kyrgyz government backed away from ordering the United States to depart its military base there, Russia also simultaneously employed its economic power by rescinding its earlier loan to Kyrgyzstan and by revoking the preferred customs duties that Kyrgyzstan had been receiving on Russian diesel and energy imports, thus raising energy tariffs on its products. These moves forced the Kyrgyz government to announce major price rises in electricity fees that were the catalyst for the demonstrations that unseated President Kurmanbek Bakiyev. Just weeks before those demonstrations, the Russian press launched a media offensive denouncing Bakiyev as corrupt and saying that Russia could not work with him as if to signal that the time had come for an uprising. All these moves suggest a concerted plan to undermine the Bakiyev government and replace it with one more amenable to and openly dependent on Moscow. Although Putin professed surprise at the demonstrations, Russian papers discussed the possibility for demonstrations in Kyrgyzstan several weeks before the actual demonstrations occurred. Stephen Blank, "Moscow's Fingerprints in Kyrgyzstan's Storm," *Central Asia Caucasus Analyst,* April 14, 2010.

37. Eli Jellenc and Kimberly Zenz, *Global Threat Research Report: Russia,* iDefense Security Report, January 2007, cited in Kara Flook, "Russia and the Cyber Threat," Critical Threats, May 13, 2009, n6, http://www.criticalthreats.org/russia/russia-and-cyber-threat.

38. Ibid.

39. Richard Weitz, "Global Insights: Russia Refines Cyber Warfare Strategies," *World Politics Review,* August 25, 2009, www.worldpoliticsreview.com/articles/4218/global -insights-russia—refines-cyber-warfare-strategies.

40. Capt. Paulo Shakarian (USA), "The 2008 Russian Cyber Campaign against Georgia," *Military Review,* November–December 2011, 65.

41. Alexander Melikishvili, "The Cyber Dimension of Russia's Attack on Georgia," *Eurasia Daily Monitor,* December 12, 2008.

42. Ibid.

43. US Cyber Consequences Unit (CCU), "Overview by the US-CCU of the Cyber Campaign against Georgia in August of 2008," 2009, http://www.registan.net/wp-content /uploads/2009/08/US-CCU-Georgia-Cyber-Campaign-Overview.pdf.

44. Brian Krebs, "Report: Russian Hacker Forums Fueled Georgia Cyber Attacks," *Washington Post,* October 18, 2008, http://voices.washingtonpost.com/securityfix/2008 /10/report_russian_hacker_forums_f.html.

45. US Cyber Consequences Unit, "Overview."

46. Ibid.

47. Melikishvili, "Cyber Dimension"; and David Hollis, "Cyberwar Case Study: Georgia 2008," *Small Wars Journal,* January 6, 2011, www.smallwarsjournal.com/jrnl/art/cyberwar -case-study-georgia-2008.

48. Ibid.; and US Cyber Consequences Unit, "Overview."

49. See Denis Volkov, "Supporting a War That Isn't: Russian Public Opinion and the Ukraine Conflict," Commentary (Washington, DC: Carnegie Endowment for International Peace, September 9, 2015), http://carnegieendowment.org/2015/09/09/supporting -war-that-isn-t-russian-public-opinion-and-ukraine-conflict/ih3x.

50. "Information Warfare," Levada Center, November 12, 2014, http://www.levada .ru/eng/information-warfare. Note: The link comes up with a warning.

51. George Gao, "Key Findings from Our Poll on the Russia-Ukraine Conflict: NATO Countries Blame Russia and Ukraine; Russians Blame West," Pew Research Center, June 9, 2015, http://www.pewresearch.org/fact-tank/2015/06/10/key-findings-from-our-poll -on-the-russia-ukraine-conflict/ft_15-06-09-nato-ukraine-russia/.

52. Geers, *Cyber War in Perspective.*

53. Ibid.

54. For more details, see Dina Evans, "Operation Armageddon: Cyber Espionage as a Strategic Component of Russian Modern Warfare" (Arlington, VA: Looking Glass, April 28, 2015), https://lgscout.com/press-release/lookingglass-cyber-threat-intelligence-group -links-russia-to-cyber-espionage-campaign-targeting-ukrainian-government-and -military-officials/.

55. "'Cyber Berkut' Hackers Target Major Ukrainian Bank," *Moscow Times,* July 4, 2014, http://www.themoscowtimes.com/business/article/cyber-berkut-hackers-target -major-ukrainian-bank/502992.html; "Pro-Russian Hackers Mug Key Ukrainian Bank," Nextgov, July 4, 2014, http://www.nextgov.com/cybersecurity/threatwatch/2014/07 /stolen-credentials-network-intrusion-data-dump-pro/1225/; and Bill Gertz, "Russian Cyber Warfare Suspected in Bank Attacks," Flash/Critic, August 30, 2014, http://flash critic.com/russian-cyber-warfare-suspected-bank-attacks-sophisticated-hackers/.

56. "Cyberberkut Hacked the Site of Ukrainian Ministry of Finance: The Country Has No Money," South Front, May 25, 2015, https://southfront.org/cyberberkut-hacked-the -site-of-ukrainian-ministry-of-finance-the-country-has-no-money/.

57. US Department of Homeland Security, "NCCIC/ICS-CERT Incident Alert: IR-Alert-H -16-043-01AP Cyber-Attack against Ukrainian Critical Infrastructure—Update A," March 7, 2016, 3, http://neih.gov.hu/sites/default/files/dlc/IR-ALERT-H-16-043-01AP.pdf.

58. Ivan Nechepurenko and Neil MacFarquhar, "As Sabotage Blacks Out Crimea, Tatars Prevent Repairs," *New York Times*, November 23, 2015, http://www.nytimes.com /2015/11/24/world/europe/crimea-tatar-power-lines-ukraine.html.

59. Ivan Nechepurenko, "Electricity Restored to Crimea after 2 Weeks of Darkness," *New York Times*, December 8, 2015, http://www.nytimes.com/2015/12/09/world/europe /electricity-restored-to-crimea-after-2-weeks-of-darkness.html.

60. *Natsional'naya Strategiya Bezopasnosti Rossii,* December 31, 2015.

61. Arquilla, "End of War," 383.

6 An Ounce of (Virtual) Prevention?

JOHN ARQUILLA

Among the most paradoxical, clouded concepts in military and security affairs are the twin notions of "preventive war" and "preventive use of force." The former is commonly associated with starting a war at a most opportune moment—for example, while the prospects of defeating an enemy's military, seizing territory, or toppling a regime are good—or at least before a growing threat worsens. The latter reflects more modest ambition and consists of shorter, sharper actions aimed at avoiding more protracted conflict or at mitigating a dangerous situation.[1] Simply put, preventive war is about fighting now, not later, and for grander aims; preventive force is about using violence now in the hope of avoiding a full-blown war or to keep the strategic situation in an ongoing confrontation from deteriorating.

Preventive war, on the one hand, has a very long pedigree. Thucydides noted that the Spartans decided to wage war against Athens, amid crises over the smaller city-states of Corcyra and Potidaea, because "they feared the growth of the power of the Athenians, seeing most of Hellas already subject to them."[2] For more than two millennia since, decisions to go to war have often been made in fear of such growing power and of the potentially dire consequences of delaying a fight until a later time.

Yet it must be noted as well that adventurer-conquerors, such as Napoleon and Hitler, have also relied on notions of preventive war to rationalize blatant acts of aggression. Indeed, both men employed the logic of preventive war in their decisions to invade Russia. Thus, the preventive warfare concept can be nebulous—resorted to either out of fear or covetousness—and clearly has been subject to abuse.

Moral philosophers, therefore, have generally disapproved of preventive war.[3] So, too, did President Dwight D. Eisenhower, who in an important speech in 1954 categorically ruled out the idea of using the nuclear supremacy the United States then enjoyed to wage preventive war against Russia or China.[4]

Preventive force, on the other hand, is not fundamentally about trying to start a fight at the most opportune moment. Rather, it is a strategic construct designed for using a modicum of violence to thwart the rise of a fresh danger, to keep an ongoing conflict from widening, or, perhaps most important, to avoid the outbreak of a large-scale conflict. A well-known example of the last motive is the Israeli air raid mounted against the Iraqi nuclear weapons facility at Osirak in 1981. This raid was a classic preventive act of violence, short in duration and applied quite sharply. In this instance it seems clear that the Israeli attack set back Saddam Hussein's plans to acquire a nuclear weapons capability, though it did not end his ambitions. Still, the "underground" nature of Saddam's reconstituted nuclear program moved slowly, guaranteeing that he would not have an ability to threaten mass destruction to deter the US-led Coalition that ejected Iraqi forces from Kuwait a decade later in 1991.[5]

In the cyber realm, the Stuxnet worm that induced a considerable number—perhaps as many as a thousand—of centrifuges in an Iranian nuclear facility to self-destruct in late 2009 and early 2010 provides another example of preventive force. It was clearly an act of prevention because its aim was to slow down a suspected nuclear weapons proliferation process. It was also considered an act of force because real physical damage—of the sort that commandos could have caused with bombs and bullets—was inflicted. But in this case the deed was done with bits and bytes.[6] The attack was not sabotage but rather what I call cybotage. This alternative to violence may have dissuaded the Israelis from mounting an Osirak-type operation—this time against the Iranians—and gained time for the diplomatic deal that followed.

The Osirak and Stuxnet examples highlight the point that in recent times, acts of preventive force have focused on counter-proliferation as a proximate goal. But clearly much serious thought has been given and continues to be directed to the idea that, in the future, preventive force, particularly in the form of cyber attacks like Stuxnet, may have the broader potential to take the place in statecraft of classic deterrence and coercive diplomacy.[7] The main point of attraction is that "cyber prevention" requires no major military field operations and may even be conducted covertly or deniably. Thus, both the costs and the risks of engaging in acts of preventive force may be sharply lowered. Taking preventive force into the virtual domain has the potential to revitalize this concept, which has significant historical roots, the analysis of which may provide immeasurable value in informing and guiding future actions.

The Classic Paradigm of Preventive Force

The archetypal historical instances that illuminate the application of preventive force are the two attacks conducted by the Royal Navy on the Danish fleet, the Danes' coastal artillery emplacements, and the city of Copenhagen in 1801 and 1807. On each occasion, in its continuing struggle against Napoleon, Britain feared that the Danes might employ their very considerable naval capabilities to close the Baltic Sea to trade in needed commercial goods and naval stores. Even worse was the dire possibility that formerly neutral fleets might actively join the French in a direct challenge to British naval mastery. In 1801 the threat to Britain took the form of an emerging League of Armed Neutrality, which comprised Denmark-Norway, Prussia, Russia, and Sweden. Rather, it was a "re-emerging league," as the first such alliance was formed against British interference with neutral trade in 1780 during the American Revolution.

To parry the threat posed by the league, Britain dispatched a naval squadron to Denmark under the command of Adm. Sir Hyde Parker, a cautious man in his sixties who had recently enjoyed softer duty in the West Indies and married a teenaged girl. His second in command, Horatio Nelson—architect of earlier naval victories at Cape St. Vincent and the Nile—had a full understanding of the need to apply vigorous preventive force; indeed, he would defeat a larger force at Trafalgar some four years later. And in a hot action on April 2, 1801, at Copenhagen, the outcome was so much in doubt that Parker ordered Nelson to disengage. The latter famously chose to disregard Parker's command, and the Danes acquiesced. The British fleet went on to cow the other members of the league into submission.

According to the great naval historian Dudley Pope, "The destruction of a considerable part of the Danish Navy was eventually to benefit Britain before the war against France ended."[8] But the achievements of 1801 at Copenhagen were fated to be put to a fresh test. By 1807 the resentful Danes had rebuilt their fleet, and Napoleon had made himself master of much of Europe after decisively defeating Prussia and Austria and pummeling the Russians sufficiently to force them into an accommodation (and a kind of wary alliance via the Treaty of Tilsit).

The war aims on each side came to focus on imposing severe economic costs on the enemy. The French relied on their so-called Continental System, which was designed to limit any trade with Britain, while the latter's Orders in Council created a countervailing naval blockade in the hope that French commerce and credit would eventually be mortally wounded. At this point Britain had no immediate hope of defeating French land forces, so maintaining a favorable balance of naval power was crucial to its ability to continue the conflict. And it was British vulnerability at sea on which Napoleon fixed, in the belief that if he could but take control of the sizable, strong navy of Portugal and the revived Danish fleet and add them to his own and other captive warships, together they would make for a winning advantage. Indeed, as the apostle of sea power Alfred Thayer Mahan noted, Bonaparte "intended to seize the navies of Europe and combine them in a direct assault upon" Britain.[9]

However, not being favorably inclined toward the French, the Portuguese sailed their fleet away from Lisbon before Napoleon could grab hold of it. The situation with Denmark, whose navy had not only been reconstituted but improved in the wake of the 1801 incident, was far more dangerous. The Danes at this point were hardly on what could be called friendly terms with Britain. They also faced an implicit French threat of land invasion if Copenhagen reached any manner of accommodation with London. Thus, a new preventive naval expedition was decided on; it was to comprise more than fifty Royal Navy ships and twenty-five thousand army troops. The latter were to be at the ready to besiege the Danish capital as and if necessary.

Admiral Lord Gambier's fleet arrived off Copenhagen in August 1807. General Lord Cathcart's forces landed, under the direct command of Sir Arthur Wellesley, later the Duke of Wellington, and swiftly routed the Danish army. Still, the

fleet had to bombard Copenhagen for nearly four days early in September, causing terrible fires in the city, before the Danes agreed to hand over their ships. As
Mahan summed up the results for the British, they "took possession of eighteen
sail-of-the-line, besides a number of frigates, stripped the dockyards of their
stores, and returned to England."[10] Napoleon's naval ambitions were thwarted.

For us today, the moral of these Napoleonic-era vignettes about the value of
preventive force is that such short, sharp actions can have profound strategic
impact. Bonaparte could never be truly secure, he knew, while Britain remained
an implacable foe. But Britain could be countered only if its naval dominance
were overturned. Only then could its support for the insurgency in Spain be cut
off and its trade with Russia curtailed. As the historical record shows, Napoleon
strove hard, over many years, to craft the kind of sea power that could achieve
these aims and perhaps even to make the threat of invading Britain credible. The
two preventive actions at Copenhagen kept sizable naval forces out of French
hands, ensuring that British sea power would remain unbroken and—for the last
decade of these wars—unchallenged.

Today, the strategic potential of preventive force remains as great, and the
lower costs and risks of cyber prevention make this option even more attractive. Perhaps. But the very attractiveness of cyber prevention may prompt
those who see themselves as potential targets to engage in policies and behaviors of an aggressive rather than an acquiescent nature. This was certainly the
case in naval affairs, when aspiring sea powers operated under the constant
fear that the Royal Navy might swoop in at any moment, as it had done twice
against the Danes.

The Rise of a "Copenhagen Complex"

While the Danes were the direct victims of British preventive force and France
had suffered the indirect strategic consequences, later in the nineteenth century
Germany seemed most traumatized by a growing fear of a potential British coup
de main mounted against its growing High Sea Fleet. In the wake of victorious
land wars, culminating with the victory over France in 1871 that cemented German unification, Berlin began to focus seriously on naval affairs. This trend
accelerated with the accession of Kaiser Wilhelm II, who was deeply taken with
Mahan's ideas. After inviting Mahan to dinner on his yacht, the *Hohenzollern,* in
1893, the kaiser ordered that a German-language translation of Mahan's *The
Influence of Sea Power upon History* be added to the onboard libraries of all German
warships.[11]

The kaiser's determination to complement his land power with a first-rate
navy led him into a key administrative alliance with the industrious Adm. Alfred
von Tirpitz. In the decades before World War I, Tirpitz built a remarkably well-
engineered fleet that was smaller than the Royal Navy but still quite substantial
and in many ways superior, ship for ship. Regarding size, for example, in 1897
the German High Sea Fleet had less than a fifth the number of capital ships that
the Royal Navy possessed, but by 1907 the Germans had closed the gap swiftly,

possessing nearly half as many all-big-gun battleships, or "dreadnoughts," as Britain's Grand Fleet.[12]

But the kaiser and Tirpitz, mindful of the British actions at Copenhagen, knew that as the fleet grew toward a point where it would have a deterrent effect on the Royal Navy—or failing that would be able to give a good account of itself in battle—the danger of provoking a preventive attack grew ever greater. And as Jonathan Steinberg once pointed out, as long as Britain "continued to expand its own fleet, the gap between the two powers, and thus the danger zone, would remain forever."[13] The omnipresent threat that the nascent Imperial German Navy might be "Copenhagened" helped to feed an antagonism toward Britain that accelerated shipbuilding. Thus, in 1912 when Winston Churchill called on the Germans to "declare a holiday" in the arms race, Kaiser Wilhelm seethed and refused even to respond.[14]

In the event, the Germans were not "Copenhagened"; indeed, the British did not use preventive force in the run-up to World War I. Further, the Battle of Jutland in 1916, in which a number of British battle cruisers blew up when hit, reflected a tactical victory for the less numerous German fleet. The British lost fourteen ships and more than six thousand sailors killed; the High Sea Fleet suffered eleven ships sunk and some two thousand men dead. Not enough damage was done to break the British blockade of Germany, but the High Sea Fleet had certainly shown itself to be a mortal threat to Britain—so much so that when serving as First Lord of the Admiralty, Winston Churchill held that the Grand Fleet commander was "the only man on either side who could lose the war in an afternoon."[15]

So here is an instance where fear of a preventive attack spurred a "proliferator" in an arms race to go fast, to take significant defensive steps—like mining sea approaches to home waters and widening an interior canal for secure big-ship movement—and to succeed in building up to a point where its capabilities posed a true and very dangerous challenge to British naval mastery.

What does the Copenhagen complex mean in the cyber era? Can one really draw an analogy between fleet bombardments and the use of cyber prevention? The answer is yes, if one accepts that cyber attacks will remain hard to detect and defend against. Certainly the aforementioned expert opinions surveyed in this chapter suggest that leak-proof defenses are unlikely to arise. Much as the Royal Navy could not be kept from sailing to Copenhagen, it will likely prove very difficult to seal out computer worms, viruses, Trojan horses, and the wide range of other malicious software. In the United States alone, the high-profile hacks of commercial and government sites suggest that cyber attack will remain a tool of choice for statesmen, insurgents, criminals, and others for many years to come.

Thus, in the wake of the Stuxnet attack, the Copenhagen complex remains relevant. Its principal implication is that the potential for the preventive use of "virtual force" could impel those who feel threatened to take significant steps both to mitigate the risk of being on the receiving end of such an action and to accelerate, expand, and more diligently secure their own efforts, especially in the realm of developing weapons of mass destruction. For example, Iran, after the Stuxnet attack on its centrifuge program, quadrupled the number of centrifuges it deployed, bringing a burgeoning proliferation crisis with Tehran to a boil.[16] Was the alleged American-Israeli preventive use of force in this instance a boon or a bane? To be sure, the action gained some time for negotiation, but the proliferator's resolve to continue to possess an enrichment capacity was reaffirmed.

Whether the threat of another use of preventive force, either virtual or physical, will contribute usefully to sustaining the diplomatic solution now in place remains a vexing unknown at this point. This is particularly a risk for the use of advanced cyber techniques, which are to some extent "wasting assets." Once used, such exquisitely precise, targetable tools are unlikely to work as effectively when applied again. Patches will cover up specific vulnerabilities, and generally increased security awareness will stiffen defenses. These efforts will not eliminate the possibility of another use of cyber preventive force—new tools are always under development—but it does raise the cost of this form of intervention.

Another downside of the situation created by the Stuxnet exploit is that the very use of cybotage for preventive purposes, which may have gained time for diplomacy, opened a door to more general uses of cyber attack for retaliatory or signal-sending purposes. If the Shamoon virus that severely disrupted Saudi oil industry data was an Iranian attack, as many experts say, then it would be a logical follow-on to the Stuxnet operation's preventive use of cyber capabilities.[17] Thus, cyber as a mode of preventive force may seem more usable than violence, but it may also spawn more cyber wars. The matter is worth weighing in the balance as acts of cyber prevention are considered.

Other Lessons to Be Learned from the British: From Oran to Vemork

British leaders fully embraced preventive force in World War II. The most notable episode occurred after France fell in June 1940, when the Royal Navy's Force H sailed to the Algerian coast and bombarded the heavy French naval squadron in port at Mers-el-Kébir, near Oran. A brief, lopsided action saw one French warship sunk, others damaged, and over a thousand French sailors killed. Only one of the major combatant vessels escaped to Toulon, as did some of the lighter French ships in other parts of North Africa. Concurrent with the Oran strike (Operation Catapult), Operation Grasp undertook the seizure of other French ships and submarines that had made their way to British-controlled ports.

Britain took these steps to prevent the weak-kneed Vichy government from transferring the large French Navy to Germany. Winston Churchill, Britain's prime minister at the time, was deeply ambivalent but aware also of historical precedent and pressing need. In his history of the Second World War, he described the matter with sad eloquence: "This was a hateful decision, the most unnatural and painful in which I have ever been concerned. It recalled the episode of the destruction of the Danish Fleet at Copenhagen in 1801; but now the French had only yesterday been our dear allies, and our sympathy for the misery of France was sincere. On the other hand, the life of the State and the salvation of our cause were at stake. It was Greek tragedy."[18]

No less a personage than Adm. Sir Andrew Cunningham, commander in chief of the Mediterranean Fleet, strongly opposed the preventive use of force, arguing that it would "alienate the French throughout the French Empire."[19] But action at Mers-el-Kébir did not lead to greater conflict, nor did a later attack at Dakar, where Vichy naval forces lost two submarines sunk and suffered serious damage to the battleship *Richelieu*. The French fought as hard as they were able against the British in these actions—much as the Danes had resisted so vigorously against Nelson in 1801—but as was the case with the Napoleonic-era preventive actions, these operations during World War II stayed contained too. To be sure, relations with the Vichy government were soured in part because of Oran and Dakar; this was a nontrivial cost of preventive action. But the gain—keeping the naval balance in Britain's favor—by far outweighed this political cost.

Beyond naval actions, the British used preventive force in other ways as well during World War II. Concerns arose about German progress toward building a nuclear weapons capability. While the Allies had the Manhattan Project under way, the Nazis also wanted to develop an atomic bomb. One of the principal components in the proliferation process was (and still is) heavy water, or deuterium oxide, which, very simply, slows down neutrons to the point where they can sustain a nuclear chain reaction using uranium-235 or other fissile material. In German-occupied Norway, the Norsk Hydro plant was able to produce heavy water and so became a serious concern of the Allies.[20]

Several attempts were made to destroy the plant. In the context of an ongoing conflict, repeated and protracted uses of preventive force may be needed and, thus, expected. Such was certainly the case when it came to slowing or stopping the German nuclear proliferation effort. First, British commandos tried an airborne assault, but it failed. Norwegian resistance fighters had their innings next and did some serious damage to production that took months to repair. Air raids followed. Hundreds of bombs were dropped on the plant, and the damage disrupted but did not cripple production. Nevertheless, all this attention prompted the Germans to try to relocate the heavy water stocks to relative safety closer to home. Thus, a big opportunity came when the Germans started to move the heavy water from Vemork in February 1944, with the first leg of the transit being by ferry across Lake Tinn. One Norwegian commando planted a bomb below the waterline of the vessel, and it detonated while the material was being ferried, sinking the boat and its cargo of heavy water. More than a dozen innocent Norwegian passengers died. Almost none of the heavy water was salvaged, dooming whatever small hopes the Germans had of bringing a nuclear reactor on line in time to build an atom bomb that might have changed their failing fortunes in the war.

The British use of preventive force during World War II reflected a clear willingness to move beyond single actions of relatively short duration and widely separated in time—like the strikes at Copenhagen in 1801 and 1807. To some extent this was evident in the Royal Navy's actions against the French in 1940. As noted earlier, they were geographically quite widespread and waged over a period of nearly three months when one includes the action at Dakar. But looking beyond naval affairs to the counter-proliferation efforts against the heavy water facility at Vemork, one can glimpse the outline of an entire campaign of preventive force. This campaign took years; included several different forms of action, ranging from commando raids to aerial bombing; and concluded with the successful sabotaging of the Germans' attempt to move the heavy water to a safer location. The campaign was well worth the costs of the sustained effort. In the opinion of Kurt Diebner, one of the leading German atomic scientists of the wartime period, "When one considers that right up to the end of the war . . . there was virtually no increase in our heavy-water stocks in Germany, and that . . . there were in fact only two-and-a-half tons of heavy water available, it will be seen that it was the elimination of German heavy-water production in Norway that was the main factor in our failure to achieve a self-sustaining atomic reactor before the war ended."[21]

The key insight for the cyber era that can be drawn from British preventive practices during World War II thus may be to think about using cyber means of prevention in protracted campaigns—for example, against rogue proliferators, terrorists, and perhaps other adversaries—rather than restrict such actions to one-off events such as Stuxnet. Given that cyber measures can often be taken covertly—that is, with plausible deniability as to the identity of the perpetrator—and sometimes even clandestinely, with the target's being wholly unaware of the action taken, this notion of conducting protracted preventive campaigns grows more attractive. Indeed, the concept seems well suited to an era of perpetual irregular warfare. Whereas Winston Churchill found his choice to take preventive action against the French Navy in the summer of 1940 a "hateful decision," the decision maker today—and tomorrow—armed with cyber options may find fewer practical or ethical constraints standing in his or her way, whether the choice made is for a one-time coup de main or to pursue a more protracted preventive course.

While its potential for covert, deniable action may make protracted cyber preventive campaigns attractive, particularly in a time of open-ended conflicts with terrorists and proliferators, costs and risks are associated with such a longer-term approach. A strategic factor of concern is the likely loss of the veil of anonymity over time. When actions are aimed at clearly malevolent actors already being opposed openly, and perhaps even militarily, the costs and risks are acceptable. But a protracted cyber preventive campaign to counter a competing nation's aims in some theater of operations, or to curtail its continuing theft of intellectual property from one's own commercial sector, might come completely undone or lead to conflict escalation when the cyber exploits are "outed." A further though less critical risk is that in any protracted cyber prevention campaign, the tools being used will eventually lose their potency as the adversary's defenses improve. This problem can be mitigated by developing more tools, but it does impose a cost that needs to be considered.

Assessing the Prospects for Cyberspace-Based Preventive Force

Clearly much insight can be derived for our time, and the future, from the earlier history of preventive uses of force. Whether the intent is to stem the tide of proliferation or to preserve a favorable balance of power, the preventive use of force has proved a valuable tool of statecraft and strategy. Early in the cyber age, already one well-known example (Stuxnet) involves a computer worm inserted into a system to cause centrifuges employed in the nuclear enrichment process to self-destruct. One can only think that this covert, very low-risk means of intervention will continue to provide an attractive option to decision makers who are trying to cope with the potential threat of an adversary's "trading up" to nuclear-power status.

While preventing or delaying proliferation by means of cybotage highlights one aspect of maintaining a favorable balance of power, trying to shore up one's more traditional military edge over a competitor by cyber preventive means will likely be a daunting challenge. Certainly the pursuit of such an aim requires thinking in terms of more protracted preventive measures, which will involve coming back with cyber strikes again and again as needed—similar to the Royal Navy's return to Copenhagen in 1807, six years after the first preventive attack there. But the target set today, and on into the future, will be far more complex and will undoubtedly require taking aim at those civilian industries providing the advanced technologies on which their militaries depend. While a hard task, surely, it will be far from impossible. Indeed, the very porousness of cyber defenses of US high-technology firms has led to considerable hemorrhages of their intellectual property. And the same exploits that have led to such theft could just as easily be used to destroy or corrupt data in ways that slow or per-haps even reverse progress.

Whatever its ultimate limitations, cyber prevention's covert nature can still enable and empower protracted campaigns as opposed to limited, short-duration strikes against particular targets. If British strategists were committed to con-ducting attacks on French naval assets for months and then on the German nuclear program for years during World War II, there is little reason to believe that counter-proliferation via cyber means will be delimited from doing the same.

Of course, the demands of a more protracted approach to preventive action based on cyber capabilities will have some unique aspects of their own, with the principal one being that a method employed in one instance may not obtain over the longer term. This is because, once known, a cyber exploit may be straightfor-wardly defended against. Thus, a kind of wasting-asset feature to cyber methods must be considered, and any protracted campaign of prevention will have to be waged with an arsenal of unique exploits ready to be used one after the other. The value of a successful preventive action must be weighed against the cost of the future loss of use of any particular cyber weapon.[22]

Aside from defensive measures that might be taken by the adversary—as were so apparent in the Germans' attempts to protect their budding High Sea Fleet prior to World War I—the further problem is that a persistent fear of preventive attack may spark very aggressive action, particularly in the form of arms racing. Again this can be seen in the case of Wilhelmine Germany, where the Copenhagen complex led to an accelerated pace of building all-big-gun battleships—considered the strategic weapon nonpareil of that time—in the hope that the "danger zone" might be escaped by the sheer speed of development and production.

Other forms of threatening action may be of a more covert nature—that is, designed to hide illicit or prohibited activity from view. One case in point of this latter type of (passive?) aggressive behavior is seen in the North Koreans' nuclear weapons program, which continued secretly in the wake of the 1994 Carter Accord and the much more fleshed-out Agreed Framework that, it seemed, had handled the problem.[23] It is now openly acknowledged that the United States considered the possibility of using preventive military force against North Korea and widely believed that Pyongyang had real fears of such an action being taken.[24] Yet, as in the case of the German High Sea Fleet, production efforts went ahead. In 2006 the North Koreans successfully tested a nuclear weapon and more recently have demonstrated their long-range ballistic missile capabilities. It is a cautionary tale for those who think the existential threat of preventive action alone might achieve the ends desired in any given interaction.

Cyber prevention mitigates this problem, at least to some degree. Striking with bits and bytes is, above all, a more usable option than attacking with bombs and bullets, especially in peacetime. Stuxnet's use may have been an act of war, but the identity of the perpetrator was never proved beyond a reasonable doubt.[25] Further, Iranian and international reactions were likely far more muted than would have been the case in the wake of an air raid, a missile strike, or a commando attack.

The next application of cyber preventive force may prove to be far less showy than Stuxnet, taking instead the form of deeply embedded malicious software that can, from time to time, conduct acts of cybotage or corrupt critical data in unseen, and unnoticed, ways. Both types of action—sporadic cybotage and manipulation of key data—might be used in protracted cyber prevention campaigns and could prove effective in the disruption of, say, an illicit weapons proliferation process. For more general purposes, these modes of cyber prevention could do a great deal to undermine the military effectiveness of potential adversaries, particularly those advanced enough to have developed dependencies on secure, ubiquitous flows of information in support of field operations. Indeed, it seems that although advanced information technology can do much to empower, at the same time it imperils. As Martin Libicki has observed, "The complexity of today's information systems is a central factor in making them vulnerable."[26] This makes for very fertile ground when it comes to taking a cyber approach to prevention.

Cyber prevention might also prove an ideal means for detecting and disrupting terrorist networks, for slowing their recruitment processes, and for generally undermining trust and morale. Dark networks are hard to deter or coerce, so preventive action may be the only way to keep their operatives from signing on, linking up, and then pursuing their murderous ways. Much as the British acted to keep Napoleon from co-opting the navies of minor powers for conquest, so today cyber prevention may keep terrorist networks from rising to even more dangerous levels. Perhaps cyber prevention can even reverse their flow, sending terrorist networks down the path to ultimate defeat.

On balance, one can see considerable room for the application of cyber preventive force in the future, whether against rogue nations or terrorist networks. Will it lead to the kind of dystopian "cool war" world envisioned by the novelist Frederik Pohl, where a neo-Hobbesian war of all against all is waged but covertly?[27] Perhaps so. Perhaps such a development is unavoidable, as the merits of cyber prevention come to be more widely appreciated. Thus, the concept of preventive force may migrate from the physical to the virtual world and come back again.

Notes

1. "Danger," if to the point of an imminent threat of violent action, moves the issue from prevention to preemptive calculations about launching a "spoiling attack."

2. Thucydides, *The Peloponnesian War*, trans. Richard Crawley (New York: Modern Library, 1951), 50.

3. For a good survey of the ethical problems that attend preventive wars, see Deen K. Chatterjee, ed., *The Ethics of Preventive War* (Cambridge: Cambridge University Press, 2013). A more benign assessment of preventive war can be found in John Yoo, *Point of Attack: Preventive War, International Law, and Global Warfare* (Oxford: Oxford University Press, 2014). Yoo was a senior adviser to President George W. Bush on legal and ethical matters relating to the decision to invade Iraq in 2003, an act that supporters and detractors viewed as a quintessential case of preventive war.

4. Eisenhower's rejection of preventive war is discussed in John Lewis Gaddis, *Strategies of Containment: A Critical Appraisal of Postwar American National Security Policy* (Oxford: Oxford University Press, 1982), esp. 170.

5. The seminal study of this preventive action is Rodger W. Claire's *Raid on the Sun: Inside Israel's Secret Campaign that Denied Saddam the Bomb* (New York: Random House, 2004).

6. See Kim Zetter, *Countdown to Zero Day: Stuxnet and the Launch of the World's First Digital Weapon* (New York: Crown, 2014).

7. See, for example, Abraham Sofaer, *The Best Defense? Legitimacy and Preventive Force* (Stanford, CA: Hoover Institution Press, 2010).

8. Dudley Pope, *The Great Gamble: Nelson at Copenhagen* (New York: Simon & Schuster, 1972), 512.

9. Alfred Thayer Mahan, *The Influence of Sea Power upon the French Revolution and Empire*, vol. 2 (Boston: Little, Brown, 1898), 276.

10. Ibid., 277.

11. Barbara Tuchman, *The Proud Tower: A Portrait of the World before the War, 1890–1914* (New York: Macmillan, 1966), 152.

12. George Modelski and William Thompson, *Sea Power in Global Politics, 1494–1993* (Seattle: University of Washington Press, 1988), 76.

13. Jonathan Steinberg, "Germany and the Russo-Japanese War," *American Historical Review* 75 (1970): 1965–86. Steinberg also wrote on this theme in a key article, "The Copenhagen Complex," *Journal of Contemporary History* 1 (1966): 23–46; and in his other study of this era, *Yesterday's Deterrent: Tirpitz and the Birth of the German Battle Fleet* (Oxford: Oxford University Press, 1965).

14. See A. J. P. Taylor, *The Struggle for Mastery in Europe, 1848–1918* (Oxford: Oxford University Press, 1954), 501. Taylor goes on to note that Churchill repeated his call in 1913, and again the Germans refused to respond.

15. Geoffrey Bennett, *The Battle of Jutland* (London: B. T. Batsford, 1964), 155 and 42 for statistics on the battle losses and the Churchill quote, respectively.

16. See Ariane Tabatabai, "Hitting the Sweet Spot: How Many Iranian Centrifuges?," *Bulletin of the Atomic Scientists*, October 2014.

17. "U.S. Officials Believe Iran behind Recent Cyber Attacks," *CNN*, October 2012. See also Nicole Perlroth, "Cyberattack on Saudi Firm Disquiets U.S.," *New York Times*, October 24, 2012.

18. Winston S. Churchill, *The Second World War*, vol. 2, *Their Finest Hour* (Boston: Houghton Mifflin, 1949), 234.

19. Cited in Correlli Barnett, *Engage the Enemy More Closely: The Royal Navy in the Second World War* (New York: W. W. Norton, 1991), 175.

20. Even before the Germans invaded Norway, French intelligence had contrived a way to move Norsk Hydro's supply of heavy water (a little less than two hundred kilos) to France. But they left the plant untouched, so it remained a persistent threat.

21. Cited in Thomas M. Gallagher, *Assault in Norway* (New York: Harcourt Brace Jovanovich, 1975), 229. See also Dan Kurzman, *Blood and Water: Sabotaging Hitler's Bomb* (New York: Henry Holt, 1997), 238. He asserts that the preventive campaign "constituted the coup de grâce that finally doomed the German nuclear program."

22. Calculations of this sort are explored in a thoughtful study by Robert Axelrod and Rumen Iliev: "Timing of Cyber Conflict," *Proceedings of the National Academy of Sciences* 111, no. 4 (January 2014): 1298–303.

23. Former president Jimmy Carter, in his negotiations with the North Koreans, had gotten somewhat ahead of the Clinton administration in terms of commitments made, but the agreed framework ultimately followed his recommendations fairly closely. See International Atomic Energy Agency, "Agreed Framework of 21 October 1994 between the United States of America and the Democratic People's Republic of Korea" (Vienna: IAEA, November 2, 1994).

24. This is made clear in Joel S. Wit, Daniel B. Poneman, and Robert L. Gallucci, *Going Critical: The First North Korean Nuclear Crisis* (Washington, DC: Brookings, 2004).

25. However, reportage at the time suggested that Stuxnet was the product of an American and Israeli joint venture. See, for example, William Broad, John Markoff, and David Sanger, "Stuxnet Worm Used against Iran Was Tested in Israel," *New York Times*, January 15, 2011.

26. Martin C. Libicki, *Conquest in Cyberspace: National Security and Information Warfare* (Cambridge: Cambridge University Press, 2007), 240.

27. See Frederik Pohl, *The Cool War* (New York: Ballantine, 1982).

7 Crisis Instability and Preemption

THE 1914 RAILROAD ANALOGY

FRANCIS J. GAVIN

If the first historical analogy American policymakers and pundits reach for during a foreign policy crisis is the 1938 Munich Agreement, international relations scholars are more likely to cite the July crisis of 1914.[1] Many of our most powerful concepts—the offense-defense balance and the security dilemma, misperception and inadvertent escalation, the cult of the offensive and preemptive and preventive war, to name a few—draw heavily on what is believed to be the historical lessons of the European political crisis that exploded into the First World War more than a hundred years ago.[2]

The role of new technologies, especially the massive expansion of rail lines throughout Europe and their ability to move huge numbers of men and weapons more quickly to the battlefield, is often seen as a key element of how the July 1914 Crisis began and played itself out in a catastrophic world war. Can we generate insights from this history into how emerging cyber capabilities might affect great power crises in the future? What, if anything, can the story of railroads and their effect on international stability tell us about cyber's influence on crisis stability today?

To answer these questions, we must first explore both what actually happened during the July crisis and what the consumers of this analogy *believe* happened. Over time, many aspects of the July crisis analogy have worn thin as historical scholarship has provided a more nuanced view of the origins of the First World War.[3] Older notions of the war being inadvertent, driven by miscalculation, or caused by strict adherence to mobilization schedules—in which the function of railroads was crucial—have been challenged.[4] That said, comparing the two new technologies and assessing their influence on crisis management and stability are revealing, and the work may provide ideas for how to minimize the dangers posed by cyber capabilities in a conflict.

Historical Analogies and the July Crisis

Historians have mixed feelings about analogies. First, historians are often skeptical of the methods other social scientists use to define, identify, cumulate, and explain past phenomena. Even if events can be coded correctly, we are wary of making predictions. There are far too many omitted variables and confounding factors to meaningfully compare contemporary events, to say nothing of technologies, separated by over a century. Historians would be thrilled if they could explain important single point events or even aspects of bigger questions (such as how railroads influenced decision-making during the July crisis). They point out that the effort to derive generalizations often sacrifices complexity and context.[5] Another reason historians are ambivalent about analogies is they update their understanding of past events. Many analogies that international relations scholars use, especially surrounding the First World War, are based on long-since contested accounts of what happened during the July crisis. Many international relations scholars still base their analogies on the work of the West German historian Fritz Fischer, despite that professional historians have contested and even discredited many of his arguments.[6]

Recognizing these shortcomings, using well-thought-out historical analogies can still be worthwhile. Human beings reason through analogies, and policymakers often reach for analogies from the past to make sense of the present. Ernest May and Richard Neustadt once suggested that it was like teenagers and sex education: teens are going to do it, so why not help them do it better and more safely?[7]

While there is little consensus on the short- and long-term causes driving World War I, the facts behind the July crisis are well understood. On June 28, 1914, Archduke Franz Ferdinand, the heir to the Habsburg throne, and his wife, Sophie, were assassinated by Gavrilo Princip in Sarajevo, Bosnia.[8] Princip and his accomplices were part of a secret, pan-Serb organization that sought to expand Serbia's territory and pry Bosnia away from Austria-Hungary. Soon it became clear the attack was undertaken with the knowledge and complicity of high-ranking members of the Serbian government, especially its notorious head of intelligence, Dragutin Dimitrijević, otherwise known as Apis.

A faction led by the Austria-Hungarian military chief of the general staff Count Franz Conrad von Hötzendorf believed the appropriate response was to crush Serbia once and for all. Serbia's territory—and, many thought, its irredentist ambitions—had increased after winning the Balkan Wars in 1912 and 1913. Others, particularly Prime Minister of Hungary Count István Tisza, wanted to avoid a war. In the end, after a drawn-out debate (over three weeks) but with strong backing from Germany assured, the dual monarchy issued a harsh ultimatum to Serbia that any sovereign state would have found difficult, if not impossible, to comply with.

Throughout July each of the major European powers engaged in intense deliberations, diplomacy, and signaling within their governments, among their allies, and with their adversaries. Austria-Hungary would not move without Germany's support, which it received. Both hoped to keep the crisis localized to the Balkans. Russia, however, saw itself as the protector of Slav interests and was wary of Habsburg designs in the region. Still stung by Austria-Hungary's 1908 annexation of Bosnia-Herzegovina, Russia refused to stand by and allow its client, Serbia, to be humiliated. Russia understood, however, that a clash with Austria-Hungary likely meant a war with Germany. France, worried about Germany's economic and military rise and always seeking an opportunity to reclaim Alsace and Lorraine, backed Russia. Germany swung between aggressive rhetoric and desires to launch a war to fears and concerns about the consequences of a global conflict. Great Britain remained uncertain until the end, fearing German power and intentions, yet at times unenthusiastic over its commitment to alliance partners France and Russia.

It is almost impossible to sort out the vast array of short- and long-term drivers and how they combined to turn the crisis into a world war. Each of the major players was dealing with sharp domestic-political crises that both distracted the government and may have provided a reason to see the July crisis as a welcome diversion. Longer term, the Anglo-German naval race, the imperial competition, the rise of nationalism, the decline of the Ottoman and perhaps the Austria-Hungarian Empires, and the perceived increases in German and Russian economic and military capabilities all generated great instability, as did demographic pressures and ideological clashes.

Europe, however, had weathered almost constant crises and instability in the decade before the outbreak of war in 1914. In addition to arms races between the powers, the First Moroccan Crisis of 1905–6, the Bosnian Crisis of 1908–9, the Second Moroccan Crisis of 1911, and the Balkan Wars in 1912 and 1913 had been very dangerous affairs but had not resulted in a world war between the powers. Tensions were high, but diplomacy worked in each. Cooler heads prevailed, and a global conflagration was avoided. Why was the July crisis different?

Understandably, scholars have focused on the dynamics of the July crisis itself to determine an explanation. Perhaps no country wanted a war, it has been suggested, but perhaps did something about the military environment make escalation more likely and world war unavoidable?

This is where railroads come in. Railroads had first been developed in Great Britain in the early nineteenth century when steam power and innovations in materials used for wheels, wagons, and rails combined to make rail transportation possible. Rail transport was soon competitive with and quickly overtook horse-drawn wagons and canals. Its innovations spread quickly to Western Europe and North America (as well as some European colonies and Latin America) and were key drivers of massive industrialization, urbanization, and economic growth. A century after they were first developed, tens of thousands of miles of rail sprawled throughout Europe, with the capacity to move massive amounts of people and goods in relatively short times. They were widely welcomed as a transformative technology that revolutionized transportation and, with it, society and the global economy.

The military application of railroads was likely first understood and exploited by Germany (as early as in Prussia in the 1840s). Planning for railways and developing war plans against potential adversaries became viewed as connected. Prussia's successful use of rail during its surprisingly quick victories in the wars of German unification between 1864 and 1871 convinced other European countries, especially France and Russia, of the need to better utilize this technology to secure their own national security interests. After 1871 the pace of rail construction intensified, and the technology of rail improved, all while European nations became increasingly cognizant of rail's potential military uses. Railways were also seen as a way to connect sprawling colonial possessions and increase national influence over wide territories. Germany's construction of the Berlin-to-Baghdad railway, for example, was considered a blatant effort to exercise influence in the Near and Middle East and to threaten especially Russian and British interests.[9]

As European tensions increased in the first decade of the twentieth century, Germany designed and adapted a war plan initiated by the chief of the Imperial German General Staff, Field Marshall Alfred von Schlieffen. Developed for numerous scenarios, the most interesting and ultimately relevant part of the plan crafted in 1905 and 1906 (and updated several times before 1914) envisioned moving quickly with most of the German army and knocking France out of the war before turning against Russia. The plan's success would be predicated on Germany's ability to mobilize and move its armies quicker than France and much faster than Russia could. A smaller German force in the east, cooperating with Austria-Hungary, would stay on the defensive against Russia until France could be defeated and forces moved by rail from the western to the eastern front. France and Russia, however, had their own plans to increase both the size and speed of their respective mobilizations. Russia also planned to massively increase the number and quality of its railways.

Perhaps coincidentally, after 1870 the railroads became more important at the same time political tensions and geopolitical competition increased. The competing military plans and the role of railroads in them also had potential short-term and longer-term consequences for crisis stability. First, if Germany were to prevail in a two-front war with France and Russia, it would have to mobilize rapidly vis-à-vis its potential adversaries. In a crisis, each country would have powerful incentives to mobilize its troops and railways first. If Germany waited too long and France and Russia gained enough of a head start on mobilizing, the former's plans for victory would be undermined. Given that these plans were not a surprise in 1914—Germany's war plans were an open secret—all sides had great incentives to launch their forces preemptively and gain advantage or nullify the advantage over adversaries. This situation had the potential to escalate a middling political crisis into a full-blown clash of arms. The pressure on the European powers to mobilize would be enormous, thereby shortening the time horizons for diplomacy and negotiations to work. Once one side or the other thought conflict likely, it had little incentive to hold back. Worse still, these mobilization plans relied on very rigid, tightly planned movements over railways. Once implemented, hundreds of thousands of troops would be moved forward in ways that would be hard to reverse or alter. The plans of different states appeared to be interlocking; that is, once one country had mobilized its armies and sent them over rail, others had to implement their own mobilization plans lest they be open to defeat.

A longer-term strategic issue also involved railroads. Germany's faith in its military plans was based on its comparative advantages in the size of its army and the speed with which it could move it to the fronts, and the latter was based in no small part on the quantity and quality of its railways in 1914. German planners recognized that their lead, however, might not last forever. France and especially Russia had ambitious plans to increase and improve their railways and thus nullify Germany's mobilization advantages. German leaders feared that in a crisis several years later, after France and Russia had implemented their plans, the mobilization edge they possessed in 1914 would vanish.

In other words, scholars have suggested that Germany saw a closing "window of opportunity" to exploit its mobilization-railroad advantages, which were a "wasting asset." If war between Germany and France and Russia were inevitable, the Germans might reason, wouldn't it be better to have it take place when Germany still possessed comparative advantages in mobilization power and speed, on which its whole plan for victory was based? If war was sure to come, wouldn't now be better than later? Such thoughts, scholars have suggested, would certainly have influenced German thinking during the crisis, thus making Kaiser Wilhelm II's regime far more willing to take political actions that risked war. The pressures behind mobilization were further intensified by what was known as the "cult of the offensive."[10] Many (though by no means all) decision makers believed there were military advantages to going first and striking a knock-out blow. The spirit of the offensive was also seen as an important part of building a passionate national identity.

In sum, the rapid mobilization and movement to the front of mass armies were made possible in large part by railroads. Railroads, according to the analogy, were destabilizing technologies that made a crisis more likely to escalate toward war. By playing into nationalist ideologies about the ease of the offensive, by decreasing the time and motivation to engage in long, drawn-out crisis management and diplomacy, and by providing powerful incentives to create and implement preemptive military strategies, railroads helped undermine efforts to localize the Balkan crisis and avoid world war. Are lessons here for considering cyber and its influence on great power competition, crisis dynamics, and the outbreak of war?

Similarities and Differences between Cyber and Rail

How are emerging cyber technologies similar to and different from railroads in influencing questions of war and peace? Four similarities and several differences stand out.

First, both rail and cyber are often more commonly understood as "facilitating" technologies. It is accurate that cyber capabilities can be weaponized in ways that can be massively disruptive—for example, by disabling military platforms on the ground and in the air, as well as by potentially crippling an adversary's command-and-control operations. That said, cyber's true strength likely manifests itself when combined with other kinetic capabilities. Alone, neither rail nor cyber are the most powerful, destructive instruments of violence to ensure success on the battlefield, especially in a long, drawn-out conflict. Even when cyber attacks are especially disruptive, their consequences may be temporary or reversible in ways that conventional attacks often are not. Thus, many cyber capabilities are unlike other technological breakthroughs—for instance, the battleship, the bomber, the tank, or nuclear weapons—that directly and dramatically influenced the intensity and effectiveness of killing the enemy (both his military and civilian populations) and physically destroying his military assets. Rail and cyber technologies may be most effective when they improve military operations in conjunction with more traditional weapons.

An adversary with powerful cyber technologies but impoverished kinetic capabilities can certainly cause damage and create complications for the United States or other cyber-dependent powers. It may also impede, for a time, US progress on a battlefield; however, in the end, it is unlikely to determine outcomes on the battlefield without other technologies or military forces. Cyber alone, for instance, cannot invade or occupy a country. While comparative differences in rail capabilities (namely, speed and volume) certainly mattered, what and whom the railways delivered to the battlefield—the quantity and quality of the soldiers and their weaponry—proved decisive. As with railways, any assessment of potential adversaries' cyber capabilities must be done in a holistic way and consider connections to other assets.

Second, both rail and cyber have thrived as revolutionary civilian technologies that were motivated by and then transformed the economic landscape. Railways are part of what might be considered as the second transportation revolution (with long-distance navigable European ships being the first and aircraft being the third). Domestically, rail replaced horse-drawn transport and canals as the primary means of moving commercial goods. In the process, the cost of shipping goods fell markedly as the distance and volume of goods shipped increased dramatically. This led to tremendous economic growth and increased prosperity in Europe and North America. Similarly, cyber is the key part of what has been recognized as a profound revolution in telecommunications. The ability to move massive amounts of information quickly and at a fraction of previous costs has generated enormous wealth throughout the world.

Why does this matter? Rail and cyber are dual-use technologies with both civilian and military applications that are sometimes hard to distinguish. Undoubtedly the more destabilizing aspects of both rail and cyber technologies were underplayed or underestimated during the early years of each technology revolution. In particular, the military applications of these tools were poorly understood. The opposite worry, however, might be greater cause for concern. Certainly some cyber capabilities are meant purely for coercive or military purposes; for example, the weaponized payload of a virus like Stuxnet obviously has no civilian purposes. Distinguishing civilian from nefarious cyber capabilities ahead of time, however, can be challenging. Many cyber capabilities fall into a murky area, and it may be hard to identify them as "weapons" prior to their use. While measures to limit the dangers and vulnerabilities presented by cyber are eminently sensible, these measures would be ill advised if they undermined or dramatically impeded the enormous economic benefits brought by the information revolution. More work is needed to effectively distinguish and understand where on what might be considered a spectrum of malevolence—from entirely benign uses to primarily cyber weapons—a potential adversary's cyber capabilities lay.

Third, both technologies are products and enhancers of the process of globalization. Railroads and cyber have shortened distances, both physical and nonphysical, and compressed time by eliminating or reducing intermediary processes. In other words, goods, ideas, and intelligence could be delivered far faster. Political institutions that had developed to deal with more slowly evolving movements may have been challenged by reductions in time. Their decision-making could be compromised by new, unexpected realities, leading more easily to mistakes, accidents, misunderstandings, and misperceptions.

With railways and cyber, this globalizing process led to increased connections, drawing states and societies into closer contact and often obscuring long-held borders and boundaries. Thus, in both cases the issue to focus on is not the technology per se but rather the consequences of the globalizing process on international stability and crisis dynamics. There are two schools of thought on globalization and war. Many believe that the greater interdependence brought by globalizing technologies increases the possibilities of peace, as nations have a greater economic stake in each other. Furthermore, disappearing borders displace entrenched social, ethnic, and economic groups and create constituencies whose identities and interests transcend the prejudices of nationalism. A darker view posits that dislocations attendant to disruptive technologies within states, plus increased exposure and interaction between national groups, can generate greater opportunities for friction between states and increase the chances for conflict.[11] The same factors that drive growth and interdependence also expose critical vulnerabilities and weaknesses. Intuitively, the first, more optimistic view of technologically driven globalization holds great appeal. But greater interdependence did not prevent the First World War, and globalizing information technologies today often empower illiberal and destabilizing forces such as the Islamic State.

Fourth, both rail and cyber are compressive technologies. Each, in their own way, condenses the effects of space and time. This is not to say one can neglect the enormous lead time needed to design and construct both rail and cyber platforms, and to plan for their use. When deployed for battle, however, both rail and cyber can dramatically intensify the pace of battle. The speed and carrying capacity of rail moved people farther in far shorter times than in the past. The world became smaller and faster. Cyber has a similar effect. Massive amounts of information and communications can be moved instantaneously with no regard for distance or geography. Space and time are key variables in military conflicts, and this rapid compression might dramatically increase the pressures on decision-making during a crisis.

What are some differences between cyber and rail and their influence on crisis stability and war? The biggest difference surrounds the question of mobilization. With railroads, it is fairly clear who mobilizes, how they mobilize, and for what purpose. Furthermore, what mobilization looks like is obvious: railroads carry massive numbers of soldiers and matériel to the front according to a strict time line. With cyber, however, the details of who, how, for what purpose, and appearances are far less clear.

Railways were built and are operated out in the open for all to see. A nation's rail capabilities are impossible to hide. To the extent they are part of a military balance of power, they are relatively easy to measure and compare. Rail lines and rolling stock are overt and expensive assets, have large physical footprints, and are relatively transparent and predictable in how they can be deployed. Smart intelligence agencies can study and evaluate them to learn an adversaries' capabilities and intentions. When, where, and how railways are built, for example, may provide important clues to what a state is interested in and what its intentions are. A massive buildup of rail capacity to a border, for example, would be an obvious sign and would allow a state to prepare and perhaps initiate defensive countermeasures.

Railways are also a relatively rigid, binary, linear capability. The direction and size of railroads cannot be changed quickly, easily, or secretly. Once a rail-based strategy is launched, it is hard to adapt or change. In many ways, it is a quite predictable capability; railroads, once understood and measured, rarely surprise. Railroads also do not have an attribution problem; when they aid a military action, it is clear where the train originates and where it is going. While a surprise attack may still be possible, an anonymous attack by rail is not.

Finally, railroads are the ultimate manifestation of state power. The rise of the modern nation-state went hand in hand with the rise of rail, which also reflected the ability of the state to generate and mobilize significant resources.

Cyber technologies, in contrast, can easily become tools employed by non-state actors and in fact may reflect the relatively decreased importance of the nation-state in world politics. Furthermore, cyber capabilities exhibit far less observable physical footprints than railroads do. Computer hardware, or physical assets, are obviously involved in cyber activities; however, this technology is usually far smaller, more portable, and more easily hidden than rail technology is. Cyber capabilities can be developed and implemented covertly, unlike railroads. Even if one could locate, identify, and properly evaluate the hardware capabilities of an adversary, the conclusion would largely miss the mark. The key determinant of cyber capabilities is software, both in terms of the computer programs developed and the human talent that produces and operates it. It is extraordinarily difficult to accurately measure cyber capabilities before they are used and even harder to evaluate the balance of cyber capabilities among actors.

Cyber technologies are also flexible and adaptable, ever changing in both peacetime and war, thus making a priori assessments of an adversary's cyber strength very challenging. Cyber attacks can be finely calibrated, making it hard to know how much of a capability is being revealed and making escalation dynamics trickier to predict. As mentioned, both railroads and cyber are dual-use capabilities with civilian and military purposes, and it may be difficult to assess with 100 percent accuracy when they are being used for good or ill. Once railroads are converted to military purposes, however, the shift is clear and, as noted, tough to reverse. Cyber capabilities exist in a more liminal space, where they can shift quickly, easily, and without detection back and forth between military and civilian uses. Adversaries have powerful incentives to hide the true intent of their capabilities and to make the line between cyber capabilities and cyber weapons murky. The origin of a cyber attack is far easier to hide; thus, one can imagine cyber attacks where the perpetrators are never identified.

The analogy between railroads and cyber capabilities and their influence on crisis stability is, at best, an imperfect fit. A recent high-level study produced by Booz Allen instead explored historical analogies that focused on transnational actors and problems of the global commons. Nuclear nonproliferation, infectious disease outbreaks, food safety in the United States, wildfire suppression, and the response to the 2004 tsunami disaster—all were seen as appropriate cases to mine for historical lessons to deal with cyber attacks.[12] The July crisis was nowhere to be found.

Furthermore, unanswered questions still surround the analogy between rail and cyber and their relationship to conflict. How high are the barriers to entry for both rail and cyber, and how hard is it for states to catch up? Railways, once possessed only by great powers, were soon within the reach of almost every state in the world. Will the same eventually prove true for cyber?

Both technologies are integrative; in other words, they allow people within a nation and between countries to come into closer economic, cultural, social, and political contact with each other. What will the consequences of increasing interdependence be?

Scholars also debate how such technologies influence the so-called offense-defense balance. Do these technologies provide a first-mover advantage that makes conflict more likely? Or, like tanks or aircraft, are they also effective at improving a state's capacity to improve its defensive capabilities? Arms control—based on counting and verifying equipment—helped manage fears of offensives by conventional and nuclear forces. But cyber capabilities do not allow such quantification and verification.

Finally, how do these technologies affect geographical calculations? Railways are located in and affect specific physical spaces, and they presumably have greater influence on conflicts between states that share borders. Still, they also helped connect and deepen ties among sprawling global empires. Cyber may be just the latest manifestation of shifting world politics from local and regional to global concerns, a movement that began with the transportation revolution (naval, rail, air, and, more recently, space) and that intensifies with the more recent revolution in telecommunications and computing.

Exploring answers to these questions can be useful in devising policies to manage offensive and defensive cyber warfare capabilities.

Conclusion

There is a danger in focusing on technology to the exclusion of underlying political factors. Railways did not cause World War I, and it is unlikely cyber threats will create a great power conflict in the years to come. One of the failings of international relations theory has been to focus too much on a particular military technology, assessing whether it makes offense or defense easier and attributing that characteristic to increases or decreases in the chances of war. The great powers were driven to conflict in 1914, however, by underlying political tensions. German ambitions, both at sea and in continental Europe, aroused suspicion all around. France wanted to recapture its lost provinces of Alsace and Lorraine. Austria-Hungary worried about the threat of Serbian nationalism. Russia had ambitions of its own, especially as the Ottoman Empire continued to recede. Combined with a lethal mix of imperialism, nationalism, economic volatility, demographic pressures, and social Darwinism, conflict and crises were constant in the decade before the First World War. Great power war, though obviously regrettable and avoidable, was thus not completely surprising. No doubt the idea that new military capabilities, including railroads, might aid the offensive and make a war short and decisive created a more permissive environment for states to gamble and risk war. Still, the underlying political tensions and rivalries were the cause of the war and should always be the focus of study. The United States, Russia, Ukraine, China, North Korea, Iran, Saudi Arabia, and any other possible adversaries would do well to understand each other's ideological and geopolitical dispositions first, before assessing how certain technologies would make conflict more or less likely.

That said, cyber does possess characteristics that, similar to other technologies in the past, might be especially destabilizing during a political crisis. Three worrisome characteristics stand out.

First, it is often difficult to identify the sources of a cyber event and even more so to measure cyber capabilities before they've been used. Second, cyber capabilities—even as they take time to develop and deploy—may increase the speed of a conflict once started. By compressing the time available to make decisions, cyber can overwhelm institutions, organizations, and individuals who are used to a more deliberate battlefield. Third, cyber capabilities are neither static nor linear. They can adapt as a battle goes on and, in conjunction with other military capabilities, may have multiplier effects in conflict. This can rapidly shift how the battlefield looks. Furthermore, cyber attacks may be oriented in comprehensive ways at the participants' command, control, communications, and intelligence capabilities, blinding either one or all sides to what is actually happening on the battlefield. These qualities may increase the incentive to use cyber preemptively, as there may be large first-mover advantages. These characteristics may also impede war termination or efforts to prevent escalation, as one side or another may lose the capability to assess the battlefield and might assume the worst.

What lessons might the First World War provide? One is struck, looking about over a century ago, at how under-institutionalized Europe was. While massive numbers of soldiers and military equipment could be moved far more quickly, information and intelligence—and the ability to properly assess, share, and deliberate on them—did not seem to keep pace. Not only was there little opportunity for adversaries to discuss conflicts, reveal their intentions, and negotiate stand-downs but also there was often complete opacity among allies and even within governments. In other words, railways had compressed the amount of time needed to make good decisions, but the political and diplomatic institutions that were part of this process had not advanced. One can imagine a world where the leaders of those great powers had a place and a reason to discuss their differences (i.e., a United Nations), where the allies had a place to better understand and synchronize their political and military strategies (i.e., a North Atlantic Treaty Organization), and where the civilian and military wings of government could better consult and share their plans and could better coordinate and offer a unified national strategy (i.e., a National Security Council). A more deeply institutionalized Europe in July 1914 might not have resolved the deep underlying political tensions driving tension, but it might have prevented their escalating into a catastrophic great power war.

With cyber capabilities, information and intelligence can now move quite quickly, yet the ability to process, assess, share, and deliberate them may not exist. The making of cyber strategy and policy in most countries today appears to be divided up between different groups and disaggregated, still lagging the innovation of capabilities. Cyber represents a technology that, once again, compresses the time available during a crisis to make decisions. Furthermore, cyber capabilities may actually degrade the ability to make such decisions. At the very least, leaders and experts should think about the institutional capacities of states to deal with massively increased amounts of information coming from a variety of different sources and in an environment where cyber attacks might be oriented toward degrading and blinding their capabilities. In other words, while little can be done in the cyber realm to shape larger political dynamics, steps can be taken to lessen the dangers that a cyber attack during a crisis will make war more likely or deadlier.

Notes

1. For the best historical work on how policymakers use and misuse analogies, see Yuen Foong Khong, *Analogies at War: Korea, Munich, Dien Bien Phu, and the Vietnam Decisions of 1965* (Princeton: Princeton University Press, 1992); and Ernest May, *"Lessons" of the Past: The Use and Misuse of History in American Foreign Policy* (Oxford: Oxford University Press, 1975). While most policymakers do reach for Munich, the July crisis haunted President John F. Kennedy during the Cuban Missile Crisis. See Robert Zaretsky, "Struggling with Destiny: Barbara Tuchman's Legacy as an Historian," *TLS: Times Literary Supplement*, February 24, 2012.

2. See, for example, the essays in Steven E. Miller, Sean M. Lynn-Jones, and Stephen Van Evera, *Military Strategy and the Origins of the First World War* (Princeton: Princeton University Press, 1991).

3. For an extraordinary, in-depth review of recent historiography before the recent slate of books, see Samuel R. Williamson Jr. and Ernest R. May, "An Identity of Opinion: Historians and July 1914," *The Journal of Modern History* 79 (June 2007): 335–87.

4. The centennial of the July crisis of 1914 saw many new works of history on the origins of the First World War. For an overview of the key new works, see Francis J.

Gavin, "History, Security Studies, and the July Crisis," *Journal of Strategic Studies* 37, no. 2 (2014): 319–31.

5. For an excellent example of this view, see John Lewis Gaddis, *The Landscape of History: How Historians Map the Past* (Oxford: Oxford University Press, 2002).

6. Marc Trachtenberg, "New Light on 1914?," July 29, 2015, unpublished paper.

7. Richard E. Neustadt and Ernest R. May, *Thinking in Time: The Uses of History for Decision-Makers* (New York: Free Press, 1988).

8. This overview is from Gavin, "History, Security Studies."

9. Sean McMeekin, *The Berlin–Baghdad Express: The Ottoman Empire and Germany's Bid for World Power* (Cambridge: Belknap, 2012).

10. Stephen Van Evera, "The Cult of the Offensive and the Origins of the First World War," in Miller, Lynn-Jones, and Van Evera, *Military Strategy*, 59–108.

11. For a comprehensive view of the scholarship on whether, when, and how economic interdependence causes war or peace, see Dale C. Copeland, *Economic Interdependence and War* (Princeton: Princeton University Press, 2014). For a study that focuses on the cultural and civilizational tensions generated by interdependence, see Samuel P. Huntington, *The Clash of Civilizations and the Remaking of World Order* (New York: Simon & Schuster, 1996).

12. Booz Allen Hamilton, *Cyber Operations Maturity Framework: A Model for Collaborative, Dynamic Cybersecurity* (McLean, VA: Booz Allen Hamilton, 2011).

Wake Tech. Libraries
9101 Fayetteville Road
Raleigh, NC 27603-5696

8 Brits-Krieg

THE STRATEGY OF ECONOMIC WARFARE

NICHOLAS A. LAMBERT

> To bring the pressure of war to bear upon the whole population, and not merely upon the armies in the field, is the very spirit of modern warfare.
> ALFRED THAYER MAHAN, November 1910

During the last quarter of the nineteenth century, for statesmen with an interest in national security, understanding the strategic implications of *globalization*—the phenomenon by which national economies became entwined with and progressively subsumed within the international economy—was one of the thorniest and most mentally challenging problems they faced. Today, as national economies become increasingly dependent on and intertwined with cyberspace, the topic of cybersecurity moves steadily up the defense agenda. Now critical to the global economy on which societies depend, cyber systems are a major factor in national defense and international stability.

Like globalization, cyber warfare is a multifaceted yet amorphous subject: barbed, hard to define, and difficult to conceptualize. The paucity of tangible examples of cyber warfare does not help matters, because it is difficult to theorize about a subject when one does not understand the parameters of the possible. Until very recently there was but one reasonably well-known instance of cyber warfare, Stuxnet. For a period, indeed, its name became almost synonymous with the term "cyber warfare." Yet to frame an understanding of a subject on a single manifestation would clearly be unwise. In conceptual terms, moreover, Stuxnet was the cyber equivalent of a precision tactical weapon, whereas it is possible to think of other forms of cyber warfare. For instance, it is generally acknowledged that cyber weapons have been developed for use in conjunction with combat forces. Similarly, the possibility of attacking an enemy's critical economic infrastructure to degrade their military or civilian capabilities has now become widely known.

Yet even these uses do not exhaust the possibilities. The employment of cyber warfare to assist combat forces is operational, while targeting critical economic infrastructure is a precision attack on physical assets. But could not a state use cyber means as a weapon of mass destruction or disruption, targeting an enemy's confidence as well as its infrastructure, with the aim of causing enemy civilians to put political pressure on their government?

To consider such scenarios it is helpful to seek an analogy. As it happens, recent history affords several possible examples of strategies to which this type of cyber warfare—we might call it strategic cyber warfare—might profitably be compared. These attacks are often described as forms of economic warfare. The most commonly employed historical example is the Allies' strategic bombing campaign in World War II; less common are the German U-boat campaigns of the First and Second World Wars and the US submarine campaign against Japan during 1942–45. A better historical analogy for thinking about cyber warfare is Britain's economic warfare plan implemented at the outbreak of the First World War.[1] For several reasons this analogy is especially attractive.

First, the international economy of today bears a closer resemblance to that of the three or four decades preceding the First World War era than to the more recent era encompassing the two world wars. The world economy was relatively more globalized (less autarkic) during the fifty-year period prior to the outbreak of the First World War than it was during the fifty years afterward. During the first era of globalization, as in the second (i.e., today), the stability of the national economies and the international economy rested on the free movement around the globe of goods, money, knowledge, and information. The flow of physical goods over the seas also hinged on a parallel yet separate flow of real-time information via undersea cables. Accurate and instantaneous information relaying details of supply, demand, and prices was essential to all businesses and especially to the financial services industry that facilitated the movement of commerce with ever-increasing velocity. The flow of information, paralleling the international flow of goods and services, became integral to economic systems.

Second, then as now, defense policymakers seeking to forecast the nature of future wars found themselves in a very new, almost alien, strategic environment—and with good reason. The advent of new military technologies changed the ways in which wars could be fought, but more fundamentally the transformation of the world economic system introduced changes in the nature of war itself. In particular, the development of the cable network impacted the structure of the world economy in ways that presented multiple strategic challenges and opportunities. Not only could militaries use the cable network to achieve unprecedented speeds of communication but, more important, businessmen and consumers around the world also came to depend on the smooth functioning of the cable network. Interrupting the network could therefore impact civilians—not just their governments or armed forces—more directly and more rapidly than had previously been possible. This interruption need not even be achieved by the armed forces. The very parameters of warfare were changing.

Third, before 1914 the British government had devised an economic warfare strategy that included the targeted disruption of the aforementioned, complex global communications network. In fact, the economic warfare strategy as implemented in August 1914 aimed at more than disrupting specific industries or elements of national critical infrastructure. Here, the term "economic warfare" is not referencing bombing ball-bearing plants or oil refineries (done with precision or otherwise), as in the Second World War, nor even the interdiction of global supply chains, as in the German and US submarine campaigns. These forms of economic attack were all comparatively limited in scope, intended to create bottlenecks and choke points in critical-path supply chains in the hope of producing knock-on systemic consequences. In 1914 the British aim was far higher: to "derange" the enemy's entire national economy, thereby delivering an incapacitating knock-down blow that would obviate the need for less intense but more prolonged types of war. Put another way, economic warfare transcended specific systems; it was not intended to be systems specific but society specific. Indeed, Britain's plan for economic warfare may well have been the first attempt in history to seek victory by deliberately targeting the enemy's society (through the economy) rather than the state. To be more precise, the target was the systems supporting the society's lifestyle rather than the society itself. This was a novel approach to waging war.

To be clear, economic warfare, as envisioned in 1914 and as defined here, was not analogous to the Allies' strategic bombing campaign from 1942. The differences are fundamental. Whereas strategic bombing targeted the ability of the state to make war and could work only through attrition, economic warfare targeted the enemy's society by deranging its national economy with the object of rapidly undermining the legitimacy of and domestic support for the enemy state. Similarly, whereas in strategic bombing civilian casualties were typically viewed as collateral damage, in economic warfare civilians were the target. These differences are summarized in table 8.1.

In positing this analogy, I do not mean to suggest that there exist direct parallels down to every last detail between the British strategic thinking before 1914 and the cyber problems of today. Nor do I mean to suggest that the nature of the technological problems and possibilities are similar, for in fact they are quite different. Rather, I seek to offer a different way of thinking about the possibilities of cyber warfare from what seem to me to be the most common approaches. The points that I wish to emphasize and the questions that I raise, therefore, pertain to the economic, political, and legal implications of waging warfare within a globalized trading system and to the difficulties and dangers of trying to weaponize any of the underpinning infrastructure. The analysis should serve also as a reminder of how serious the stakes can be when warfare—cyber or

Table 8.1. Differences between economic warfare and strategic bombing

Economic warfare	Strategic bombing
Targets the society	Targets the state
Fast acting	Slow acting
Psychological damage	Matériel damage
Shock strategy	Attritional strategy

otherwise—disrupts the global trading system and thereby causes significant economic collateral damage. As the British discovered in 1914, employing an economic warfare strategy is easier said than done.

There are four basic parts to the story. First, why would one choose to weaponize the international trading system in the first place? We must understand how Britain came up with the strategy of economic warfare and why some British planners thought it would work and others thought it too dangerous. This question pertains to the strategic environment created by globalization, which must be described in some detail to set up the analogy to our cyber era. Second, how did British strategists intend to implement their strategy? Clearly the technologies were different from those of today, but if we accept that cyberspace might include a psychological aspect, and not just electronic and virtual dimensions, then we can begin to see how the British conceptualized their offensive and think about some functional requirements or opportunities. Third, we must look at the consequences of implementation, both unexpected and underestimated. Last, our final basic questions are, can one prepare to defend as well as to attack? What are some inherent risks in and opportunities for defense against economic warfare? How does a state prepare to endure economic warfare as opposed to preparing to wage economic warfare?

Globalization and Its Strategic Implications—Then and Now

Historians have long marveled at the tremendous expansion in world trade during the long nineteenth century and concomitant dramatic rise in the ratio of foreign trade to global economic output. Between 1800 and 1913, world output per head doubled; over the same period the volume of world trade per capita multiplied by a factor of eleven.[2] By far the greatest upward leap occurred during the last third of the nineteenth century. Led by Great Britain, between 1870 and 1896 the volume of world trade doubled, and by 1914, in the space of just seventeen years, it had doubled again. All nations, especially the industrialized European powers, saw a steady rise in the ratio of foreign trade to economic output.

The late-nineteenth-century growth in international trade has been attributed mainly to the remarkable fall in the cost of long-distance transportation, with nods to the parallel communications revolution and developments in financial services.[3] A series of innovations in steam technology led to a steady drop in the cost of carriage by land (railways) and by sea (steamships). These changes made it economically practicable to transport bulk commodities, or staples with a low value-to-weight ratio, over great distances. Between 1868 and 1902, for instance, the cost of transporting wheat across the Atlantic fell by more than three-quarters.[4] Delivery was not only cheaper, moreover, but also quicker and more reliable.

The four or five decades before the outbreak of the First World War are now regarded as the first "golden age" of globalization. Although in many respects the facts are not new—historians were long aware that the volume of world trade had majorly increased during this period—the conceptual shift in interpreting those facts is. This shift is significant. In the words of the Cambridge historian Martin Daunton, "The context for thinking and writing about British economic growth has changed: the late nineteenth century can now be interpreted less as a period of decline and more as an era of globalization." Similarly, the tremendous increase in the volume of world trade is now more viewed as "the consequences of the new steam technology of the industrial revolution."[5] Here Daunton uses the concept of globalization to rethink the story of British power at the turn of the century and to relate the increase of global trade to industrialization.

Although the work by historians of globalization is extremely valuable, it does not fully capture all the macroeconomic aspects of the phenomenon. The process of globalization involved much more than an increase in trade driven by the application of industrial technology to long-distance transportation. It included also the development of a truly global commodities market, which was primarily in agricultural products and not, as the focus on industrialization might suggest, in manufactured goods. While the transportation revolution gets pride of place in most accounts of globalization, the communications revolution was at least as important.[6] The creation of the network that permitted instantaneous communication between almost any two points on the globe profoundly changed the ways in which business was conducted and commerce transacted on a day-to-day level. Within just twenty-five years, most international (and much domestic) commerce became reliant on access to cable communications to allow buyers and vendors to find each other in the first place; to negotiate contracts; to determine a fair market price; to arrange credit financing (a bill of exchange drawn on a London bank), insurance, and shipping; and to schedule payment and final delivery.

The development of the financial services industry, in conjunction with the communications revolution, is another crucial though underappreciated part of the story. As the cost of transportation fell and the global communications network spread, merchants looked farther afield for produce to buy and resources to exploit. As they did so, one by one, distant local markets became subsumed into the single world market. This process was particularly clear in the international grain trade, which before 1914 was the single most traded commodity. During the second half of the nineteenth century, the price of wheat in each distant locality increasingly came to be determined by the conditions of demand and supply in all parts of the world. News of a drought in India, for instance, or the expectation of a bumper crop in the Ukraine had an immediate effect on the price of wheat quoted in Liverpool and Chicago. The creation of a world market—and world price—was reflected in the general convergence of global prices. In 1870 the spot price of wheat in Liverpool exceeded Chicago prices by 57.6 percent. By 1895, however, the gap was down to 17.8 percent and in 1913 to just 15.6 percent. Price convergence was equally evident within national markets: in 1870 the wheat price spread between New York City and Iowa was 69 percent; by 1910 it had fallen to just 19 percent.[7] In short, globalization represented a fundamental shift in the structure and shape of the world economic system.

While it is important to understand the macroeconomic aspects of globaliza-tion, we must not lose sight of its microeconomic and social aspects, which were arguably of greater strategic importance. Globalization was more than a disem-bodied large-scale economic phenomenon. Like the development of cyber at the end of the twentieth century, with incredible rapidity it penetrated every nook and cranny of life. Writing in 1961 about the phenomenal growth in British trade during the late nineteenth century, before the term "globalization" had even been coined, the economic historian William Ashworth discerned the national and local impact of these global changes: "The country had moved away from self-sufficiency farther and more rapidly than before. A bigger proportion of the fundamental necessities of life and industrial livelihood was brought from abroad; a wider range of the commodities produced at home incorporated, directly or indirectly, a certain amount of irreplaceable imports; and the com-munities of more and more localities found in their midst some export industry whose fortunes appreciably affected the amount of their sales and income."[8]

In their seminal volume on *Globalization and History*, Kevin O'Rourke and Jef-frey Williamson agreed that "by 1914, there was hardly a village or town any-where on the globe whose prices were not influenced by distant foreign markets, whose infrastructure was not financed by foreign capital, whose engineering, manufacturing, and even business skills were not imported from abroad, or whose labor markets were not influenced by the absence of those who had emi-grated or the presence of strangers had immigrated."[9] In effect, commercial supply chains had begun to stretch around the world, with national and local economies growing ever more dependent on each other and on the global trad-ing system.

While most contemporary commentators applauded globalization for the host of benefits it brought in its train—unparalleled levels of economic prosper-ity, generally higher standards of living, and cheaper food for populations around the world—others became apprehensive at the potential detriments. From the standpoint of national security, for example, dependencies translated into vulnerabilities. As in the cyber world of today, moreover, the fear was that the critical economic (and social) systems of the newly globalized world seemed not only intrinsically fragile and susceptible to disruption but also extraordi-narily difficult to protect. Writing in 1902, the noted geostrategist and pundit Capt. Alfred Thayer Mahan echoed the worries of many when he observed that "the vast increase in the rapidity of commutations has multiplied and strength-ened the bonds knitting the interests of nations to one another, till the whole now forms an articulated system, not only of prodigious size and activity, but of an excessive sensitiveness, unequalled in former ages. National nerves are exas-perated by the delicacy of financial situations and national resistance to hard-ship is sapped."[10]

The deleterious strategic implications of globalization were most visible at the microeconomic level. For instance, in the grain market, which before 1914 remained the premier internationally traded commodity, the microeconomic behavior of merchants and farmers changed. For merchants engaged in the storage and handling side of the business, for instance, who bought and stockpiled wheat for resale at some future date, their business was now significantly riskier. An unexpected piece of news from a place far away could change overnight the worth of the wheat stored in their elevators and silos. Since they had purchased this stored grain with borrowed money, they might have to default on their loans.

Put another way, in the grain business, as the single market came to encompass more and more of the global supply, the number of variables in the pricing matrix exponentially increased. As a result, anyone who bought and held wheat for any period (including millers and bakers) was exposed to potentially catastrophic financial risk. If the price dropped between a merchant's purchase and sale, a competitor was sure to buy cheaper and cut prices, thereby compelling the merchant to follow suit and sell at a loss. Farmers too suffered from price risk. From sowing to flowering, crops were in the ground for six months, during which time the price could dramatically change. Obviously, when farmers chose what grain to sow, be it wheat or barley or whatever, they could not know with any certainty what price they would receive for their crop in six months' time.

The solution to the increase in business risk that accompanied globalization was found in a new financial instrument called a futures contract, also known as a *derivative*. The primary purpose of derivatives is to mitigate risk caused by likely price fluctuations; it is only their secondary purpose to facilitate commercial trade, though in fact they do and indeed are necessary to the conduct of business in a globalized economic world. Derivatives allowed merchants to project their future costs and revenues with much greater certainty. Very simply, whenever a grain merchant purchased grain, he could at the same time sell an equivalent amount at an equivalent price to a "futures" broker—that is, a professional financier who specialized in analyzing global market information (crop forecasts and the like) to predict future movements in prices—thus "hedging" himself against the risk of a significant shift in market prices during the period between his purchase of the grain from the farmer and his sale of the grain to the miller. In effect, in willingly shouldering the risk that merchants in the grain trade found intolerable, the broker anticipated that he could exploit his superior market intelligence to correctly predict the future price of wheat and turn a profit on the "future" he purchased. Thus, the derivatives market was the result of microeconomic calculations by merchants about the risk—itself a result of the increased complexity of a globalized market—that they were willing to tolerate.

Despite intense opposition from farmers, futures markets remained in operation because no credible alternative system for managing risk could be devised. The excesses of the system, such as rampant corruption and market manipulation, were undeniable.[11] But the fact remained that without such a system to mitigate risk, merchants who traded commodities in a global marketplace could not safely conduct their business without prohibitive risk of bankruptcy.[12] Futures trading became—and remains—fundamental to the entire international commodities market system.[13] From the strategic standpoint, the development of futures trading was as significant as it was illustrative of the increased complexity in the global economic system. It solved the problem of complexity (the risks created by a more complex and thus less predictable business environment) by introducing an additional layer of complexity (a market in grain futures as well as grain), thus making the entire system even more vulnerable to failure in the event of a major system shock.

Another set of microeconomic adjustments to globalization with macroeconomic and strategic implications was the sharply growing practice of "just-in-time" ordering. Contemporaries more colorfully called it living from hand to mouth. Traditionally, because of so many uncertainties in the market—due especially to the inability of vendors and customers to communicate in real time over any distance or to exchange market information concerning supply, demand, and prices—at all stages in the supply chain there was a need to protect against the unexpected by maintaining significant buffer stocks, a practice that incurred storage and other costs. With the advent of the cable, the emergence of the continuous market, the ready availability of supply, and the expectation that it would be possible to communicate with sellers at the last minute, however, merchants assumed they could safely reduce or eliminate their buffer stocks and trim their costs, thereby reducing their exposure to losses due to changes in price. As the chairman of the Baltic Exchange remarked in 1904 to a British government inquiry,

> The whole course of [the grain] trade is altering in order to save warehouse and other charges. When millers and others want grain, the merchant sells it to them on cost, freight and insurance terms, or, in the case of Liverpool, ex quay. That grain goes direct to the mills, and the charges for warehousing and other things are escaped. Therefore our trade is getting every day into one of cost, freight, and insurance, or of selling ex-ship, without incurring any of the other charges.[14]

Again, these changes in microeconomic behavior had macroeconomic and strategic effects. British defense planners were startled to discover, for instance, that between 1893 and 1903, average stocks of wheat held in the United Kingdom declined by no less than 40 percent, coming to be measured in terms of weeks rather than months of supply. At the most basic biophysical level, then, globalization had created a major new strategic vulnerability. Thanks to the advent of just-in-time ordering, cities contained no stockpiles beyond what was on the shelves, at most enough to last for a few weeks, and therefore were dependent on systems that brought them a steady supply of food.

But the new strategic vulnerabilities went beyond the biophysical. The drive for improved efficiencies touched commodities other than food. As the need to hold reserves of commodities declined, the global economic system became increasingly optimized for profit, and lengthier chains of dependencies developed that often extended over the seas. Whether they were aware of it or not, the microeconomic behavior of individuals in industrial, urbanized societies had

adjusted to globalization. Their economic well-being now required uninter-
rupted access to, and a steady flow of goods and staples through, the global trading system, which in turn required high levels of global economic prosperity.[15]
Not only did consumers need imports but also producers depended on selling
their goods in a constant stream of commerce. If it piled up on the wharves, they
would be in deep trouble, as would the banks that had extended loans to them.
National economies were dependent on each other and on the globalized economy, while the social and political stability of nations was dependent to a considerable degree on everyone's economic well-being.

These changes generated two related but distinct types of fragility. One was
the fragility of the economic system itself, a product of increasing optimization
and of correspondingly declining resilience. The other related to the fragility of
politically aware industrial societies, whose socioeconomic stability at the
national level increasingly depended on the smooth functioning of an optimized
but fragile global economic system. This fragility was rooted in the microeconomic, day-to-day changes wrought by globalization. Vast numbers of ordinary
people had come to depend on the smooth functioning of the global economy
quite literally for their daily bread, to say nothing of all the other goods with
supply chains now stretching around the world. For them, a shock to the system
would not happen at the abstract level of the state or society; instead, it would
happen in their daily lives by having to pay much more for items they needed, if
they could procure them at all. Given time, they could return to the old ways or
otherwise adapt, but time was quite literally of the essence. The key questions
that troubled pre-1914 statesmen across Europe were, how much time was
needed for businesses and economies to adapt? And, of course, what would happen in the interim?

If one intersection of microeconomics and strategy caused by globalization
was futures trading and another was the drive for newly possible efficiencies, a
third may be found in international law. The entire structure of international
maritime law pertaining to war at sea, and specifically the law of contraband, was
based on an understanding of centuries-old commercial practices. For instance,
international law assumed that a belligerent—its navy and its prize courts—could
determine from papers found on board a merchant vessel its ultimate destination and the ownership of its cargo. In the age of the cable and express mail
steamers, however, these assumptions were no longer valid because of fundamental changes in the day-to-day conduct of international trade: Ownership
papers no longer accompanied cargoes but were held as collateral by the (London) bank that financed the cargoes, cargoes in transit frequently changed ownership during the voyage, and even at the port of unloading a cargo might not
have a clear owner.

Already at the beginning of the twentieth century, vendors in the United
States commonly dispatched from New York wheat-laden merchantmen without an ultimate intended destination. Even the master of the vessel in question
did not know his ultimate destination until late in the journey. Only after cross-
ing the Atlantic, a voyage that took approximately ten days, and touching at

Falmouth (UK) to refuel would the master of the ship obtain his instructions on where to discharge the cargo (be it London, Rotterdam, or Hamburg). In 1903 one authority estimated that 60 percent of all wheat discharged in British ports had been purchased through an exchange while already on the ocean in transit.[16]

Again, these microeconomic changes had strategic implications. The day-to-day conduct of international trade had changed so fast, as a result of globalization, that no means or mechanism—national, municipal, or international—existed anywhere to verify the ownership or destination of merchant ships' cargoes. Even in peacetime, financial trails that paralleled each international transaction were notoriously difficult to follow. In wartime, merchants bent on contraband running in the pursuit of fabulous profits found it all too easy to shroud the ownership question in a tangle of paperwork. Existing international law offered no guidance to those judges attempting to decide whether cargoes were contraband without tangible proof that a specific parcel of goods was destined for an enemy. The implications were stark. In time of war, the immutable rights of neutrals under international law to maintain their legitimate trade had become fundamentally irreconcilable with the equally immutable rights of belligerents to prevent illegitimate contraband from reaching their enemies. Quite simply, even with the best will in the world, disputes could not be settled by applying international law. That the square peg of modern commercial practices would have to be banged into the round hole of laws designed for the age of sail did not mean that no one would try, however.

The Idea of Economic Warfare

Toward the end of the nineteenth century, various commentators with an interest in strategic affairs and political economy began to speculate that the ever-growing interdependencies and interconnections between the great industrial powers must reduce the likelihood of war between them. Such thoughts sprang not from idealism but from widespread perceptions of brittleness within urban-industrial societies and the belief that the new global economic system was inherently fragile and susceptible to shock. In the extreme, some argued warfare that dislocated global trade, and the world economic system, would be so catastrophic as to raise the specter of social collapse. Theoretically warfare (certainly protracted warfare) would then be an existentially self-defeating proposition for anyone contemplating it. "The future of war," the Polish railway tycoon turned military theorist Ivan Bloch wrote in his famous treatise on warfare, "is not fighting, but famine, not the slaying of men, but the bankruptcy of nations and the break-up of the whole social organization."[17] The idea that globalization made protracted, large-scale warfare unlikely, if not impossible, was popularized by Norman Angell in his 1911 edition of *The Great Illusion*, which sold more than two million copies.

Although military planners (and theorists like Alfred Thayer Mahan) balked at this extreme viewpoint, many (including Mahan) nevertheless seem to have admitted the plausibility of the central argument: a major war would severely dislocate the world economic system, resulting in severe economic, political, and social consequences that would have strategic implications. More than any other, this idea stimulated the belief that the next war "must"—of economic and social necessity—be short in duration. The most important variable in any future war, in other words, would not be the relative military prowess of the combatants but the relative economic, social, and political resilience of the warring societies amid an economic Armageddon. If victory could not quickly be achieved, then a prompt negotiated peace would be necessary to avert socioeconomic collapse. Hence, the widespread conviction in 1914 was that the troops would be home before the leaves fell or that the war would be over by Christmas.

Standing at the epicenter of the global trading system, at the hub of the global communications network, Great Britain appeared to have more to lose than most from a war-induced meltdown of the global economy. Yet from about 1901, Admiralty planners began toying with the strategic possibilities of deliberately deranging the global economy to undermine an enemy's socioeconomic stability. In effect, they contemplated weaponizing the global trading system. They believed that in such an eventuality Britain would suffer relatively less than other powers, especially Germany, because Britain's dependence on the smooth functioning of that system was matched by its considerable control over the levers of the system, whereas Germany's was not. In effect, Britain was in a position to deny Germany access to world markets while retaining access for itself.

This assessment was predicated on several factors. First, the Royal Navy was the most powerful navy in the world, with an unrivaled capability to exert direct control over seaborne trade. Second, the Admiralty possessed by far the most sophisticated information- and intelligence-gathering network in the world, as well as an understanding of how to leverage this relative advantage into global situational awareness. Third, British economic institutions generally appeared to the Admiralty to be better placed than those of other nations to weather the financial and economic storm that was expected at the outbreak of war. Alone among the great powers, the British state possessed impeccable creditworthiness. In time of war, the state's ability to borrow and spend freely could not be overstated.

Most prominently, British companies dominated the physical and virtual infrastructure of the global trading system. The cable networks strung across the globe, the maritime insurance and reinsurance industry, the banks financing international sales, the discount market for bills of exchange, and, of course, the companies that owned the merchantmen that transported goods and staples across the oceans—all were based in London. In the eyes of British defense planners, this dominance meant that the British government could potentially wield effective monopoly control over the critical infrastructure on which the global trading system depended. The close connections between Westminster and the city of London are well documented. Less well known are the close links between the Admiralty and key international companies such as Lloyds of London (insurance), the Baltic Exchange (freight forwarding), and the Eastern

Telegraph Company (cables). In modern parlance, there were numerous private-public partnerships in the sharing of information.

Above all else, however, was a conceptual breakthrough: the Admiralty realized that the strategic environment within which navies must operate was substantively defined by the structure and character of the world economic system. Naval planners recognized that the nervous and circulatory system of the global economy increasingly depended on the sea and in ways that were not entirely obvious to others. Whereas sea communications—traditionally the target of naval pressure—had once been limited to merchant ships carrying goods and letters, by the early twentieth century they also encompassed the networked international financial services industry, which was built on the global undersea cable communications grid. It can be easy to miss the novelty and significance of these developments. From the strategic perspective, this expanded definition of "communications" opened the door to recognizing that an array of new vulnerabilities and opportunities now existed.

What is more, very little of this new, expanded strategic environment was governed or regulated by internationally agreed rules and laws. Whereas plenty of precedents governed the interdiction of ships and goods in wartime, almost none governed the interdiction of electronic information; yet seaborne trade could now be interdicted just as effectively through non-naval as well as through naval means. At the same time, however, older laws governing maritime economic warfare retained a superficial applicability (to observers then and to most historians since) even though the commercial practices they governed had fundamentally changed. As explained before, the entire structure of international maritime law pertaining to war at sea, and specifically the law of contraband, was based on an outdated understanding of trading practices. For better or for worse, the inapplicability of international maritime law to modern commercial practices would make it difficult to judge with confidence that Britain's (or anyone else's) wartime conduct was illegal.

Against this background, to exploit the changed strategic environment to Britain's benefit, Royal Navy planners conceived the strategy of economic warfare. The essence of the proposed strategy was for Britain to exploit the natural economic and financial forces set in motion by the outbreak of war, forces that were expected to cascade though the economies of all nations and leave widespread chaos in their wake. In other words, Britain would take certain naval (and non-naval) measures calculated to channel and intensify the magnitude of the inevitable and inescapable economic shock expected to strike the global economy.

The aim of this form of economic warfare was fundamentally different from others that sometimes receive the same moniker. The Admiralty's means were not to pressure choke points through simply restricting an enemy's maritime trade (as in submarine warfare) or precisely attacking specific individual industrial or military targets (as in strategic bombing) but to undertake a wide range of actions designed to undermine confidence in the commercial access and financial systems underpinning Germany's economy. Britain's strategic aim was not merely to interrupt enemy military operations but rather to quickly destabilize and disorganize civilian economic systems, to create chaos and panic, and ultimately to generate social upheaval and political unrest.

It needs to be understood, furthermore, that British actions were calculated to target both the physical and the psychological. Weaponizing the infrastructure of global trade would translate into a shock—not attritional—attack on an enemy society. The means and ends of this plan were also very different from a traditional *blockade* (the term most often used to describe Britain's wartime economic warfare campaign). Blockade predated the globalized world economy and was based on older trading practices and international law. Moreover, it worked by targeting an enemy state's revenues through the interdiction of physical goods, and it could work only slowly. The new economic warfare, by contrast, targeted an enemy's society (not state) psychologically (not physically), and it could work quickly. This conceptual model was fundamentally novel.

It should be further noted that contemporary planners, grasping for the vocabulary to describe their new ideas, sometimes resorted to the old phrase "blockade" when they meant something quite different. For instance, when in September 1913 First Lord of the Admiralty Winston S. Churchill referred to the strategy in a letter to the prime minister, he used the word "blockade," yet the context in which he used it made clear that he envisioned something new and very different. "The one thing that really matters to the Admiralty is the power of effective blockade. We want to be able to cut off and arrest completely the sea-borne trade of Germany, and by this means to injure and dislocate her *economic system* so as to compel a peace."[18] Traditional blockade did not seek to dislocate an enemy's economic system; the new economic warfare did.

It is perhaps not surprising, in view of the novelty of the Royal Navy's thinking, that its new strategy faced criticism from more traditionally inclined thinkers. When the navy first mooted its ideas in 1905, during discussions between the army and navy general staffs concerning British war policy in the event of war with France, economic warfare was ridiculed as an "invertebrate measure of offence" (!). The director of military operations (the head of the army's planning staff) dismissed the navy's proposals as nonsensical, holding that they reflected "a very grave divergence of opinion . . . not so much on the general question of strategy as upon the whole question of war policy, if not indeed upon the question of what war means."[19] The general, of course, was quite correct: the conceptual implications of the navy's thinking was much more than a new approach to the application of naval force; its economic warfare plan involved a wholesale rethinking of what war meant. No longer would war consist only of armed forces seeking to impose their will through physical violence, a paradigm to which even the revolutionary *levée en masse* (mass conscription) of the late eighteenth century could be accommodated; now it might be waged without physical violence by public-private partnerships. And these "invertebrate" means, its advocates insisted, could potentially collapse an enemy's ability and will to fight more certainly and cheaply than "vertebrate" means ever could.

As they struggled to turn theory into a workable strategy and at the same time gain a more sympathetic audience for their new approach to warfare, naval planners made and acted on several additional conceptual breakthroughs. First, they realized that they needed advice from the people who conducted and studied international trade if they wanted to understand the global economy in the necessary detail. As a result, they began speaking to leading economists, bankers, shippers, and businessmen. Second, they recognized that very significant legal implications would arise with British interference with global communications. Already Admiralty officials were conducting what would now be called lawfare, as they were doing their best to ensure that British negotiators at international legal conferences such as the Hague Conference of 1907 favorably shaped the international maritime legal environment (though with limited success, it must be said, because the British civilian plenipotentiaries would not cooperate). Third, they realized that interfering with communications would affect the interests of multiple foreign and domestic stakeholders who had significant political influence. Implementation, therefore, would require the highest political approval. Theirs was not simply a naval strategy, in other words, but a national strategy. As a result, they encouraged and participated energetically in interdepartmental discussions.

Victory in these interdepartmental discussions was far from assured. From the perspective of the political executive (represented by the prime minister and eight other senior cabinet ministers), the Admiralty's plan for economic warfare required revolutionary innovations in the strategic policy process and the assumption of enormous political risk. Both requirements derived from the extensive array of stakeholders whose interests would be affected by a campaign of economic warfare. They included British consumers, British businesses (especially in the shipping, communications, and financial services industries), and foreign neutrals. In the British government, these stakeholders were represented chiefly by the Board of Trade, the Treasury, and the Foreign Office.

The mere act of including the Board of Trade in strategic defense discussions was itself revolutionary by the standards of the day. Traditionally, strategy had been a matter for the Admiralty, the War Office, and perhaps the Foreign Office. In January 1912 Arthur James Balfour, a former prime minister (1902–5) who remained a permanent member of the Committee of Imperial Defence (CID), observed in a letter to another CID member "that war is no longer carried on solely by the Admiralty and War Office, and that every branch of the Public Service is concerned, are truths which have become more and more clear in consequence of the investigations of the CID."[20] By the same token, the mere act of trying to enlist the support of British business interests for the strategy—to say nothing of actually adopting or implementing the strategy—required substantial expenditures of political capital. The British government in the years before the First World War was Liberal, which then meant it was ideologically committed to free markets and free trade. Quite simply, seeking businesses' support for wartime government control over the three pillars of global communications—cables, merchant ships, and financial services—risked alienating the government's core constituency.

For more than a year, between February 1911 and May 1912, a group of senior government officials sat as members of a committee chaired by Lord Desart to assess the relative risks of economic warfare. The establishment of the Desart Committee reflected the political executive's recognition that economic warfare was too important a matter to be left to the admirals. Adopting their strategy would be a matter of national strategic policy—or grand strategy—and weighing its merits and drawbacks required input from multiple governmental and nongovernmental stakeholders. The Desart Committee's investigation, which included testimony from leading bankers, shippers, and insurers, made clear that implementing economic warfare would meet powerful and significant resistance from British business.

The Board of Trade and the Foreign Office voiced their concerns about the domestic and foreign costs of the strategy very plainly. They argued, correctly, that economic warfare would entail large-scale state intervention in the workings of both the domestic and international economy, starkly challenging traditional ideas about the role of government. Moreover, the military's adoption of economic warfare, and the unprecedented state intervention into the economy that would ensue, would far exceed established boundaries of what constituted national strategy and indeed the very nature of war. The domestic and diplomatic backlash, the Board of Trade and Foreign Office predicted, would be massive.

Though disconcerted, the political executive discounted these warnings. Impressed by the Desart Committee's assessments of the potentiality of economic warfare, the political executive (acting in conjunction with the CID) gave the defense establishment permission to forge ahead with preparations for offensive warfare. The government resolved that in the event of war it would assert its right to intervene in the economy. In secret, the government drafted a set of regulations and penalties to govern the activities of British companies in wartime that would prevent their trading with the enemy or on the enemy's behalf. They were articulated in a series of royal proclamations, drafted before the war, that forbade British merchants, financiers, and shippers—any imperial subject—to trade or conduct business with the enemy. The naval authorities were further granted "pre-delegated authority" (a truly extraordinary innovation in defense arrangements with huge constitutional implications) to implement immediately upon declaration of war stringent controls over a wide range of commercial enterprises connected with international trade.

Implementation and Abandonment, August–October 1914

When Britain declared war on Germany at 11:00 p.m. on August 4, 1914, the global economy was already in disarray. The mere expectation of war during the previous week caused a virtual cessation of world trade, an impact more dramatic than even the most pessimistic commentator had forecast. By July 31, every stock exchange around the world (including Wall Street) had shut its doors. There was a global liquidity crisis. Banks recalled their loans. Foreign exchange was simply unavailable though on the gray markets in New York, sterling was selling for $6.00 (up from par $4.86). In London, meanwhile, the accepting houses that funded international trade were unable to meet their obligations and technically were bankrupt. The British government was compelled to step in and underwrite the entire stock of outstanding bills of exchange (in the world), thus increasing the national debt obligation *overnight* by approximately three-quarters.[21]

The British government went to war extremely nervous about the health of its domestic economy and worried by the specter of large-scale unemployment. "The chief fear that haunts ministers," the well-connected Lord Esher noted in his journal on August 3, "appears to be not the naval or the military situation, but the inevitable pressure of want of employment and starvation upon the operatives in the North and Midlands; this may lead to a highly dangerous condition of affairs."[22] "Distress will come upon us very swiftly," Herbert Samuel, the government minister responsible for unemployment, wrote to his wife on August 4. "In a fortnight's time," he immodestly anticipated, "mine will be the heaviest task of all, except the Admiralty's."[23] A further, more tangible measure of concern was the cabinet's insistence on retaining at home one-third of the available infantry divisions to meet anticipated civil disorder in the industrial north instead of sending them to France to help stem the German invasion.[24]

Such worries notwithstanding, the cabinet tacitly approved the implementation of economic warfare. In so doing, it compounded the global economic chaos, as had been intended. Then came the backlash, which the Desart Committee had foreseen, but it was far swifter and more intense than expected. As the scale of the economic devastation became increasingly apparent, domestic interest groups became ever more vocal in clamoring for relief and lobbying for special exceptions, and neutrals howled in outrage at collateral damage to their interests. In the government, their protests received a sympathetic hearing from officials at the Treasury, the Board of Trade, and the Foreign Office who had never fully approved of economic warfare in the first place. Compounding the problem, inadequate economic data clouded understanding and spawned uncertainty, leading to hesitation. Political commitment to the strategy began to crumble; more and more exceptions to the published rules were granted, thereby further undermining the effectiveness of economic warfare; and implementation stalled.

In October 1914, aware of evasions and growing outright defiance by domestic interests, combined with mounting pressure from powerful neutrals (especially the Woodrow Wilson administration), the economic warfare strategy was aborted. As a result, the British were compelled to wage war in ways they had previously agreed were undesirable, unthinkable, unworkable, and even fatal. The reasons bear consideration by any nation contemplating a similar strategic policy.

What Went Wrong?

The war exposed the limits of prewar planning, and of the political will to engage in prewar planning, in several ways.[25] One was the failure to understand fully how relatively narrow technical details in economic warfare could have large political consequences. For example, in 1914, to prevent trading with the enemy, the British cable censors required that all messages be transmitted in plain English (i.e., no shorthand or abbreviations or code) and that each telegram include the recipient's full name and address of recipient (i.e., no internationally registered abbreviations such as LAMP32, which is akin to a dot-com address like "shoes.com"). In so doing, however, they either forgot or did not appreciate the finite limitation in cable capacity or bandwidth, to use the modern term. The net effect of the new regulations governing telegram content doubled or even tripled the length of each message; consequently, a communications logjam grew, exacerbating the global commercial paralysis. Overruling objections from the military censor, the government quickly relented, dialed down the regulations, and agreed to share communications resources with corporations (both nominally British and foreign).

Another way in which the war exposed the limits of prewar planning concerned the behavior of British businesses. Before the war, faced with abundant evidence that they would resist government regulation but seeking to avoid a politically damaging confrontation, the government defaulted to blithe hopes about private sector conduct. The government expected that moral suasion would translate into effective control, that businesses would cooperate with regulations, and that capitalists would forgo enormous opportunities to make a profit on the black market out of patriotism. Such an assumption ignores the reality that capitalistic economies are built on a reward system that encourages firms (and individual businessmen) to deviate from the conventional and to pioneer new methods. Those who succeed earn disproportionate rewards; those who fail risk bankruptcy. Put crudely, the instinctive and essentially rational behavior of businessmen is to make money through innovative means. It might be said that conforming to government expectations is antithetical to business mentality.

Aside from the political costs of confrontation, the prewar structure of British business made measuring its compliance with regulations extremely difficult. Tracking large corporations was one thing; tracking small businesses, through which an enormous amount of economic activity flowed, was another. Generally speaking, then as now, an inherent conceptual bias when talking about the problem is reflected in envisioning the economy only in terms of large corporations, big systems, and big data. In reality a vast (unquantifiable) amount of economic activity flows through the enormous base of small businesses. In any case, the prewar British government never set up sufficient detection and enforcement mechanisms to ensure compliance with the announced prohibitions on trade.

As a result, certainly within six months, perhaps within three, certain British businesses were conducting a roaring trade with the notional "enemy," and most of it was financed though British banks, which also financed most contraband from Americans to Germany via neutrals. To add insult to injury, much of this trade was transacted over British cables and the goods carried to the enemy in British ships. Although these violations were apparent to some degree, the military authorities responsible for waging economic warfare found themselves powerless to prevent these violations and perhaps often lacked the commercial expertise even to recognize what was happening. Early attempts to improvise a better organization were resisted by other government departments (whose assistance was needed), while political leaders turned a blind eye. In the meantime, British trade with "previously unknown" corporate entities located in countries contiguous to Germany grew exponentially. For more than a year the British government remained unaware of the scale of the problem, as it lacked the means to gauge it and did not want to believe the worst.

The government's ability to impose effective control over the economy developed only gradually and not because British businessmen suddenly discovered a hidden reservoir of patriotism. By 1916 many in the private sector were sufficiently worried by the prospect of a social revolution that they were seemingly willing to tolerate relatively moderate state interference as a preferable alternative to arbitrary confiscation of private property by a radicalized socialistic society. The government and businesses had different understandings of what constituted a security emergency: for the government, the national security emergency was the prospect of military defeat; for businesses, the corporate security emergency was the prospect of social revolution. In other words, when businesses finally began to cooperate with the government, they did so not because the government's prewar expectations about corporate patriotism were correct but because they came to fear something more than government regulation.

The government's failure to anticipate the behavior of British businesses reflected its even more fundamental failure to reach consensus with key stakeholders about the proper relationship between the state and society in wartime. While the authority of the state to conscript its citizens was well enough established during the nineteenth century, cemented by Prussia's victory over France in 1870, the state's right to conscript never extended to private property. Social cooperation with a strategy that affected property interests had to be voluntary; it could not be legally compelled (and still cannot?). For the government, voluntarism was necessary not only to avoid legal challenge but to acquire the information—in effect, the "targeting data"—needed to prosecute the strategy of economic warfare. National security imperatives required society to reconceptualize its relationship to the state, but neither party realized the degree to which reconceptualization was necessary.

The Commons Strike Back: Implications for Cyber Warfare

In seeking to disrupt global communications, broadly defined, via economic warfare, Britain enjoyed many advantages:

1. A near monopoly over the communications infrastructure of international trade
2. Naval officials with the imagination to understand that the character of the global economic system had redefined the navy's operating environment and to spot resulting new strategic opportunities
3. Government officials who acknowledged that they lacked expertise on economic behavior and day-to-day business practices and were willing to seek civilian assistance
4. Other officials who realized that any attempt to interfere with maritime communications posed serious legal problems and attempted to shape the legal terrain accordingly
5. The broad recognition that any attempt to interfere with maritime communications was a grand strategic rather than an operational problem that required input from multiple stakeholders both inside and outside government
6. A political executive willing to conduct strategic discussions with multiple stakeholders, even at the risk of alienating its core political constituency
7. A strong prewar political commitment to the strategy of economic warfare, manifested concretely in the pre-delegation of authority

Even with all these advantages, however, Britain's strategy of economic warfare still failed. Indeed, it was barely tried.

The planners of cyber warfare could use this story to assure themselves that they would not make the same mistakes today. Moreover, they could use it as an opportunity to ask whether the United States (as the present hegemon) or any other actor enjoys the same advantages that Britain enjoyed and to think through some of the difficulties they might face and the risks they would be running.

One obstacle is simply to define cyberspace. Just as definitions of maritime communications were not self-evident before World War I, so definitions of cyberspace are not self-evident today. How does one distinguish private from public cyberspace, or one state's cyberspace from foreign cyberspace? Will it suffice to defend just one's own military cyber systems and critical infrastructure and key resources? Further, given that the health of the national economies of the United States, most European states, China, Japan, and others depends so very greatly on a healthy world economy, should national security measures encompass this too? Where does one draw the line?

It may be helpful to consider potential parallels between maritime space and cyberspace. The maritime space most readily identified as the *global commons* are the high seas, or oceans, but they are contiguous with progressively smaller and more sovereign (i.e., non-common) waters: gulfs, bays, deltas, ports and harbors, rivers, inland seas, lakes, and so on. Some of these waters may be reachable by continuous voyage, while some may be more isolated. Determining exactly where the commons turns into a sovereign area is not easy. By analogy, the private sovereign areas of cyberspace are the cyber equivalents of inland seas, ports and harbors, great lakes, and so forth. Permission from their owners, or sovereigns, may be required to enter these "places." Alternatively, entrance to these spaces without permission is tantamount to use of force and espionage. The idea of cyberspace as a commons coexists uneasily with various private, sovereign claims.

As if the challenge of defining interests in cyberspace is not enough, it also involves very difficult legal questions—just as the challenge of defining maritime communications posed very difficult legal questions. The fact is US firms dominate cyberspace, with very large portions of global Internet traffic passing through the United States, but other states host data and traffic too. China, for example, has taken numerous steps to wall off its critical information infrastructure from the outside world and subject it to state control.[26] Will societies so readily accept claims by their own or a foreign government of the right to commandeer or withhold bandwidth by which one gains access to cyberspace—in other words, the radio frequencies used by the mobile communications systems through which users increasingly interact with cyberspace—as mobile platforms and related applications proliferate? Similarly, the US government, without necessarily owning cyberspace, presumes to control access to portions known as dot-gov and dot-mil because it has control over the servers and the content. Does it own the interaction space, or the protocols and gateways by which people gain access to those servers and data? In the Western world at least, does any state's exercise of sovereignty amount to legal ownership? Does ownership per se confer the right to control access? Does the United States or any other government have an international legal or moral right to defend, regulate, or control access to cyberspace in ways that will very likely impinge on others' interests? What would be the political costs in asserting such claims?

In wartime these and related questions will push their way to the front of political awareness. The exact answers are less important here than the recognition that these questions must be asked. The British experience with economic warfare suggests that it would be very dangerous for the United States or any other government to assume that it could readily translate national dominance of cyberspace—however defined—into legal or effective state control.

If a state does decide to wage cyber warfare, how will it insulate itself from the collateral damage caused by deranging the global commons, whether the state's own law authorizes the action or not? How will the United States or any other government gain the cooperation and monitor the compliance of its private companies, many of which are multinational in their ownership and operations? How will countries respond when their consumers complain about rising prices and when businesses protest heavy state regulation, unfair foreign competition, and falling profits? What will states do when allies and neutrals complain that actions targeted at belligerents are hurting them? The issue is not necessarily that the United States or any other state will be unable or unwilling to act unilaterally within cyberspace. Rather, the issue is that *if* the United States or any other globally connected state acts unilaterally, *then*, for a variety of reasons—primarily economic but also political and diplomatic, not to mention legal—such action will impinge on the critical interests of others and risk a backlash. Effective measures that a highly globalized state might take in cyberspace could hurt its own and foreign interests so much that it might be compelled to call off its attack, just as Britain had to do in October 1914.

To illustrate this potential dynamic, consider a state-authorized cyber attack that corrupted the integrity of data or algorithms in a major international bank or stock exchange. Such an attack could be intended to damage and thereby coerce a particular country. But what if the effects undermined trust in wider international financial systems, thereby jeopardizing the stability of the international economy? Even if the effects were successfully localized—perhaps because the attacking state was relatively unconnected (which seems implausible) or unimportant to the international financial system—should other states attribute the attack to their satisfaction, then they would be expected to take retaliatory action. Such action could then impose economic pain on the attacking state and its population, including through sanctions. Depending on the ensuing political-economic developments in the attacking state, the consequences could resemble those that Britain experienced in 1914.

Britain's campaign of economic warfare is a parable of unintended consequences. If the British experience has a single lesson, it is that the infrastructure of a globalized economic system makes for a weapon of mass destruction rather than a precision strike weapon. Accordingly, weaponizing it entails pervasively political problems. It is not a problem for computer experts or national security agencies to tackle alone.

To have any hope of success, a strategy to weaponize critical economic infrastructure requires acknowledging the multiple stakeholders involved—foreign and domestic, inside and outside the government—and gaining their cooperation. Its formulation demands direction from the highest political authority and the assumption of substantial political risk by elected officials even to seek cooperation from powerful constituencies, let alone to alienate them by actually implementing the strategy. The more aggressive the weaponization of the global economic infrastructure, the more severe the damage it will cause not only to its intended target but also to collateral stakeholders, including neutral nations, domestic business interests, and domestic consumers who vote.

For the strategy to survive the likely backlash, or for the intensity of the backlash to be reduced, a case must be made to stakeholders before the strategy is implemented that the costs of an alternative strategy, or no strategy, would be even worse—say, a war that drags on for four years, costs millions of lives, and raises the specter of revolution at home. These stakeholders would include not only the citizens and the businesses of the state contemplating cyber-economic warfare but also the allies, the friends, and the major trading partners as well as the multinational financial institutions whose stability is vital to the international system. It may be impossible to secure the cooperation of all interested parties, but it is certainly impossible to do so without realizing that their cooperation is necessary.

In the event of a future major conflict, waging economic warfare within the context of a very different global economic structure would be, as it was a century ago, quite different in its character from anything experienced before. It thus behooves us now to devote serious and persistent thinking to the subject.

Notes

I thank Dr. John Arquilla of the Naval Postgraduate School for the term "Brits-Krieg." An earlier version of this chapter appeared in Emily O. Goldman and John Arquilla, eds., *Cyber Analogies* (Monterey, CA: Naval Postgraduate School, 2014) as "The Strategy of Economic Warfare: A Historical Case Study and Possible Analogy to Contemporary Cyber Warfare."

1. Nicholas A. Lambert, *Planning Armageddon: British Economic Warfare and the First World War* (Cambridge, MA: Harvard University Press, 2012).

2. A. G. Kenwood and A. L. Lougheed, *The Growth of the International Economy, 1820-1960* (Albany: SUNY Press, 1971), 91; and Peter Cain and Anthony Hopkins, *British Imperialism: Innovation and Expansion, 1688-1914* (London, Longmans, 1993), 161ff.

3. Kevin O'Rourke and Jeffrey Williamson, *Globalization and History: The Evolution of a Nineteenth-Century Atlantic Economy* (Cambridge, MA: MIT Press, 1999), 5, 33–36.

4. Ibid.; and James Belich, *Replenishing the Earth: The Settler Revolution and the Rise of the Angloworld* (London: Oxford University Press, 2009), 449. See also Graham L. Rees, *Britain's Commodity Markets* (London: Paul Elek Books, 1972), 133.

5. Ronald Findlay and Kevin O'Rourke, *Power and Plenty: Trade, War, and the World Economy in the Second Millennium* (Princeton: Princeton University Press, 2007), xxiv; O'Rourke and Williamson, *Globalization and History*, 14; and Martin Daunton, *Wealth and Welfare: An Economic and Social History of Britain, 1851-1951* (London: Oxford University Press, 2007), 166, 201ff.

6. Useful overviews may be found in Daniel Headrick, *The Invisible Weapon: Telecommunications and International Politics, 1851-1945* (London: Oxford University Press, 1991); and David Nickles, *Under the Wire: How the Telegraph Changed Diplomacy* (Cambridge, MA: Harvard University Press, 2003).

7. Figures extracted from O'Rourke and Williamson, *Globalization and History*, ch. 3.

8. William Ashworth, *An Economic History of England, 1870-1939* (London: Methuen, 1960), 138. The origins of the term "globalization" are usually traced to a 1983 article by Professor Theodore Levitt.

9. O'Rourke and Williamson, *Globalization and History*, 2.

10. Capt. Alfred T. Mahan, "Considerations Governing the Dispositions of Navies," *National Review*, July 1902, reprinted in Alfred Thayer Mahan, *Retrospect and Prospect: Studies in International Relations, Naval and Political* (Boston: Little, Brown, 1902), 143–44.

11. Jonathan Levy, *Freaks of Fortune: The Emerging World of Capitalism and Risk in America* (Cambridge, MA: Harvard University Press, 2012), 232–51.

12. Federal Trade Commission, *Report on the Grain Trade: Terminal Grain Marketing* (Washington, DC: Government Printing Office, 1922), 3:162. See also vol. 5, 1920, 23ff.

13. Morton Rothstein, "America in the International Rivalry for the British Wheat Market, 1860–1914," *Mississippi Valley Historical Review* 47 (December 3, 1960): 401–18, 411;

Morton Rothstein, "Centralizing Firms and Spreading Markets: The World of International Grain Traders, 1846–1914," *Business and Economic History* 17 (1988); and Morton Rothstein, "Multinationals in the Grain Trade, 1850–1914," *Business and Economic History* 12 (1983): 85–93.

14. Royal Commission on the Supply of Food and Raw Material in Time of War (Balfour of Burleigh Commission) Cd.2643 (London: His Majesty's Stationary Office, 1905), 74, Q. 2049ff.

15. Cain and Hopkins, *British Imperialism,* 449–51.

16. Royal Commission on the Supply of Food, Cd.2643, transcript of testimony, Mr. Stanley Woods, Q.2419–25.

17. Ivan S. Bloch, *The Future of War in Its Technical, Economic, and Political Relations: Is War Now Impossible?,* ed. William Stead (New York: Doubleday & McClure, 1899), xvii. See also xxxviii, xlix, 179–82, 242.

18. Winston S. Churchill to H. H. Asquith, September 8, 1913, box 13, Asquith Mss., Bodleian Library, Oxford. Emphasis added.

19. Lambert, *Planning Armageddon,* 48–49.

20. Balfour to Esher, January 9, 1912, ESHR.5/40, Lord Esher papers, Churchill College, Cambridge.

21. Lambert, *Planning Armageddon,* ch. 5.

22. Lord Esher's Journal, entry for August 3, 1914, p. 4, ESHR.2/13, Lord Esher papers.

23. Herbert Samuel to wife, August 4 and 5, 1914, f.68, 71, Herbert Samuel Mss., Parliamentary Archives, London.

24. Lewis Harcourt, note of meeting at 10 Downing Street, 11:15 p.m., Tuesday, August 4, 1914, Ms.Dep.6231., Harcourt Papers, additional, Bodleian Library, Oxford.

25. What follows is based on Lambert, *Planning Armageddon,* ch. 5–7.

26. Chris Mirasola, "Understanding China's Cybersecurity Law," Lawfare, November 8, 2016, https://www.lawfareblog.com/understanding-chinas-cybersecurity-law.

9 Why a Digital Pearl Harbor Makes Sense . . . and Is Possible

EMILY O. GOLDMAN AND MICHAEL WARNER

Emerging technologies are changing how people create, share, protect, and store data; intellectual property; and wealth. New lines of operation have emerged for governments and businesses to pursue along with new weaknesses and vulnerabilities for adversaries to exploit. The unevenly governed spaces of the cyber domain have become the newest front line for military and economic confrontation because cyber attacks fit conveniently into adversarial strategies to counter superior conventional military capabilities. Cyber weapons enable even relatively unsophisticated actors to project power and operate deep within virtual and physical territory of the United States and other countries. They target the domain where the United States and other technologically advanced states are most vulnerable: our interconnected society, economy, and networked military, all of which rely on a digital architecture constructed for speed and convenience, not security.

For these reasons, a "cyber Pearl Harbor" has been one of the most prevalent and familiar analogies used by American officials, experts, and pundits to raise awareness of the dangers in this new realm of competition. The analogy conjures up grainy newsreel footage of burning battleships and the nation's entry into World War II. It evokes a devastating bolt from the blue that leaves an indelible imprint on the US psyche. In 2012 then secretary of defense Leon Panetta raised the specter of a cyber Pearl Harbor when he warned of attacks that could cripple the United States or its military. "Remember Pearl Harbor" is a call to mobilize support for increased cyber preparedness.

In spite of its critics, the Pearl Harbor analogy has endured because it usefully frames how dependence on cyberspace generates vulnerabilities that adversaries can exploit. Like all analogies, it must be applied with care. It gives less purchase when treated as a case of strategic surprise because the idea of a crippling bolt from the blue is an inaccurate characterization of historic events as well as an unlikely harbinger of future ones. But the analogy does provide insight into how an adversary could gain leverage over a conventionally superior military by avoiding areas of stronger states' military dominance and by launching cyber attacks against critical military infrastructure. It is also a warning that much of the current, tactical war-fighting capability of the United States and its allies depends on their ability to navigate and secure cyberspace, where their forces and weapons systems are linked and controlled.

What Happened at Pearl Harbor

Pearl Harbor was not a strategic, bolt-from-the-blue surprise for the United States. Diplomatic relations between Japan and America had reached their nadir in late 1941. The United States was exercising coercive power to contest Japan's occupation of China and other Asian states, and Washington expected war. Pearl Harbor was a logical, if misguided, result of Imperial Japan's long-term strategy to expand its Pacific empire and blunt the United States' effort to stop it. Japanese expansionism focused on establishing an exclusive zone of influence, the Greater East Asian Co-Prosperity Sphere. By mid-1941, however, Japan's aggression in China and its larger aims in the southwest Pacific were hampered by President Franklin D. Roosevelt's embargoes and freezing of Japanese funds in US banks, which Tokyo saw as tantamount to economic warfare. Japanese naval planners in response sought to stymie Washington's ability to frustrate Tokyo's seizure of the resources that its military and economy desperately needed. Since America had moved aggressively to frustrate Japan's military aims in China and impede its economy, the Japanese reasoned they had to hit back.

Adm. Isoroku Yamamoto, Japan's top naval strategist, understood the risks of fighting America's industrial might, which he had seen firsthand as a young officer. For Japan to win and retain the upper hand, he believed, it had to strike a decisive blow at the outset of hostilities—one that would preclude the possibility of the United States going on the offensive while Japanese forces consolidated a defensive perimeter in the western Pacific. Japan would have to eliminate US forces in the Philippine Islands (astride Japan's key supply routes) and crush the US Pacific Fleet near Hawaii to prevent its advance toward Japanese home waters before Japan was ready for the great naval clash that would decide the struggle once and for all.

Yamamoto's strategic objective was not to conquer the United States or even to seize (much) US territory but to delay the inevitable American counteroffensive. He judged that destroying the Pacific Fleet's offensive power, even temporarily, would allow Japanese forces to take control of oil supplies in the Dutch East Indies and erect a barrier chain of island bases, thereby enabling Japan to delay the Pacific Fleet's westward progression and perhaps even force negotiations from a position of strength. A model for the attack was Germany's successful blitzkrieg strategy in France: hit hard and demoralize the adversary so its people would reject a long and costly war. It would take years for the United States to recover, Yamamoto hoped, and by then the Americans would face a fait accompli, with Japan's control extending from the Indian to the Pacific Oceans.

Although the Americans possessed ample strategic warning, they had little tactical warning because Japanese operational deception worked. American leaders knew full well that Pearl Harbor was vulnerable and explicitly considered the possibility of attacks by Japanese submarines, saboteurs, or carrier-based aircraft.[1] Leaders in Washington and Hawaii did not, however, consider such attacks at Pearl Harbor to be either inevitable or imminent. US Army analysts had plentiful diplomatic signals intelligence from reading Japan's Foreign Ministry ciphers, but Japanese diplomats were not informed of the day and hour that war would begin until the last possible moment. US Navy analysts misread the intelligence clues they possessed (and failed to realize how many indicators they lacked) partly because the Japanese fleet practiced simple but effective deception and denial methods. Better management of intelligence analysts in Washington and Hawaii might well have revealed additional clues to Tokyo's intentions and spotted the Japanese deception efforts, thus providing another vital indicator of impending hostilities.

Pearl Harbor with its strong defenses proved vulnerable because the Americans lacked situational awareness and tactical control. Japanese aerial and submarine scouting of the harbor at dawn on December 7 should have prompted the base to go to battle stations, but as the US Army and Navy failed to coordinate their watches, clear indicators went unheeded. Also, had the radar system sent to guard Pearl Harbor been fully operational, the Japanese attack could have been blunted. Radar operators in training informed their chain of command of incoming planes that morning, but they were told the radar returns represented a US Army Air Forces flight from California.

The Japanese decision to attack Pearl Harbor looks logical only when one ignores its absurd premise: the island nation of Japan, already enmeshed in a war against the world's most populous country (China) and having recently fought another neighbor (the Soviet Union), could better its lot by attacking the world's foremost naval powers (the United States and Britain). Such a suspension of common sense was possible only in Tokyo's militarized political climate, in which the army dominated the prime minister, who could not form a government without the army's support.[2]

What Did Not Happen at Pearl Harbor

Japan hoped to buy time and present the United States with a fait accompli, but the Pearl Harbor operation was operationally and tactically flawed. The Imperial Japanese Navy's striking force easily reduced the Pacific Fleet's aging dreadnoughts on Battleship Row to smoking hulks, but it missed the fleet's more important aircraft carriers, heavy cruisers, and submarines. Japanese pilots had been ordered to hit these ships, but most of them, including all the carriers, were at sea to keep them away from Pearl Harbor. This represented a huge missed opportunity for Japan, as US Navy aircraft carriers and submarines would play a key role in strangling Japan's supply lines from 1942 onward and largely determine the outcome of the Pacific war.

The attack also did little damage to the Pacific Fleet's vital supplies and servicing components: fuel depots, dry docks, repair facilities, and undersea cable landings. This factor looms large in our analysis when considering the decision-making that had already occurred in Washington. Although Japanese planners saw the fleet as their main objective, the US Navy's leadership believed the facilities at Pearl Harbor were more important. Pearl Harbor was not the Pacific Fleet's main base. In May 1940 the US Navy (under firm orders from the White House) had temporarily shifted the main base of its Pacific Ocean fleet from San Diego to Pearl Harbor.[3] Admiral Yamamoto described this move as "tantamount" to a declaration of war. American admirals, in contrast, worried that the Japanese would target the harbor's "critical infrastructure" while the fleet was at sea.[4] Pearl Harbor's oil stocks, in particular, were obvious from the air and highly vulnerable. Hitting them along with the ship repair facilities might have sent the fleet back to San Diego, to which the navy already wanted to return.

Had that happened, the course of the Pacific war could have been much different. The importance of Pearl Harbor's facilities is difficult to overestimate. Oil tanks can be rebuilt and refilled, and dockyards can be repaired. But when? Not for weeks at a minimum and perhaps for months—that is, if Washington decided to rebuild and reinforce its already ruined and exposed Hawaiian base. In the event, surviving US warships had plenty of fuel to operate in the central Pacific, and indeed they were operating there and harassing the Japanese within days of the Pearl Harbor attack. Salvage operations on the sunk and damaged ships at Pearl commenced immediately. And the harbor was open for business when it mattered most, in May 1942. The aircraft carrier USS *Yorktown* had received serious damage in the Battle of the Coral Sea (May 8) but was hastily repaired at Pearl Harbor. Experts estimated a two-week spell in dry dock for *Yorktown*, but technicians did enough work in forty-eight hours for it to sail again. If those vital repair facilities had been destroyed and those workers sent back to San Diego, *Yorktown* would have been unavailable for the pivotal Battle of Midway on June 4, leaving the Imperial Japanese Navy with a much freer hand.

Japan's surprise attack might have disabled the US Pacific Fleet for a year or more, giving Japanese forces time to dig in for a lengthy conflict. But Japan's intended knockout blow didn't succeed, and it ensured the United States would fight to the end of Japanese militarism.

The Logic of a Digital Pearl Harbor

Theories of surprise attack can account for the pattern of a weaker state lashing out at a stronger opponent to gain time to consolidate its ill-gotten gains. It might involve a direct attack on the stronger party or instead a seizure of something of interest to the stronger party but not worth enough to merit a protracted conflict to regain it. The latter, less provocative fait accompli is still a half step toward war. It promises a greater chance of political victory than quiet diplomacy, but (being violent itself) it also raises the risks of escalatory warfare.

The Pearl Harbor analogy is a warning to study the calculations of adversaries. Some, like Japan in the 1930s, might consider a surprise attack a viable preemptive option to temporarily blunt superior military capabilities. Do other people actually think this way? They have and they do. Saddam Hussein of Iraq certainly did in 1990. He mounted a surprise mobilization of his best divisions and invaded neighboring Kuwait as soon as his forces were ready. Kuwait fell to Iraqi troops in hours, giving Hussein an oil-rich "nineteenth province" with a fine harbor on the Persian Gulf and changing at a stroke Iraq's strategic position with respect to Iran, its enemy in a grim eight-year war that had recently ended. A lesson for our time is that adversaries who feel their backs against the proverbial wall might lash out in new and unexpected ways.

Conditions exist today that resemble the East Asian crisis in the 1930s and could entice an adversary to strike a similar blunting attack against the United States or one of its allies in the hope of a quick victory that presents it with an undesirable strategic fait accompli. A power with a high tolerance for risk and, perhaps, a growing sense of desperation—especially one that perceives the United States or other adversaries to be seriously threatening its political and strategic fortunes—could use cyber means to shape the preconflict environment and delay or deter America's response. In such an aggressor's calculus, the United States (or other potential adversaries) might be induced to stay out of a regional conflict, in effect letting an aggressor keep his gains. Once an aggressor gained what it wanted, it might even have the ironic temerity to call on the international community to intervene and stop its opponent's pressure and retaliation. If an adversary's objective is to convince Washington or another state to leave it alone, or to allow it to pursue its aims against its neighbors, then Admiral Yamamoto's intellectual heirs in such a situation could be tempted to mount a quick strike.

But unlike Yamamoto's pilots, a contemporary adversary might either strike mobilization and logistical networks, impeding the adversary's ability to operate militarily, or manipulate information to blind and confuse it, much like China's strategy in Peter Singer's novel *Ghost Fleet*. From the adversary's perspective, a cyber attack has the virtue of damaging its opponents' ability to respond in the physical domain while not provoking public cries for retribution the way a terrorist attack on a city would.

In the 1991 Gulf War, Iraqi leaders did not fully appreciate the significance of highly advanced surveillance planes or networked computer communications. A future opponent will not likely make the same mistake.[5] Dependence on cyberspace for shared battlespace awareness may provide a decisive advantage for higher-tech militaries today, but the data infrastructure and data themselves make exceedingly valuable targets. The incentive to contaminate or disrupt the information flows on which the US military depends, for example, is enormous. A cyber-savvy adversary with outsized goals could find this type of effect attractive. Adversary countries need not even be risk acceptant to adopt this strategy. They may in fact be risk averse, but like Japan in 1941, their decision-making processes may give disproportionate weight to the most bellicose and paranoid leaders and factions. It does not take the resources of a state to mount a damaging cyber attack, and such an attack can be formulated in secrecy even from other parts of the attacking state. Or, as recent events suggest, national governments may find it challenging to exert control and oversight over all potential hacker communities within their military cyberspace apparatus, not to mention industry or patriotic hacktivists outside state control.

The Pearl Harbor analogy reinforces the maxim that while "amateurs focus on tactics, professionals study logistics." Targeting critical military cyber infrastructure today might succeed where Japan failed in 1941. This is so not because of any quality inherent in the attacker or in the political environment but because of the dependence of advanced militaries on cyberspace.

The Current Environment

These concerns are not far fetched. The adversaries of the United States, and those of other states, have invested in asymmetric means—such as anti-access and area-denial capabilities—to counter traditional US strengths and to prevent it from projecting power abroad. They are preparing the future cyber battlefield now by stealing intellectual property, conducting industrial espionage, and exploiting government networks and those of defense, financial, and communication industries. Through intelligence, surveillance, and reconnaissance against US and allied networks, they have gained penetration and established persistent access. These activities can be verified by perusing the continuous and alarming public statements made by a host of independent computer and software security experts in recent years. Of course, adversaries might believe that the United States and its allies also engage in cyber operations to gather intelligence and perhaps to conduct attacks.

US adversaries have also shown an increasing capability and intent to target its industrial control systems recently. Since 2011 known or suspected hackers in several countries have run supervisory control and data acquisition (SCADA) exploitation attempts against US critical infrastructure. In September 2015 in testimony before the House Permanent Select Committee on Intelligence, Director of National Intelligence James Clapper revealed that unknown Russian cyber actors had compromised the supply chains of at least three industrial control system vendors. He warned, "Politically motivated cyber-attacks are now a growing reality, and foreign actors are reconnoitering and developing access to U.S. critical infrastructure systems."[6] Cyberspace threats also headlined Clapper's February 2016 testimony to the Senate Select Committee on Intelligence on worldwide threats.

In subsequent hearings before the House Armed Services Subcommittee on Emerging Threats and Capabilities in March 2016, Adm. Mike Rogers, commander of US Cyber Command and director of the National Security Agency (NSA), testified that "industrial control systems and SCADA probably is the next big area for us because we've got to transition from a focus purely on the network structure."[7] He noted that the Department of Defense (DOD) has already begun looking at data concentrations and focusing more on industrial control systems and SCADA. Rob Joyce, chief of the NSA's Tailored Access Operations unit, complains that SCADA security keeps him up at night. Joyce understands how to exploit such systems;[8] he also appreciates how vulnerable the United States is, in turn, and that the "Internet of things" will multiply those vulnerabilities exponentially.[9]

Supply chain vulnerability, another dimension of the problem, was raised during Admiral Rogers's hearings before the Senate Armed Services Committee in April 2016. Specific processes exist in the US government to address these issues for some components of DOD infrastructure, particularly nuclear systems, but not for other major systems or components. The DOD's focus on network security is now expanding to focus on the risks to individual combat platforms, weapons systems, and individual data concentrations. In the National Defense Authorization Act of 2016, Congress directed the secretary of defense to complete an evaluation of the cyber vulnerabilities of every major weapons system by December 2019.

Perhaps the starkest exemplar of military vulnerability involves cyber attacks against US Transportation Command (USTRANSCOM), the command responsible for moving US troops and military equipment around the world. In September 2014 the Senate Armed Services Committee made public the results of its investigation into hacking activities targeting US military contractors. It reported that hackers sponsored by the Chinese government accessed contractors' computer systems "at least twenty times in a single year."[10] Targeted cyber attacks against USTRANSCOM persisted longer than a year with fifty cyber events documented between June 2012 and May 2013. According to the committee's report, USTRANSCOM is targeted more than any other combatant command because it is particularly vulnerable. It relies on commercial partners to deliver 70 percent of its military equipment, supplies, and personnel around the world and to keep the US military running. Ninety percent of USTRANSCOM's communications, distribution, and deployment transactions are conducted on unclassified networks because the companies it relies on cannot access the Pentagon's secured network. This truly is the Achilles' heel for US power projection.

Lessons for Today

The purpose of Japan's Pearl Harbor attack was to delay, not annihilate, US power. How long might a state take to recover from attacks on critical infrastructure systems on which its society, economy, and military depend today? Would that delay buy an adversary time to operate without fear of retaliation? Is there a contemporary analogue to the decisive blow against Pearl Harbor, or are we more secure by having a distributed infrastructure? In cyberspace a "Pearl Harbor" might consist of numerous national systems and institutions critical to the economy. They include but are not limited to undersea cables, power grids, water supplies, classified networks, electronic voting systems, banking systems and electronic funds transfer (those that enable Internet commerce, for instance), and heavy reliance on a single operating system. Recent examples of inadvertent interruptions in these systems have resulted in large consequences. The effects of intentional disruptions can only be imagined.

The first lesson is to take seriously telltale warning signs to avoid being caught tactically off guard as the US Navy and Army were in 1941. US government officials have been very public about seeing multiple cyber actors penetrating US systems. These events are not isolated but rather parts of sustained campaigns, indicating a long-term commitment to understanding systems and to ensuring the intruders possess the capability to potentially impair the country's ability to operate. Their purpose for now appears to be conducting reconnaissance and surveying systems, their vulnerabilities, and the control points that someone would want to access. But intent could change quickly. Cyber actors want to ensure they have technical options should they make the political decision to interfere with their competitors or send a message to deter them. Threat is composed of capability and intent, and multiple actors are demonstrating their ability to gain access to critical infrastructure. A premium must also be placed on watching for clues to their intent.

The December 2015 events in Ukraine, where multiple electric companies were hacked—the first power outage known to be caused by a cyber attack—should serve as a wake-up call. Hackers used malware to gain access to the Ukrainian utilities' business networks and, from there, maneuvered to their production networks and on to operator stations. The hackers then remotely disconnected the breakers of thirty substations. According to Robert M. Lee of Dragos Security, "Every bit of this is doable in the U.S. grid." Although the US grid is more hardened than Ukraine's, the former's recovery would be more difficult because if the SCADA systems are lost, the fully automated US systems cannot switch to manual control as the Ukrainians' system did.[11]

Vulnerable states and enterprises must strengthen their defensive capabilities both for government networks and for the critical infrastructure nodes that are outside government control and in the hands of the private sector. Corporate leaders in the United States and elsewhere are working hard to correct those deficiencies. The US power sector is looking at microgrids and other techniques to try and break the grid into smaller and thus potentially more defensible segments. But overcoming decades of investment in capital infrastructure—in which defensibility was never a core design characteristic—is a huge challenge.

Going forward will require a culture change as well. Expecting zero system penetrations is unrealistic because it is less a question of if than when attackers will get through the perimeter. The measure of success thus rests on how one responds when a penetration occurs. In the summer of 2015, after an intrusion into the US Joint Staff's unclassified systems, the DOD quickly disconnected the network and ensured no data was extracted or a long-term presence was established. While this action is a good model for defensive response, disconnection is not always feasible. It is necessary to learn how to retain the capability and the mission of the network while maneuvering and fighting to drive the opponent out. This work requires a different skill set and mind-set.

Culture change must extend to the entire workforce. In the United States, the DOD is working to create a culture where cyber hygiene and cybersecurity are as foundational to a DOD employee as the accountability expected of every affiliate who is issued a weapon. That weapon, of course, must be appropriately treated, appropriately used, and always secured. Traditionally, cyber and cybersecurity have been viewed as very specialized and highly technical work that only a small segment of the workforce (the information technology [IT] personnel) did. Cyber was the purview of the chief information officer or chief technologist. Senior military and defense officials, like their private sector counterparts in the C-Suite—that is, the corporations' senior executives and board members—looked to the IT experts to take care of problems they were uniquely trained to do. Increasingly they have recognized that everyone in this domain is a point of vulnerability as cyber behavior shapes the ability to defend networks. On September 28, 2015, the secretary of defense and chairman of the Joint Chiefs of Staff authorized the Department of Defense Cyber Culture and Compliance Initiative. It intends to transform DOD cybersecurity culture by improving the individual human performance and accountability of every member of the DOD cyber enterprise: leaders, service providers, cyber warriors, and users.

Another takeaway is the importance of public-private partnerships. The private sector can generate insights into what is happening online that governments cannot. The reverse is also true. Thus, the capabilities of intelligence infrastructure must be augmented by insights from the private sector to fill respective information gaps and to better understand what is happening, who the actors are, and what tactics, techniques, and procedures they are using. Defeating an enemy starts from the premise that one is aware of and understands it. Developing this knowledge reflects the power of partnership. With a legal framework that engenders confidence and enhances the free flow of information in both directions, government can push actionable information to the private sector. The US Congress in 2015 passed the Cybersecurity Information Sharing Act, which enables industry to increase its sharing of threat information with the federal government (and vice versa) without fear of losing competitive advantage or risking additional legal liability. It has established a key element in the government's efforts to improve the cybersecurity of critical infrastructure.

Connecting the dots is critical, but planning responses to attacks is also necessary. What would have been the US response if the right dots had been connected in December 1941? International law permits nations to conduct preemptive strikes in self-defense. Seeing six Japanese aircraft carriers north of Hawaii and headed for Oahu at top speed on December 6, for example, US Army bombers might well have been ordered, with clear legal justification, to launch attacks against that fleet. Currently few clear guidelines, however, exist for preemptive strikes to blunt or prevent cyber attacks against national infrastructure.

American political leaders and scholars have typically viewed cyber operations as wartime measures undertaken only during a conflict and therefore as inherently escalatory. Yet cyber conflict has been increasing for years, and cyber interference with US and other states' interests occurs daily. Cyber operations are an extension of policy and strategy, increasingly a normal part of state behavior.

Gen. Darren W. McDew, commander of USTRANSCOM, raised questions before the House Armed Services Committee's Readiness Subcommittee in March 2016 that all nations dependent on IT and computer technology networks must address: "When can I defend my network, how far out can I defend? What constitutes an attack on a commercial provider? What do they have to report as an attack, because the definition may be not as clear with every single person?"[12] After blunting an adversary attack, moreover, states must have the capability to maneuver to conduct operations that neutralize and disrupt the adversary's ability to conduct follow-on cyber operations.

Conclusion

The Internet is inherently redundant and resilient. But technologically advanced states would be foolish to rule out a lucky hit that cripples them, just as the Pearl Harbor attack would have crippled the Pacific Fleet if the Japanese pilots had hit the fleet's oil supplies and dockyards instead of its old battleships. The United States had little up-to-date experience in maneuvering large sea, air, and ground forces at the outset of World War II. The critical infrastructure of many states today likewise remains unprepared for cyber attacks.

Critics of this analogy could argue that adversaries probably would not, or could not, launch a cyber Pearl Harbor–style attack against the United States. But one could have said the same about the prospect of an air raid on Pearl Harbor in 1941. The Japanese attack was strategically foolhardy, but nonetheless it happened. The ease of applying mass and achieving surprise in the new cyber domain means that numerous competitors are in this space. Threats to national and economic security in cyberspace are increasing in complexity and destructiveness as well. Thus, given the lack of traditional warning and the absence of immediately visible consequences for malicious cyber behavior, adversaries could believe they face few costs and yet stand to reap huge benefits.

Notes

This chapter is an adaptation and extensive revision of the chapter by the coauthors (along with John Surdu) that appeared in Emily O. Goldman and John Arquilla, eds., *Cyber Analogies* (Monterey, CA: Naval Postgraduate School, 2014). Drs. Goldman and Warner serve in US Cyber Command, but the opinions expressed herein are their own and in no way represent official positions of the Department of Defense or any other US government entity.

1. Gordon W. Prange, *At Dawn We Slept: The Untold Story of Pearl Harbor*, with Donald M. Goldstein and Katherine V. Dillon (New York: McGraw-Hill, 1981).

2. Ibid., 213.

3. Ibid., 38, 93, 97.

4. Ibid., 42, 66, 97.

5. Richard J. Harknett and the JCISS Study Group, "The Risks of a Networked Military," *Orbis*, Winter 2000, 127–43, https://www.comw.org/rma/fulltext/00harknett.pdf.

6. James Clapper, director of National Intelligence, Statement for the Record, "Worldwide Cyber Threats before the House Permanent Select Committee on Intelligence," September 10, 2015, https://www.dni.gov/index.php/newsroom/testimonies/209-congressional-testimonies-2015.

7. Testimony of Michael Rogers, commander, US Cyber Command, at the hearing on US Cyber Command's FY 2017 budget request, Committee on Armed Services, Subcommittee on Emerging Threats and Capabilities, 114th Congress, March 16, 2016.

8. While the head of TAO at NSA, Joyce was tapped to become the White House cybersecurity coordinator for President Donald Trump.

9. Tom Simonite, "NSA Hacking Chief: Internet of Things Security Keeps Me Up at Night," *MIT Technology Review*, January 27, 2016.

10. Senate Armed Services Committee, *Inquiry in Cyber Intrusions Affecting US Transportation Command Contractors*, 113th Cong., 2d sess. (Washington, DC: Government Printing Office, 2014).

11. Kim Zetter, "Everything We Know about Ukraine's Power Plant Hack," *Wired*, January 2016.

12. Darren McDew, commander, US Transportation Command, testimony on the FY17 US Transportation Command FY 17 budget request, House Committee on Armed Services, Subcommittee on Readiness, 114th Congress, March 15, 2016.

PART III

What Are Preventing and Managing Cyber Conflict Like?

10 Cyber Threats, Nuclear Analogies?

DIVERGENT TRAJECTORIES IN ADAPTING TO NEW DUAL-USE TECHNOLOGIES

STEVEN E. MILLER

Alarm is mounting over large security vulnerabilities produced by the pervasive spread of cyber capabilities into vast realms of socioeconomic activity. To be sure, most cyber threats fall into the category of mischief or normal crime, but some potential cyber attacks—on nuclear power plants or other critical infrastructure or on the financial system, for example—could do enormous harm. There is a need, therefore, to seek remedies and adapt to the challenges posed by this ubiquitous dual-use technology.[1]

Other dual-use technologies have raised similar challenges of adaptation. It is natural to examine these other, possibly analogous experiences to see if there are lessons that might apply in the cyber realm. This chapter looks at the emergence of nuclear technology, examines the challenges it posed and the reactions to those challenges, explores the evolution of early thinking about the risks and benefits of nuclear technology, and considers whether the trajectories and time lines of the adaptation to nuclear technology have any resonance with the cyber issue. Is there a nuclear analogy? What elements of the response to nuclear technology, if any, have relevance for the cyber era?[2]

I attempt to answer these questions by offering a brief account of three dimensions of the nuclear experience: First, how did the nuclear age arrive and what was the response to it? Second, how did the peaceful benefits of nuclear technology fit into the picture? And, third, what answers were found to the national security threats raised by the nuclear revolution? These discussions reveal that the nuclear story is very different from the more recent experience with cyber technology in a number of fundamental respects.

One basic difference is that the nuclear tale is first and foremost about weapons with possible civilian applications rather than the other way around. The weaponized form of the technology was from the beginning, and has remained, the center of the nuclear question. In the nuclear case, civilian applications struggled to be born and in many respects were less impactful than expected. The time lines and trajectories associated with cyber and nuclear are quite different, and their areas of primary impact fall in different domains: nuclear technology is above all a geopolitical consideration, whereas cyber technology has become an enormous factor in many areas of social and economic life. Nevertheless, there are parallels as both technologies have raised the question of how to protect against an intractable threat. Some of the answers considered in the nuclear realm may find application if adapted to the cyber context.

The Nuclear Age Arrives

The nuclear age arrived with stunning suddenness. To all but the minuscule fraction of humanity that had been privy to the Manhattan Project, the unprecedented weapons employed in the bombings of Hiroshima and Nagasaki in August 1945 were completely unexpected and shockingly devastating. The world was made aware of this new development when President Harry S. Truman issued an unassuming but muscular three-page typewritten press statement on August 6, 1945.[3] "Sixteen hours ago," the statement began in almost understated plain language, "an American airplane dropped one bomb on [Hiroshima] and destroyed its usefulness to the enemy."[4]

It was necessary, of course, to explain what this meant to an unknowing world: "With this bomb we have now added a new and revolutionary increase in destruction to supplement the growing power of our armed forces. In their present form, these bombs are now in production and even more powerful forms are under development. It is an atomic bomb. It is the harnessing of the basic power of the universe. The force from which the sun draws its power has been loosed against those who brought war to the Far East."[5]

The president also spelled out in unflinching terms what this new weapon meant for Japan in the ongoing war in the Pacific: "We are now prepared to obliterate more rapidly and completely every productive enterprise the Japanese have above ground in any city. We shall destroy their docks, their factories, and their communications. Let there be no mistake: we shall completely destroy Japan's power to make war."

This language, blunt though it is, betrayed an incomplete comprehension of the destructive effects of the atomic bomb. It destroyed not docks and factories but cities, as soon became apparent when images of Hiroshima and Nagasaki were revealed to the world. But the implications for Japan's leaders were conveyed in vivid terms that left no doubt about the destructive potential of this new technology. Hoping "to spare the Japanese people from utter destruction," the president's statement called on Japan's leaders to accept the ultimatum that had been issued at the Potsdam Conference in July 1945. In perhaps his most famous passage of the statement, he continued, "If they do not now accept our terms they may expect a rain of ruin from the air the likes of which has never been seen on this earth."[6]

The extraordinary character of the atomic bomb was recognized almost instantly. In her widely read syndicated newspaper column, "My Day," for example, Eleanor Roosevelt wrote on August 8, 1945—even before Nagasaki—about the implications of the atomic bomb: "This discovery may be of great commercial value someday. If wisely used, it may serve the purposes of peace. But for the moment we are chiefly concerned with its destructive power. That power can be multiplied indefinitely, so that not only whole cities but large areas may be destroyed at one fell swoop. . . . You soon face the unpleasant fact that in the next war whole peoples may be destroyed. . . . This discovery must spell the end of war."[7]

Similarly, immediately upon hearing of the bombing of Hiroshima, Bertrand Russell wrote a small essay, published ten days later, on August 18, 1945, under the title "The Bomb and Civilization." He lamented that a historic scientific accomplishment had produced such terrible results and commented in a stunned and frightened fashion about what this development could mean for the future: "It is impossible to imagine a more dramatic and horrifying combination of scientific triumph with political and moral failure than has been shown to the world in the destruction of Hiroshima. . . . In an instant, by means of one small bomb, every vestige of life throughout four square miles of a populous city has been exterminated. . . . The prospect for the human race is somber beyond precedent."[8]

The August 20, 1945, edition of *Life* magazine, seen by millions of Americans, was devoted to the atomic bomb and included photos of Hiroshima and Nagasaki. It also contained an essay by *New York Times* military correspondent Hanson Baldwin on the implications of the atomic bomb for military power. Possibly for the first time, Baldwin raised the question of whether traditional military forces were now obsolete, thus opening a fierce debate that would haunt and damage the US military in the coming years.[9] Immediately after the bombing of Hiroshima and Nagasaki, clearly a new era had arrived that required serious rethinking of international politics and security. Above all, the sense of shock was almost universal at the scale of the destructive potential associated with this new weapon. As Paul Boyer noted in his own detailed account of reactions to the atomic revolution, "The whole world gasped."[10]

Thus was nuclear technology introduced to most of the world. The transition to the nuclear age was abrupt. From the first moments of this new era, the new technology existed as a weapon. This was not a case in which an important *civilian* technology had the potential also for malign use with wide consequences. Indeed, nuclear arrived in exactly the opposite circumstance—as a weapons technology that, it was hoped, could have civilian applications. At the birth of the nuclear age, however, no civilian uses of the new technology existed.

Nuclear technology emerged from a top-secret military program. It was in the hands of only one power that explicitly intended to keep its secret as long as possible. Further, the damage that nuclear technology could wreak was not linked to speculative scenarios or hypothetical worst cases. Two incinerated cities lay in ruins, demonstrating the gruesome destructiveness of this new technology. And in many quarters, there was immediate recognition that this new era had profound implications for international politics and security.

In its origins, the nuclear story is very different from that of cyber. No overnight passage led into a dramatically new cyber world; rather, the cyber world was built progressively across multiple decades. Though cyber history is contested, the origins of the Internet are generally traced to a relatively obscure military project in the late 1960s, the Advanced Research Projects Agency Network (ARPANET). The world did not gasp at the birth of ARPANET. It was not until the 1980s that this creation began to emerge as a public phenomenon and not until the 1990s that it began, in an accelerating fashion, to dominate communications and to penetrate wide swaths of socioeconomic life. Though it has military applications, cyber is pervasive in civilian activities and has become part of the basic infrastructure of civilian life in large portions of the world. Indeed, it is precisely the dependence of much economic and social activity on cyberspace that creates the large vulnerabilities about which we now worry. Thus, the way these two technological eras emerged and the framework of issues they raise are very different.

Developing the Peaceful Atom

The nuclear age arose out of wartime exigencies, and its initial technological manifestations were the result of a crash military program, although it was understood early on that nuclear technology might have an array of civilian applications. Even after the end of World War II, however, peaceful uses of the atom remained a subsidiary concern, especially as the Cold War rapidly emerged and an ensuing nuclear rivalry with Russia came to dominate security concerns. As Richard Hewlett and Francis Duncan observe in their still indispensable account of the early years of US nuclear policy, the postwar period was marked by "a shift from the idealistic, hopeful anticipation of the peaceful atom to the grim realization that for reasons of national security atomic energy would have to continue to bear the image of war."[11] The priorities of national policy remained fundamentally important because for more than a decade the US government retained a monopoly on the nuclear information, technology, and activity. Only after the passage of the Atomic Energy Act of 1954 was the private sector legally authorized to undertake commercial activities in the nuclear realm (and, even then, only under strict government regulation). Many of the possible peaceful uses of nuclear power experienced a protracted struggle to be realized for several reasons: Fissile material was (for a time) relatively scarce; the demand for nuclear weapons grew steadily; the weapons program retained highest priority; secrecy was highly valued; the technologies in question were expensive, challenging, and difficult to commercialize; and the number of nuclear experts was limited.[12]

Nevertheless, the apocalyptic fears of nuclear destruction were accompanied by extravagant visions of extraordinary, widespread nuclear benefits. In the initial wave of enthusiasm about the promise of the nuclear revolution, popular discussions covered everything from atomic energy vitamin tablets to atomic-propelled vehicles of all varieties (including automobiles) to breathtaking medical breakthroughs (cancer cured) to abundant and inexpensive electricity (too cheap to meter, as it was sometimes predicted). There was serious consideration of the problems that would flow from this vast nuclearization of civilian life. For example, what about the thirty million automobiles that would be rendered obsolete by the arrival of the nuclear-powered car? In the early phase of the nuclear age, the air was filled with what Boyer described as "fantasies of a techno-atomic utopia."[13] Some of this forecasting did indeed belong in the realm of science fiction, but the optimistic exploration of possible peaceful applications of nuclear technology was far from completely disconnected from policy. In 1946, for example, the Atomic Energy Commission established the Nuclear Energy for the Propulsion of Aircraft program.[14]

By the time President Dwight Eisenhower took office, however, the nuclear arms race clearly was roaring ahead, galvanized by the Soviet acquisition of nuclear weapons starting in 1949, and the peaceful atom was lagging. Moreover, Eisenhower assumed the presidency as the US nuclear arsenal was making the transition to the hydrogen bomb (H-bomb), which is vastly more powerful than the weapons employed in August 1945. Eisenhower had been briefed about the H-bomb during the presidential transition and had been "struggling with the staggering implications of a weapon that could destroy not only an entire city but perhaps civilization itself."[15] He understood the portentous implications of the H-bomb revolution and even alluded to this issue (albeit indirectly) in his inaugural address: "Science seems ready to confer upon us, as a final gift, the power to erase life from this planet."[16]

Eisenhower also had to contend with aftereffects of the nuclear testing program. By this time they raised public fears of fallout and radiation, causing international outcry and producing growing concerns about nuclear technology. After the famous Lucky Dragon incident in 1954, in which a Japanese fishing trawler was accidently showered with fallout from a US thermonuclear test, there were efforts to condemn the United States at the United Nations and there was what was described as a "worldwide expression of fear."[17] In both the private and public considerations of nuclear policy, the nuclear dangers figured prominently.

With a zeal that surprised his advisers (not least, those on the Atomic Energy Commission), Eisenhower responded to these pressures by seeking to promote the peaceful uses of nuclear power. There was "a sense of moral compulsion that drove the President to seek some redeeming value in a new technology that threatened the future of humanity."[18] Eisenhower was especially keen to see the emergence of nuclear power (that is, the generation of electricity using nuclear reactors) and was frustrated by the slow progress toward developing and commercializing that technology. His administration explored one peaceful nuclear idea after another, including the construction of nuclear-powered merchant ships (discussed in the National Security Council in 1955) and the creation of small nuclear reactors suitable for distribution as part of an "Atomic Marshall Plan."

President Eisenhower's nuclear instincts found historic expression in his remarkable "Atoms for Peace" speech, which he delivered before the UN General Assembly on December 8, 1953. In powerful and sometimes poetic language, he spoke starkly about the dangers of the nuclear arms race, noting that it was not possible to escape "the awful arithmetic of the atomic bomb." Atomic warfare, he said, would lead to "annihilation" and "desolation." He emphasized that the United States sought to avoid this destructive result: "My country's purpose is to help move out of this dark chamber of horrors into the light." He therefore proposed that the world's three nuclear powers—the United States, the Soviet Union, and the United Kingdom—divert some portion of their nuclear efforts, including fissile material, to peaceful purposes; that an international atomic energy agency be created to, among other things, control donated fissile material; and that, in general, nuclear development be pushed onto a more peaceful path. "The United States pledges before you," he concluded, "and therefore before the world, its determination to solve the fearful atomic dilemma—to devote its entire heart and mind to find the way by which the miraculous inventiveness of man shall not be dedicated to his death but consecrated to his life."[19] Eisenhower wanted to tilt the balance toward the peaceful uses of technology, hoping to achieve benefits on the same scale as the revolution wrought by nuclear weapons.

In his assessment of the Atoms for Peace program, Peter Lavoy writes that it "fundamentally altered the way the world treated nuclear energy."[20] In the United States, the promotion of nuclear power helped fuel what the Atomic Energy Commission feared was a "grandiose public vision of the nuclear age," one with an "almost unbridled enthusiasm over the potential uses of atomic power."[21] Under pressure from Eisenhower, the commission pushed the nuclear power program, and the first reactor was completed in 1957. But the program's impact internationally was even greater. The Eisenhower administration was eager to push peaceful nuclear technology out into the world and created a nuclear assistance program that shared research reactors and other nuclear technology with many other countries, contributing significantly to the spread of nuclear capability around the world. In retrospect, the administration seems to have both underestimated how it might boost the aspirations of some states for nuclear weapons and overestimated its ability to control the nuclear behavior of others. In 1955 and again in 1957, the United Nations sponsored the International Conference on the Peaceful Uses of Nuclear Energy. President Eisenhower's suggestion that an international nuclear agency be established resulted several years later in the creation of the International Atomic Energy Agency, and the inspection requirements associated with the Atoms for Peace program's assistance contained the seeds of the agency's eventual safeguards system.

With a serious push from the highest levels of the US government, the civilian applications of nuclear technology were elevated in priority, and by the late 1950s—more than a decade into the nuclear age—tangible progress was made on a number of fronts, notably in establishing a civil nuclear power industry. Moreover, in certain sectors, such as nuclear medicine and food irradiation, civil applications over the years became important, well established, and widely used. However, many nuclear dreams never came true. The atomic energy vitamin never materialized. With one exception, the nuclear propulsion programs failed. While the US Navy first succeeded in using nuclear reactors to power many of its vessels, no nuclear automobiles, aircraft, or rockets were ever achieved.[22] Perhaps most significantly, nuclear power never lived up to the expectation of providing abundant, cheap energy. A nuclear power industry was created, of course, but nuclear power proved to be costlier, riskier, less competitive, more difficult, and more unpopular than expected. As a result, it played a much more limited role in overall energy production than the optimists had foreseen. For long periods in subsequent decades, nuclear power was simply not commercially competitive compared with alternative sources of energy, and years passed without any new reactors. Time and experience revealed the costs and limits of civilian applications of nuclear technology. It seems fair to conclude that as of 2017, compared to the enthusiasms of the late 1940s and the grandiose visions of the Eisenhower years, the peaceful nuclear revolution has had disappointing results.

In the United States, the peaceful uses of nuclear technology emerged sluggishly out of a government monopoly, in part because of a top-down process. Though Eisenhower was keen to see the peaceful atom exploited, at no point was this aim the highest priority. The civilian applications were subordinate to and probably impeded by the weapons program. Indeed, the Eisenhower administration presided over a prodigious expansion in the US nuclear arsenal (some twenty thousand weapons existed by the end of Eisenhower's term); thus, the weapons program had first claim on labs, personnel, budgets, and nuclear materials. During the 1950s, a sustained effort moved civilian nuclear activities into the private sector with some success, but extreme secrecy, strict regulation, and the scarcity of nuclear expertise in the commercial world constrained progress. The unfettered market was not a powerful factor in the early civilian exploitation of the atom, and when the market did come into play, it effectively limited the expansion of nuclear power. More generally, in areas that had been expected to yield large, possibly revolutionary gains, such as power and propulsion, programs were either discouraging or unsuccessful. The "techno-atomic utopia" never arrived.

Apart from cyber's distant origins as a military program, little in the peaceful nuclear story maps well into the cyber era. Cyber did not burst on the scene as a revolutionary weapon but made itself felt as an extremely useful civilian technology that took hold gradually, then spread rapidly far and wide. The development of civilian cyberspace did not encounter disappointing limits that truncated its reach; instead, it accelerated into the computerization of everything from watches to automobiles and to the creation of vast networks of communications and commercial activity. It is doubtful that the early users of email envisioned websites supplanting retail stores, but it did happen.

The extent and diversity of cyberspace reflect not a government monopoly and policy edicts from on high but a lively, decentralized, fast-moving private sector acting in a heavily populated, highly competitive marketplace. The rise of the Internet, according to one recent account, is explained by "innovation from the edge"—that is, "multiple perspectives originating from multiple places in an industry with almost no concentrated decision-making."[23] While cybersecurity is certainly now on the agenda of governments and attracting serious attention from defense ministries, no cyber weapon overshadows and circumscribes the civilian cyberspace. In numerous fundamental ways, the nuclear experience and the cyber context are completely different.

Seeking Security in the Nuclear Age

How could security be achieved in the nuclear age, in the presence of weapons so devastating? What strategies, policies, and postures would protect state interests without provoking catastrophic war? At the beginning of the nuclear age, these questions were raised not simply in response to the emergence of a dangerous new weapon but also in the context of a growing global rivalry with the Soviet Union, a powerful adversary that was itself nuclear armed after 1949. These concerns became core issues in the foreign and defense policies of the nuclear antagonists.

In the struggle to find answers, some offered radical solutions. Given the revolutionary existence of nuclear weapons, some believed that what was required was a new world order marked by global governance or by the banishment of war or some other visionary scheme. The famous Russell-Einstein Manifesto of July 1955 warned of the perils of nuclear weapons, for example, and stated plainly that the continued existence of the human species was "in doubt." This declaration raised what Bertrand Russell and Albert Einstein called "the stark and dreadful and inescapable" problem of the nuclear age: "Shall we put an end to the human race; or shall mankind renounce war?"[24] Others believed that the only genuine answer was disarmament—the elimination of nuclear weapons—or the placing of nuclear weapons under international control. In 1946 reflecting in part the extravagant hopes for the United Nations and the large fears of nuclear technology, the United States put forward an unsuccessful plan for the international control of nuclear energy.[25] However, visions of a peaceful future in which the nuclear danger had been tamed were overwhelmed by the intractable realities of international politics. The decades after World War II were marked not by effective international government, disarmament, and the banishment of war but by ceaseless confrontation, the massive accumulation of nuclear weapons, and the division of the world into hostile blocs.

Though visions of escaping nuclear danger by reinventing international politics failed, efforts to find solutions to the problem of security in the nuclear age persisted. The first quarter century of the nuclear age was marked by intensive deliberation and debate on how to address the threat posed by nuclear weapons. By 1970 a considerable literature on nuclear strategy and policy existed, and a broad framework of concepts for minimizing or constraining the nuclear threat was in place.[26] Four large ideas shaped the evolution of the nuclear order: deterrence, damage limitation, arms control, and nonproliferation. How did these ideas work in the nuclear context, and how relevant are they to the world of cyber?

Deterrence

The central concept that emerged for managing nuclear weapons and constraining the risks of nuclear rivalry was deterrence. The core idea was to prevent nuclear attack by the threat of severe nuclear retaliation. Given the enormous destructiveness of nuclear weapons, no attack would be worth the price of absorbing a nuclear counterstrike. Retaliatory strikes, of course, required that some nuclear forces would survive a nuclear first strike. Thus, preserving a second-strike capability became the essential precondition for achieving an effective deterrent posture. To achieve this stabilizing capability, the Soviet Union and the United States deployed large numbers of forces on diverse platforms, with some protected by hardened silos, some hidden in the sea, and some held on high levels of alert.

Over the course of the Cold War, nuclear experts engaged in arcane debates about potential vulnerabilities of the nuclear forces, the required size and character of the retaliatory force, and the type and number of targets that must be threatened to achieve deterrence. But the key insight of deterrence theory was this: If each side understood that the other was capable of nuclear retaliation, then neither side would have an incentive to strike first; and if both sides are vulnerable to devastating nuclear attacks, then each will have an incentive to avoid conflict. This condition of mutual vulnerability, in which the civilian societies on both sides are regarded as hostages, was thought to provide a kind of stability. If each side heeded the dictates of deterrence theory, then a condition of mutual assured destruction (or MAD, as it was known) would prevail and prevent the use of nuclear weapons.

Is deterrence relevant to the world of cyber threats? In principle, the same concept could apply—that is, preventing attacks by threatening retaliation. And in some contexts—notably when states launch intentional cyber attacks, probably in the connection with a wider international conflict—perhaps the concept of deterrence can be adapted to the cyber world. However, several considerations circumscribe the utility of deterrence in addressing cyber threats. First, the concept of deterrence arose in a bipolar context and was aimed above all at influencing the behavior of a single coherent state, the Soviet Union. But there is no such clarity in the cyber world. The threat could emanate from anywhere. Far from being bilateral, the threat is omnidirectional. Because the barriers to entry are low and the vulnerabilities are many, just about any state could be the source of a cyber attack (as indicated by the fact that one notable case involves North Korea). However, cyber capabilities are widely distributed not just among states but among organizations and individuals as well. The attacker could be a terrorist group or a criminal gang or a crazed individual. Can a deterrence posture be effective against a diverse and multitudinous set of potential threats?

The cyber deterrence challenge is compounded by a second consideration, the problem of attribution. US intelligence would have had no doubt who was attacking if the Soviet Union had launched a nuclear strike against the United States, but identifying the source of a cyber strike may not be easy. The number of potential attackers is vast, and clever attackers have ways of hiding or camouflaging their identities. Deterrence can be undermined if it is not clear against whom retaliation should be directed. If the attribution problem becomes more tractable, then this concern will weaken. But to the extent that attribution remains a challenge, retaliatory threats lose value.

A third consideration is that the protagonists in a cyber fight may not be symmetrically vulnerable. It was clear during the Cold War that the cities, the economic infrastructures, and the militaries of both the Soviet Union and the United States were vulnerable to nuclear attack. But in the cyber context, one party may be much more dependent on cyber assets than another. The United States is vastly more dependent on the cyber world, for example, than is North Korea. Is it likely that North Korea will be deterred by the threat of cyber retaliation? The problem may be even more difficult if the attacker is a non-state actor. How does one threaten terrorist groups or criminal organizations, much less individuals, with cyber retaliation? In short, asymmetries in cyber infrastructure and reliance on cyber assets may complicate fashioning effective retaliatory threats.

Finally, nuclear deterrence rests on assured destruction of enormous magnitude, posing an unmistakable threat of unacceptable damage. Cyber attacks are more uncertain in effect. Some imaginable attacks, such as those on critical infrastructure or military command and control, could have large consequences. But their effects may be unpredictable, temporary, or even short term; they may be thwarted by a clever defender; or the disruptions may be minimized by redundancies or resilience built into the defender's systems. No doubt some attacks could be quite damaging, but it is not certain that they would result in unacceptable damage; indeed, it is not even clear what unacceptable damage *is* in the cyber context. Any use of nuclear weapons will have devastating consequences. The same is not true of cyber. As a result, mobilizing credible and effective deterrent threats in the cyber context is more complex.

Despite these difficulties, it may still be possible, at least in some contexts, to persuade adversaries that the costs of a cyber attack exceed the benefits. Joseph Nye has suggested, for example, that modern economies are so interconnected that a cyber attack by one country on another—say, by China on the United States—can be self-harming if the resulting economic damage hurts the attacker's economy. Nye's term for this is "deterrence by entanglement."[27] But the difficulties and uncertainties associated with a doctrine of cyber retaliation have led to consideration of other sorts of retaliatory measures. When a cyber attack amounts to an act of war, retaliation using conventional military forces is seen as legitimate and can be considered.[28] This logic undoubtedly applies only to a small subset of cyber attacks, many of which are too minor or too ineffective to warrant a state of war. But perhaps cyber deterrence can take the form of credible threats of retaliatory attacks by conventional forces. Further, to deter severe, large-scale cyber attacks, perhaps nuclear retaliatory threats will come into play. In its discussion of "maintaining deterrence in the cyber era," for example, the Defense Science Board has stated that "the top of that escalation ladder is the present US nuclear deterrent."[29]

Most likely, we are still in the early stages of thinking through the relationship between cyber threats and deterrence. In the nuclear realm, the basic idea of deterrence arose soon after World War II, but some of the classics of nuclear strategy—including such notable works as Bernard Brodie's *Strategy in the Missile Age* or Thomas Schelling's *Arms and Influence*—did not appear until fifteen or twenty years after the detonations at Hiroshima and Nagasaki. Moreover, the debate over the requirements for and the reliability of nuclear deterrence raged throughout the Cold War. If the nuclear experience is any indication, then we can expect that years of wrestling with the idea of cyber deterrence lie ahead. What does seem clear, however, is that it will not be simple or straightforward to adapt the core nuclear concept of deterrence to the cyber world. Cyber deterrence may prove useful in some contexts, but it will be at best a partial solution to the problem of cyber threats. Complexities abound, and as a result deterrence is not likely to play the overwhelmingly central role that it does in the nuclear context. As one prominent analysis of cybersecurity concluded, "The force that prevented nuclear war, deterrence, does not work well in cyber war."[30]

Damage Limitation

Theorists and arms controllers promoted mutual deterrence as a policy, championed it as a desirable state of affairs, and loved the stable nuclear environment that was thought to result from this approach. To the military organizations charged with managing the nuclear arsenals, and to many of the civilian authorities to whom the militaries were answerable, there existed what was generally seen as an inescapable responsibility to be prepared to fight a nuclear war if necessary. And if nuclear war came, then it seemed obvious and compelling that one of the overriding goals would be to limit the damage to one's society as much as possible. In this framework, mutual assured destruction was a condition to be resisted rather than an objective to be sought. As Robert Jervis has observed, resistance to mutual deterrence "led to a number of attempts to escape from vulnerability."[31] In terms of military doctrine and operational preparations, the notion of damage limitation has occupied a central place in thinking about nuclear weapons and the threat they pose.

In the nuclear realm, damage limitation has had both offensive and defensive components. The offensive dimension entailed the contemplation of various first-strike options. The optimal damage-limiting scenario involves a disarming first strike on an opponent's nuclear forces, but the goal of mutual deterrence, of course, with its enormous emphasis on survivable forces, was to eliminate such preventive war temptations. The argument still remained that if escalation to nuclear war seemed likely, then it was better to strike first, degrade the opponent's forces as much as possible, and deal with the "ragged retaliation" by the other side's residual forces rather than contend with a comprehensive and coherent attack by the full, undamaged arsenal of the attacker. Such preemptive incentives exist even if disarming strikes are not possible. During the Cold War, both the Soviet Union and the United States made extensive preparations for the first use of nuclear weapons despite the mutual deterrence relationship that was thought to exist between them.

There is no question that offensive cyber attacks are possible and indeed have been happening.[32] But is offensive damage limitation a useful concept in the cyber context? Preventive attacks aimed at eliminating or degrading an opponent's cyber capabilities are rendered difficult by the decentralized and widely distributed global nature of cyber infrastructure and by the ubiquity of access to the Internet. A cyber attacker need not rely on its own infrastructure, and cyber attacks need not originate from the attacker's territory. Non-state attackers, of course, will in most cases have neither cyber infrastructure nor territory. In addition, whether or how much an opponent's capability is degraded, and for how long, will be very difficult to assess. With respect to preemptive cyber attack—that is, striking first in response to an opponent's preparations to strike—the options are limited by the opaqueness of the cyber world. Lacking any visible mobilization prior to a cyber attack, and hence having no warning of attack, makes a preemptive strike impossible.

States will have multiple reasons for conducting offensive cyber operations: to seek information, to punish adversaries, to undermine WMD programs, and to support conventional military operations. No doubt offensive damage limitation will be among them. But as with deterrence, the notion of damage limitation fits imperfectly with some realities of the cyber threat. Nevertheless, the goal of disrupting an opponent's capabilities as much as possible is likely to remain enticing. And whereas nuclear weapons came to be regarded as unusable in all but the most extreme circumstances, cyber attacks are a routine occurrence.

Nuclear damage limitation also has a defensive dimension. The vast destructiveness of nuclear weapons, the huge numbers of weapons that the Cold War superpowers amassed in their arsenals, the fact that deploying offensive rather than defensive capabilities is easier and cheaper, and the impossibility of developing perfect missile and air defenses make it difficult to envision achieving meaningful levels of defense. Indeed, concerns that an offense-defense arms race would provoke ever higher offensive deployments without providing significant protection led to the 1972 Antiballistic Missile Treaty, which placed severe limits on the deployment of missile defenses by the Soviet Union and the United States. Nevertheless, interest in defenses persisted, substantial investments in research and development on defenses were sustained, and the wisdom of remaining defenseless while relying on deterrence was recurrently challenged. Most memorably, President Ronald Reagan in his famous 1983 "Star Wars" speech announced a program aimed at achieving high levels of defense. And in the early 2000s, the United States exercised its legal right to withdraw from the Antiballistic Missile Treaty and began to deploy missile defenses. This move was prompted in part by the emergence of smaller nuclear threats, such as North Korea, against whom some level of defense is more feasible. The instinct to defend is a powerful one even in the nuclear context, where costs are high, progress is slow, and benefits are circumscribed.

This same instinct is evident in the cyber domain. Here, damage limitation is at the heart of much of the discussion concerning how to use defensive measures to respond effectively to cyber threats. A central concern is to protect critical infrastructure—such as electricity grids, nuclear power plants, the financial system, and key industrial facilities—from cyber attack. Similarly, today's most powerful states place major priority on preventing opponents from disrupting or degrading the military cyber capabilities on which these states rely. Because cyber is central to a wide array of economic, social, and military activity, a huge range of disruptive or destructive attacks is possible, and the cyber terrain to be defended is quite extensive. Many of what might be called damage-limitation measures figure in thinking about cyber defense. For example, critical infrastructure can be insulated as much as possible from the cyber world. Cyber assets can be "hardened"—that is, made more difficult to penetrate. Key cyber functions can be hidden or shifted frequently around the cyber infrastructure. Relying on redundancy can complicate an attack and possibly prevent an attacker from achieving his or her objectives. Similarly, investment in rapid recovery capabilities may allow a defender to ride out an attack and still function afterward. Much can be done to limit and neutralize the threat of cyber attack.

However, cyber shares with nuclear one fundamentally important attribute: effective defense is very hard to achieve because vulnerabilities are endemic to the technology. In its 2013 report, for example, the Defense Science Board observes that US cyber networks are based on "inherently insecure architecture" and concludes that "with present capabilities and technology it is not possible to defend with confidence against the most sophisticated cyber attacks ."[33] Thus, as in the nuclear case, defenses in the cyber world are desirable but difficult. While they will surely be pursued, for the foreseeable future there will be limits to what can be achieved. The nuclear revolution has meant living with an inescapable level of vulnerability despite our best efforts; the cyber revolution may mean the same.

Arms Control

Another approach to taming the danger of nuclear weapons began to emerge around 1960. In the Soviet-American context, the key insight was that nuclear war posed a massive, existential threat to both antagonists; hence, both had a profound interest in avoiding it. The idea of arms control was to mitigate nuclear danger by constructing a managed competition via negotiated constraints. Though the rivalry remained intense, the two antagonists could nevertheless collaborate in the joint pursuit of their shared interest in preventing a nuclear catastrophe. Beginning around 1970, nuclear arms control became a regular and central, if occasionally interrupted, feature of Soviet-American relations. It has remained so for nearly fifty years, even after the demise of the Soviet Union.

Arms control generally falls into one of three categories:

- *Limits on forces and force postures.* Many of the major strategic arms control agreements focused on placing limits on the size, character, and modernization of nuclear forces.
- *Crisis management measures.* Some arms control arrangements put in place institutions and procedures for containing the danger of crises, principally through communication and consultation.

- *Confidence-building measures.* These steps are aimed at dampening the intensity of the competition and preventing misunderstandings through such measures as information-sharing, prenotification of military exercises or missile tests, and regular consultations.

Over a period of decades, the Cold War protagonists built up an extensive web of treaties and arrangements (not all of them nuclear) that shaped their relationship and governed their nuclear competition.

Could negotiated arms control help manage the cyber environment? The answer is mixed. Some aspects of the Cold War's arms control experience do not translate into the cyber world. Strategic arms control treaties, for example, were preoccupied with observable objects and activities and were centered on things that could be counted. The parties generally believed that it was only possible to limit what could be verified. It was, however, possible to verify nuclear arms control agreements, including by remote surveillance using what were labeled national technical means.

The cyber world does not have a discrete force posture that can be constrained by numerical limits. Further, it is hard to see how sufficient levels of transparency and verifiability can be attained; hence, cyber arms control would be limited in scope. Moreover, cyber arms control will need to encompass a huge universe of actors if it is to fully address the potential sources of threat. Multilateral arms control is possible, of course, and some significant multilateral nuclear treaties, signed by large numbers of states, do exist. But in the cyber arena, states are not the only actors and, in the eyes of some, are not even the most important actors. As P. W. Singer and Allan Friedman comment in their study of cybersecurity, "There is a notion that the Internet is a place without boundaries, where governments do not matter and therefore do not belong."[34] It will not be easy to fashion multilateral cyber arms control in an environment in which states are not necessarily the dominant players and in which serious threats can emerge from an infinite mélange of individuals, corporations, criminal organizations, and terrorist groups, as well as from states.

For these reasons, neither traditional nuclear arms control as practiced between the United States and the Soviet Union nor multilateral nuclear arms control as it has existed in the past seem a promising model for cyber. There are too many relevant actors, too few countable objects, and too little verifiability for these approaches to be effective shapers of the cyber environment. However, there may still be room for other types of arms control measures—that is, crisis management and confidence-building measures. Given the opacity of the cyber realm, the potential difficulty in identifying potential attackers, and the lack of time for assessment, deliberation, and decision-making (because cyber attacks will happen in an instant), there is great potential for confusion, uncertainty, misperception, mistaken judgments, and misdirected retaliations. Hence, some states are interested in measures that facilitate consultation, rapid and reliable communication, and cooperation in addressing shared threats (such as criminal or terrorist exploitation of cyber vulnerabilities or attacks that disrupt the cyber architecture on which all depend). Some such measures already exist and many others have been proposed.[35]

Moreover, while constructing a comprehensive global regime for cyber management may not be possible, cyber governance measures can be established in important bilateral relationships or in significant groupings of states. In May 2015, for example, Russia and China signed a cybersecurity agreement. During President Xi Jinping's September 2015 visit to Washington, the United States and China issued a joint statement that addressed an array of cyber issues. Various groupings of states have agreed to measures aimed at addressing one piece or another of the cyber problem: The G20 has tackled cyber theft, the Shanghai Cooperation Organization has condemned information war, and a group of forty-seven states has accepted the Budapest Convention on Cybercrime.[36] Thus, though some forms of arms control as practiced in the nuclear realm seem unsuitable in the cyber context, negotiated rules, procedures, and constraints evidently will influence the emerging cyber order.

Nonproliferation

In the unconstrained early years of the nuclear age, the expectation was that the number of states possessing nuclear weapons would grow steadily in the future as more states developed the technical capacity to build them. This expectation was accompanied by a fear that the dangers associated with nuclear weapons would multiply as they spread into more hands. As Albert Wohlstetter suggested in an influential study, "life in a nuclear armed crowd" seemed perilous and extremely unattractive. Accordingly, efforts to prevent the spread of nuclear weapons have been one of the main hallmarks of the nuclear order and have figured prominently in the foreign policies of the major powers. Francis Gavin argues, for example, that nuclear nonproliferation has been a core imperative of US grand strategy since the end of World War II.[37]

The legal foundation of the nonproliferation regime is the 1968 Nuclear Nonproliferation Treaty (NPT), which now encompasses nearly every state in the international system. All NPT signatories without nuclear weapons have agreed not to acquire nuclear weapons. But the nonproliferation regime does not rely on this legal instrument alone. In the nuclear realm, technological choke points impede the path to acquiring nuclear weapons. Without enriched uranium or plutonium, for instance, it is impossible to manufacture them. These materials and the technologies to produce them are in relatively few hands, and access to them is limited. In effect, an elaborate system of technology denial is in place that consists of national export control regulations and increasingly harmonized international guidelines for restricting the sale of sensitive, weapons-related dual-use items. The major suppliers of nuclear technology have also institutionalized their collaboration in the Nuclear Suppliers Group.[38] Worrisome recipients can be and are denied access to dual-use items, and all exports of some sensitive technologies (such as plutonium reprocessing) are universally discouraged. In addition, the NPT system is monitored. Any peaceful civilian facility that handles nuclear materials is subject to inspection by the International Atomic Energy

Agency (IAEA). States can circumvent the nonproliferation regime by developing indigenous technology, by acquiring dual-use items illicitly on the international black market, or by misusing existing permitted facilities (though in this latter case, inspections might detect the cheating). But on the whole, with the notable exception of North Korea, the nonproliferation system of a legal regime, combined with technology denial and inspection, has been remarkably effective at preventing the spread of nuclear weapons and the technologies to make them.

In the cyber context, nonproliferation is a nonstarter. This area is where the divergence between nuclear and cyber is clearest and most stark. For one thing, cyber technology has already spread. Globally, billions of devices are connected to the Web. Individuals commonly possess multiple devices that give them access to the Internet. The only barrier to the spread of cyber technology appears to be poverty; in the wealthier parts of the world it is ubiquitous. Second, where the nonproliferation regime is built substantially on technological choke points, no such choke points exist in the cyber arena. Rather, cyber is a market of many suppliers, rapid innovation, and widespread adoption with little leverage for restraining the spread of this technology. Finally, the nonproliferation regime is a monitored system. IAEA safeguards are applied to all facilities that handle nuclear material. No equivalent system exists for cyber, and it is hard to imagine what international inspection scheme could offer assurance against the hostile use of cyber technology. The nuclear nonproliferation experience holds little relevance for cyber.

Conclusion

Nuclear and cyber technology both raise the challenge of coping with threats of enormous potential consequence. Any use of nuclear weapons, of course, would be devastating. The same is not true of most cyber attacks, but in their most dangerous incarnation, they can cause what the Defense Science Board described as "existential" levels of damage.[39] The scale of the most threatening cyber attacks invites invocation of the nuclear analogy. The board put it plainly: "The cyber threat is serious, with potential consequences similar in some ways to the nuclear threat of the Cold War. . . . The Task Force believes that the integrated impact of a cyber attack has the potential of existential consequence. While the manifestation of a nuclear and cyber attack are very different, in the end, the existential impact to the United States is the same."[40] There is a certain symmetry here: two technological revolutions, two large and potentially existential threats, two difficult but unavoidable challenges to security policy.

The analogy, however, is imperfect. The trajectories and time lines of these two technologies have been quite different. With nuclear technology, the weapons side has been preeminent while the civilian side has been government dominated, sluggish, and less extensive than expected. For cyber, market-driven civilian applications have spread like wildfire, and concerns about security vulnerabilities have followed in the wake of its penetration into most walks of economic and social life. With nuclear, the number of relevant actors is few, the sensitive technologies are relatively inaccessible, and the weapons are generally regarded as unusable. With cyber, the number of relevant actors is enormous, the technology is widely distributed and widely accessible, and attacks are frequent (though generally low impact). Though serious worries about nuclear terrorism exist, nuclear technology is still overwhelmingly the province of states, and nuclear weapons are in the hands of only a small number of states. In striking contrast, the pace and direction of the cyber world are driven by the private sector, innovation flows from companies and individuals, the state struggles to be relevant, and cyber weapons are potentially in the hands of anyone with a laptop.

Given these differences in the ecosystems of the two technologies, it is not surprising the conceptual framework that developed to cope with the nuclear threat applies only imperfectly to the cyber world. A mix of deterrence, preparations for damage limitation, arms control, and nonproliferation has managed to keep the nuclear peace for more than seven decades. As we have seen, some of these concepts will be adaptable to the cyber world, but the nuclear framework is not directly transferrable to the cyber context. The distinctive character of the cyber threat will require a distinctive set of answers.

Notes

1. For a particularly thoughtful analysis of the cyber challenge, see, for example, Lucas Kello, "The Meaning of the Cyber Revolution: Perils to Theory and Statecraft," *International Security* 30, no. 2 (Fall 2013): 7–40.

2. On this theme, see also Joseph S. Nye Jr., "Nuclear Lessons for Cyber Security?," *Strategic Studies Quarterly* 5, no. 4 (2011): 18–38.

3. Because of time zone differences, this was August 7 in Japan.

4. The identity of the location is blanked out in the original document. Truman seems to have been under the impression that Hiroshima was a purely military target, which may account for the rather elliptical language. See Alex Wellerstein, "The Kyoto Misconception," *Restricted Data*, August 8, 2014, http://blog.nuclearsecrecy.com/2014/08 /08/kyoto-misconception/.

5. Harry S. Truman, "Statement by the President of the United States," White House, August 6, 1945. The document is available in Ayers Papers, subject file Army US, the Harry S. Truman Library & Museum Archives, Independence, MO, http://www.trumanlibrary.org /whistlestop/study_collections/bomb/large/documents/index.php?pagenumber=1&docu mentdate=1945–08–06&documentid=59&studycollectionid=abomb.

6. Ibid.

7. Eleanor Roosevelt, "My Day," August 8, 1945, Eleanor Roosevelt Papers Project, digital edition, 2008, George Washington University, https://www.gwu.edu/~erpapers /myday/displaydoc.cfm?_y=1945&_f=md000097.

8. Bertrand Russell, written just as news of Nagasaki arrived, probably on August 9, 1945, originally under the title "The Atomic Bomb." Available in Russell's collected papers at Russell Editorial Project, vol. 24 of the Collected Papers of Bertrand Russell, McMaster University, Ontario, http://www.humanities.mcmaster.ca/%7Erussell/brbomb.htm.

9. Hanson W. Baldwin, "The Atomic Bomb and Future War," *Life*, August 20, 1945, 17–20.

10. Paul Boyer, *By the Bomb's Early Light: American Thought and Culture at the Dawn of the Atomic Age* (Chapel Hill: University of North Carolina Press, 1994), 3.

11. Richard G. Hewlett and Francis Duncan, *Atomic Shield: A History of the United States Atomic Energy Commission*, vol. 2, *1947-1952* (University Park: Pennsylvania State University Press, 1969), xiv.

12. This is one of the themes, in fact, of the Hewlett and Duncan histories of the Atomic Energy Commission, especially with respect to nuclear power.

13. Boyer, *By the Bomb's Early Light*, 107. Boyer's chapter covering popular visions of the atomic future provides an arresting picture of these nuclear enthusiasms.

14. Hewlett and Duncan, *Atomic Shield*, 72.

15. Richard G. Hewlett and Jack M. Holl, *Atoms for Peace and War, 1953-1961: Eisenhower and the Atomic Energy Commission* (Berkeley: University of California Press, 1989), 41.

16. Quoted in ibid., 34.

17. Ibid., 275.

18. Ibid., 239.

19. All quotes in this paragraph are from "Text of the Address Delivered by the President of the United States before the General Assembly of the United Nations in New York City, Tuesday Afternoon, December 8, 1953," Dwight D. Eisenhower Library, Abilene, KS, http://www.eisenhower.archives.gov/research/online_documents/atoms_for_peace /Binder13.pdf.

20. Peter Lavoy, "The Enduring Effects of Atoms for Peace," *Arms Control Today*, December 2003, https://www.armscontrol.org/act/2003_12/Lavoy.

21. Hewlett and Holl, *Atoms for Peace and War*, 208, 239.

22. The success of the nuclear navy is an interesting story and important more broadly because the navy program contributed significantly to the development of power reactors. The history is detailed in Richard G. Hewlett and Francis Duncan, *Nuclear Navy, 1946-1952* (Chicago: University of Chicago Press, 1974).

23. David Warsh, "Will 'Innovation from the Edge' Help with Global Warming?," *Economic Principals*, November 29, 2015, http://www.economicprincipals.com/issues/2015 .11.29/1833.html. The quote is from Warsh, from his review of Shane Greenstein, *How the Internet Became Commercial: Innovation, Privatization, and the Birth of a New Network* (Princeton: Princeton University Press, 2015). The concept of "innovation from the edge" is Greenstein's.

24. "The Russell-Einstein Manifesto, July 9, 1955," First Pugwash Conference on Science and World Affairs, Pugwash, Nova Scotia, https://pugwash.org/1955/07/09/statement -manifesto/.

25. See, for example, Barton Bernstein, "The Quest for Security: American Foreign Policy and the International Control of Atomic Energy, 1942–1946," *Journal of American History* 60, no. 4 (March 1974).

26. A vast literature is related to the nuclear debate. For an excellent overview, see Robert Jervis, *The Meaning of the Nuclear Revolution: Statecraft and the Prospect of Armageddon* (Ithaca: Cornell University Press, 1989).

27. Joseph S. Nye, "Can Cyber Warfare Be Deterred?," Project Syndicate, December 10, 2015, https://www.project-syndicate.org/commentary/cyber-warfare-deterrence -by-joseph-s—nye-2015-12. Nye has developed these ideas in his essay, "Deterrence and Dissuasion in Cyberspace," *International Security* 41, no. 3 (Winter 2016/17): 58-60.

28. For a detailed discussion of the relationship between cyber attack and the laws of war, see Oona Hathaway and Rebecca Crootof, "The Law of Cyber Attack," *California Law*

Review 100 (2012): 817–85. They point out that the law of war applies only to a small percentage of cyber attacks.

29. Defense Science Board, *Resilient Military Systems and the Advanced Cyber Threat* (Washington, DC: Office of the Undersecretary of Defense for Acquisition, Technology and Logistics, US Department of Defense, January 2013), 40.

30. Richard A. Clarke and Robert K. Knake, *Cyber War: The Next Threat to National Security and What to Do about It* (New York: HarperCollins, 2010), introduction.

31. Jervis, *Meaning of the Nuclear Revolution*, 50.

32. Numerous illustrations can be found in Fred Kaplan, *Dark Territory: The Secret History of Cyber War* (New York: Simon & Schuster, 2016).

33. Defense Science Board, *Resilient Military Systems*, 1.

34. P. W. Singer and Allan Friedman, *Cybersecurity and Cyberwar: What Everyone Needs to Know* (Oxford: Oxford University Press, 2014), 181.

35. For a thorough and very useful survey of existing and proposed crisis management and confidence-building measures in the cyber context, see Herbert Lin, "Governance of Information Technology and Cyber Weapons," in *Governance of Dual-Use Technologies: Theory and Practice*, ed. Elisa D. Harris (Cambridge, MA: American Academy of Arts and Sciences, 2016), 141–48.

36. Ibid., 129–32, surveys some of these developments.

37. Francis Gavin, "Strategies of Inhibition: US Grand Strategy, the Nuclear Revolution, and Nonproliferation," *International Security* 40, no. 1 (Summer 2015).

38. For a full discussion of these national and international constraints, see James M. Acton, "On the Regulation of Dual-Use Nuclear Technology," in Harris, *Governance of Dual-Use Technologies*, 8–59.

39. "*Existential Cyber Attack* is defined as an attack that is capable of causing sufficient wide scale damage for the government potentially to lose control of the country, including loss or damage to significant portions of military and critical infrastructure: power generation, communications, fuel and transportation, emergency services, financial services, etc." See Defense Science Board, *Resilient Military Systems*, 2. Emphasis in original.

40. Ibid., 1, 5.

11 From Pearl Harbor to the "Harbor Lights"

JOHN ARQUILLA

The specter of a looming digital "Pearl Harbor"–style attack has been and remains a central element in the American discourse on cybersecurity. Clearly, the iconic example of a disabling surprise attack on an unsuspecting fleet, more than seventy-five years after the event, still speaks powerfully to the fresh threat posed by a cyberspace-based attack on a technology-dependent society and its equally vulnerable military. Given the deep emotional effect evoked by memories of that "day of infamy," one would expect significant steps would be taken to mitigate such a risk.

Yet, over the past twenty years, a time during which the notion of a digital Pearl Harbor has proved a useful analogy, little visible effective preventive public policy has been made. Writing in 2013, cyber experts P. W. Singer and Allan Friedman noted that in the United States "some fifty cybersecurity bills [are] under consideration," representing just a small portion of the total number that had been proposed in a decade. They also observed, "Despite all these bills, no substantive cybersecurity legislation was passed."[1] Since then only one bill, the Cybersecurity Information Sharing Act of 2015, which encourages voluntary information sharing with the government about cyber incidents in the private sector, has been enacted into law.

Among the difficulties experienced in efforts to pass good cybersecurity legislation, privacy concerns, ranging across the Right–Left political spectrum, have sparked and sustained very strong, steady resistance.[2] In the wake of the massive breach of the Office of Personnel Management's "secure" files that began (apparently) in March 2014, confidence in the government's ability to solve the riddles of cybersecurity remains quite low. Over twenty million members of the military and the civil service have been affected. The US government's dispute with Apple, Inc., in which the Federal Bureau of Investigation sought Apple's assistance in decrypting the information on a smartphone seized in the investigation of the 2015 domestic terrorism incident in San Bernardino, further strained public-private amity in cybersecurity-related matters.

Even worse, in addition to its inability to protect vital information under its control, the US government is also seen as obstructionist. Policymakers worry that more secure products will make it harder for law enforcement and intelligence agencies to employ "cyber taps" on criminal and terrorist organizations. This concern has led, beyond the issues raised by the San Bernardino matter, to the US government's desire to be able to access any and all private communications via cyberspace as and when deemed necessary. The Cybersecurity Information Sharing Act, in the view of some leading cyber analysts, is thus seen as entailing two deleterious effects: it would make not only "any future data breach . . . far more catastrophic" but also "everything you do and say online less safe and more susceptible to government eavesdropping."[3]

As for market-driven solutions, consumers have a record of not demanding very secure products. For decades producers did not seriously try to make their systems more "hack proof." But given the string of costly attacks across a wide range of enterprises over the past few years—Anthem Blue Cross, Sony Pictures Entertainment, Target, and Yahoo are just a few of the most high-profile victims—US and other manufacturers have become determined to craft computers, cell phones, and other systems that are ever more secure.[4] They have done so while going against the wishes of some in government—with the notable exception of former president Barack Obama—to be able to outflank strong encryption by means of "backdoor" keys, which allow intrusion into anyone's system.[5]

Thus, it seems that Washington, which has trumpeted the Pearl Harbor metaphor, has failed to act in a helpful manner as defenses are developed against such a virtual bolt from the blue. As to Silicon Valley, and throughout the commercial information technology (IT) sector, the decades of neglect to produce more secure products have contributed to leaving cyberspace and its countless users quite vulnerable to hackers. If the government were somehow to cease its efforts to impede the launch of far more secure products, the situation would surely improve, at least at the margins. But much more is needed, as the United States and many other countries remain far from having a true ability to prevent, preempt, or counter the effects of a digital Pearl Harbor.[6]

An Alternative Analogy: "The Harbor Lights"

Despite its deep resonance in military and intelligence communities, the foregoing analysis of the Pearl Harbor analogy lacks traction in the political and economic arenas. Perhaps this is because military analysts do not speak directly to the commercial consequences of a major cyber attack. The Pearl Harbor imagery easily conjures visions of a stunned military, but it does little to illustrate how a surprise attack could affect the economy or could sway votes in an election. Thus, it is worthwhile to consider adding a cyber analogy that can engage decision makers in government and business—and the mass public—in political and economic ways. One does not have to look far beyond Pearl Harbor to find an analogy that serves this purpose very well.

Four days after the December 7, 1941, attack on Pearl Harbor, Axis partners Germany and Italy declared war on the United States. Strategic analysts criticized the move, given that Congress had authorized war against Japan only. This precipitate move by Adolf Hitler and Benito Mussolini brought America actively into the European conflict much earlier than it might otherwise have done—if at all. On November 21, only a few weeks before the Japanese attack, nearly two-thirds of respondents to a Gallup poll opposed the very idea of war against Germany and Italy. But as the diplomatic historian Thomas Bailey once noted, with equal eloquence and irony, thanks to Hitler's taking the matter into his own hands and to Mussolini's following his lead, "American opinion was spared the confusion of a debate over fighting the European Axis."[7] Thus, the Allied coalition was forged by the aggressor.

The fight against Germany and Italy was for the most part conducted in theaters thousands of miles from the United States. Any sense of immediate danger was, to say the very least, lacking. To be sure, many Americans of Japanese descent—and some of German and Italian ancestry as well—were soon put in camps due to war paranoia. There was also belt-tightening and rationing, but for the most part, Americans' life patterns retained much of their normalcy. On the East Coast, this fact was manifested in how cities and harbors continued to light up at night, and most coastal maritime traffic sailed unescorted. All this occurred despite the German U-boats—the one enemy weapons system that could reach the United States—having done grievous harm to Britain's shipping since September 1939.[8]

Five weeks after Pearl Harbor, in mid-January 1942, Karl Doenitz, the commander of U-boat forces, had a handful of his submarines operating off the East Coast of the United States. Dispatched under the grand name *Paukenschlag* (Drumbeat), these U-boats—at most a dozen at any given moment—more than lived up to the operation's title, inflicting steady, heavy losses on coastal shipping. One member of the U-boat service recalled the time as the "American Shooting Season," during which the Germans could quietly lurk off "open anchorages and undefended harbors . . . a veritable Eldorado."[9]

For three long months, coastal cities refused even to dim their lights at night. The illumination helped U-boat skippers immensely. The eminent naval historian Samuel Eliot Morison labeled this inaction America's "most reprehensible failure."

In Morison's analysis of these events, "the massacre enjoyed by the U-boats along our Atlantic coast in 1942 was as much a national disaster as if saboteurs had destroyed half a dozen of our biggest war plants." Indeed, during this U-boat "happy time," Germany sank 2.5 million tons of shipping, or about half the total losses inflicted by German submarines in the first two years of the war. Morison's unsettling bottom-line assessment is quite biting: "Ships were sunk and seamen drowned in order that the citizenry might enjoy pleasure as usual."[10] And all this came at a minimal cost to the U-boat arm. Though the US Navy claimed twenty-eight kills of enemy submarines from January to March, in actuality these were all false claims made by overenthuisastic American skippers. *No* U-boats were sunk during this period, and only half a dozen were lost by July 1942.

How could it be that the harbor lights stayed on so long during this absolute crisis? The only reasonable reply to this question is to explain that very powerful political and economic interests trumped sound strategy. All along the Eastern

Seaboard, big-city mayors and local business leaders objected that blackouts would cause catastrophic economic losses. Florida, which was experiencing the height of its flow of winter visitors from the North, strongly resisted pressures to darken its coastal cities' lights, even though U-boat sinkings along this section of the coast were devastating. Naval historian Henry Adams has noted, for example, that "Miami especially was urged to employ a dimout to reduce the deadly glow, but its Chamber of Commerce refused, saying it would ruin the tourist season."[11]

By the spring of 1942, losses had grown heavy enough that President Franklin D. Roosevelt (FDR) ordered blackouts all along the seaboard and for miles inland. This action was accompanied by the US Navy's grudging willingness to start moving coastal ship traffic in convoys. From the start of the war, Adm. Adolphus Andrews, who oversaw the antisubmarine campaign on the East Coast, took the position that, thanks to air patrols, ships could "seek the protection daylight affords" and "break their passage by lying over in sheltered harbors at night."[12]

This approach failed miserably. Only when escorts could strike back at the U-boats did the latter begin to suffer growing losses. As Michael Gannon, another historian of submarine warfare, has summed the matter up: "What really broke the back of the U-boat campaign in U.S. waters was the coastal convoy."[13]

Clearly, blackouts, dimouts, and convoys helped solve the problem, but they did not really break the back of the U-boats. Between January and August 1942, only seven German submarines were sunk in US waters. But this number is the wrong metric by which to judge the outcome of the antisubmarine campaign; instead, the number of U-boat *attacks* should be considered. By March–April they had risen to ninety-eight; by July–August they had fallen to twenty-six, or a drop of 73 percent.[14] Attacks fell in part because Admiral Doenitz simply decided to stop investing in the long-transit, short-dwell time of U-boats in American waters.

As US defenses improved, it made little sense for a U-boat to spend two-thirds of its patrol time in transit to and from the target zone. Thus, protected convoy targets were to be found and attacked far closer to home, requiring much less transit time. One of Admiral Doenitz's aides, Wolfgang Frank, summed up the principal reason for ending *Paukenschlag*: "This was not because the A/S [antisubmarine] defenses off the American coast had grown too strong, but because with the end of independent sailings and the introduction of convoys it was no longer worthwhile to send boats so far out."[15]

Assessing the Harbor Lights Analogy

Perhaps the first and most important point to derive from the American experience on the Eastern Seaboard in the months after entry into the war is that just as far too few civil defense measures were taken then, the same is true today in the virtual domain. Throughout all too much of cyberspace, the "harbor lights" remain on and illuminate the commercial sector's intellectual property, sensitive data held by government and the military, and the personal information of individuals. All have been exposed, providing rich targets for attack by the latter-day counterparts of the World War II–era U-boat raiders, hackers. Indeed, former US cyber czar Richard Clarke, along with his colleague Robert Knake, have rated US cyber defenses as worst among leading nations. They also note that the "senior government official charged with coordinating cybersecurity was . . . in an office buried several layers down in what was turning into the most dysfunctional department in government, DHS [Department of Homeland Security]."[16]

When he entered office in 2009, President Obama did try to elevate the cyber czar's role, affirming that the "cyberthreat is one of the most serious economic and national security challenges we face as a nation."[17] But his choice for leadership, Howard Schmidt, turned out to be skeptical about the gravity of security affairs in the virtual domain. As he once opined: "Anytime someone commits a denial-of-service attack or someone intrudes into a system to steal intellectual property, it's not a cyber war. This kind of hype is beneficial to no one."[18]

Schmidt left government service in 2012 to join a cybersecurity firm, partnering with, in a moment of true irony, the former head of DHS governor Tom Ridge. Michael Daniel, Schmidt's successor as cyber czar and a former congressional staffer with little actual expertise in computing or IT, suffered sharp criticism for being missing in action as waves of costly, debilitating hacks swept over US commercial and governmental sites.[19]

While part of the reason why the harbor lights are still on throughout the many sectors of cyberspace has to do with organizational dysfunction in Washington, some blame can also be placed on IT manufacturers, whom the consumer markets did not press to craft more secure products until recently. Then when manufacturers made efforts to produce far more secure products, the government—law enforcement in particular—expressed its concern that secure smartphones and other communications devices might impede their investigations.

A third culprit hearkens to the harbor lights metaphor as well—that is, the strategic paradigm employed by those charged with the defense of cyberspace. Central elements of this security paradigm are antiviral software and firewalls; together, it is much hoped, they are able to keep the cyber barbarians from breaking in. But they do not, at least not often enough. Good hackers break right through firewalls, which stop only the viruses, worms, and malware that they can already recognize.

Faith in firewalls has led to failure to adopt the most effective tool of cybersecurity, widespread use of very strong encryption. The reluctance to make end-to-end encryption the norm in cyber communications is analogous to the stubborn unwillingness to use convoys to protect vessel traffic on the Eastern Seaboard during the early months of 1942. And the unwillingness to keep stored data strongly encrypted is very much akin to keeping port cities illuminated.

Interestingly, the harbor lights analogy cuts both ways, providing lessons for the attacker as well. As noted previously, Doenitz had to calculate the factors of time, space, and force in determining the optimal use of his U-boats, and his analysis ultimately led him to shift his forces away from the US Eastern Seaboard when its defenses improved. It may well be that cyber attackers will have similar incentives to redirect their own operations if security improves thanks, say, to the ubiquitous employment of strong encryption or the dispersal of targets in the Cloud, that place of places outside one's own system.[20]

Attackers' first inclinations under such circumstances—that is, when data is strongly encrypted or harder to find and exploit by virtue of being secreted in the Cloud—would likely be to search for "softer" targets elsewhere. Doenitz showed a penchant for doing exactly this; as shipping defenses firmed up along the East Coast, he shifted his subs to the Gulf of Mexico, the Caribbean, even to Panama. When defenses in these areas improved, becoming virtually equal to convoy protection in the North Atlantic, he pulled back and concentrated the U-boats on targets that took less time to reach from their home bases. Eventually, with the major advances in Allied radio direction-finding equipment, the substantial increases in escort vessels, and the breaking of the Nazi Enigma codes by the boffins working at Bletchley Park, the U-boats were defeated.[21]

But this happy ending is not necessarily going to be repeated in the case of cyberspace for two fundamental reasons. First, the possibility that cyber malefactors will simply decide to switch to softer targets when one's defenses improve may be slim, as the targets of greatest value may not be available in places that are easy to breach. The intellectual property of leading American firms is not to be found in soft targets in other places around the world. Doenitz could move his U-boats to the Gulf of Mexico and the Caribbean knowing full well that ample oil tanker targets were in each maritime zone; thus, his payoff was still good. The world's softer cyber targets do not come replete with such high-value assets of their own.

The second problem with the hope that improved defenses will send attackers away in search of easier prey is that it is not enough to provide better security relative to the starting point. Rather, improvements must be substantial in *absolute* terms. The absolute capabilities for cyber defense in the United States, for example, have been quite poor for decades. Indeed, as noted earlier, former cyber czar Clarke and his colleague Knake rate them the worst defenses among all the major cyber powers.[22]

Their judgment has been affirmed by the long trail of high-profile hacks of major US commercial firms, as well as of sensitive government sites, and even the personal account of the director of central intelligence. Thus, making substantial improvements in the relative level of cybersecurity will likely not do enough to drive away the intruders.

Clearly, what is needed at this point is a paradigm shift in the whole way of thinking about cybersecurity. Ample evidence shows firewalls, the latter-day equivalents of Admiral Andrews's "sheltered harbors," are as ineffective as were his faulty remedies back then. Instead, stored data can and should be "blacked out," and information flows can be well "escorted" by the employment of very strong encryption and evasively routed via the Cloud.

These sorts of steps are only now beginning to be regularly taken in the United States, but they are serious indicators of real progress. Still, the habits of mind of those who rely on massive data flows and their ready availability remain steeped in the old paradigm, one on which the existing cybersecurity consulting industry is itself all too dependent. New defensive methods simply must be considered.

Comparing the Pearl Harbor and Harbor Lights Analogies

In light of the abovementioned concerns, how are good cybersecurity legislation and regulation to be enacted and pursued? In the United States, the Obama administration relied heavily on the Pearl Harbor analogy; indeed, it was a main line of argument advanced by former secretary of defense Leon Panetta when he was in office.[23]

But as this chapter has argued, this analogy has a fundamental problem: Pearl Harbor speaks primarily to the strategic and military aspects of cybersecurity. Defending the virtual domain from costly, disruptive hacks, however, has profound economic and political dimensions. With these factors in mind, I propose adding "harbor lights" to Pearl Harbor in making an operative analogy.

In December 1941 a great deal of US naval power was concentrated at Pearl Harbor, and a sharp blow to it was inflicted, enabling Japan to pursue its expansionist aims for a while. Of the eight US Navy battleships berthed there, four were sunk and another four seriously damaged. And if the *Kido Butai,* the Japanese carrier strike force, had caught the three American aircraft carriers deployed to the Pacific in port—they were out to sea at the time of the attack—or had blown up the base's massive fuel storage tanks, the damage would have been catastrophic. Pearl Harbor was a true single point of failure. And if the Japanese had not been outfoxed and outfought at Midway, even those surviving aircraft carriers would have been of little moment in the Pacific war's strategic balance.[24]

Nothing quite like the sort of concentration of power in a battleship row now exists in cyberspace. Indeed, part of the logic behind the creation over forty years ago of an Advanced Research Projects Agency Network, which would prove to be a key building block of the Internet, was to ensure continued communications even in the wake of a nuclear war. Redundancy and resilience are the key notions that lie at the heart of the structure of cyberspace.[25] Yes, there are very important, even "critical" nodes here and there, but work-arounds and fallbacks abound too. Thus, cyberspace is similar to the oceans that cover two-thirds of the world in that it has its various choke points, but there are always alternate routes.

If the Pearl Harbor analogy is somewhat limited, perhaps even misleading, it is because it encourages the dangerous belief that defenses can be concentrated in one or a few major areas to provide strategic protection to most, if not the vast majority, of threatened spaces. The harbor lights analogy is both more expansive conceptually—in that it speaks to military, economic, and political factors—and more accurately depicts the widely distributed defensive challenge that characterizes efforts to secure cyberspace.

This new analogy speaks in an interesting way to military matters, but its true value lies in engaging the range of politico-economic challenges. They are delineated in the harbor lights analogy by the costly failure of President Roosevelt to order a blackout along the East Coast, despite the growing depredations of the U-boat skippers, who were having their "happy time" teeing up targets for night attacks because they were so well illuminated.

Clearly, the harbor and other coastal lights stayed on far too long. In search of causation, this illustration leads us to the point that the failure, though ultimately FDR's, was driven by local political pressures, which were themselves the product of economic considerations. For several months in 1942, mayors of coastal cities resisted pressure to enforce blackouts because they feared a loss of business would ensue and plunge their economies, still not yet fully recovered from the Depression, into fresh downward spirals. It was only when the shipping losses grew to dangerously high levels that the blackout was finally put in place and merchant ships began to move in escorted convoys. This tactic didn't put an end to the U-boat menace, but it did bring it under control and encouraged Admiral Doenitz to send his submarines elsewhere in search of prey.

Today, the harbor lights are on—all over cyberspace. A wide range of targets is well illuminated and highly vulnerable to all manner of cyber mischief. Technologically advanced armed forces, all of which are increasingly dependent on their connectivity to operate effectively in battle, can be virtually crippled in the field, at sea, or in the air by disruptive attacks on the infrastructure on which they depend but that are often not even government owned.

As to leading commercial enterprises, they hemorrhage intellectual property to cyber snoops every day—a point Governor Mitt Romney made twice in debates with President Obama during the 2012 election campaign. Regarding mass publics, countless millions of people in the United States and around the world have had their personal security hacked and now serve unwittingly as drones, or zombies, impressed into service in the robot networks, or botnets, of master hackers. As do billions of their Internet-connected smart home appliances.

Why do the harbor lights remain on in cyberspace? Because rather than focusing on security, for decades IT manufacturers and software developers have been driven by market forces impelling them to seek greater speed and efficiency in their products—all at highly competitive consumer prices. In short, the virtual harbor lights have stayed on because the perceived economic costs of improved security—that is, of enforcing a virtual blackout—have been seen as too high. And, just as FDR did, American political leaders today have shied away from forcing the private sector's hand. In this current case, however, the motivations of those in government are a bit more mixed. Their reluctance to champion, or even to require, production of the most secure cyber products extends far beyond fealty to market forces. Instead, the government's intelligence and, even more, law enforcement departments fear that the improved security afforded by the ubiquitous use of, say, strong encryption will curtail their own information-gathering capabilities.

Clearly, the harbor lights analogy speaks very powerfully to the economic and political dimensions of the cybersecurity challenge. But it has its limitations, as no analogy can address every aspect of a problem. One way the analogy breaks down is in its inability to speak to the "invisible" nature of many of today's cyber depredations. The mass ship sinkings of the early months of 1942 were tangible events that (eventually) horrified the nation and its civilian and military leaders. Today the ongoing compromise of highly sensitive military information systems, the theft of intellectual property, and the unwitting recruitment of men, women, and children into zombie armies all pass largely beneath our levels of awareness. Cyber warfare is a lot like Carl Sandburg's fog coming in on "little cat feet."[26]

Another problem is that whereas FDR had the authority to compel the darkening of coastal regions, it is not at all clear today that the president, or "government" more generally, has the same ability. Can the ubiquitous use of encryption, cloud computing, or other measures be dictated? Legislated? Likely not. Still, the presidency is a bully pulpit. If the chief executive were to use what presidential scholars such as Samuel Kernell and Richard Neustadt believe is the true power of the office—the power to persuade—then there would be a greater likelihood of gaining significant voluntary compliance.[27]

To be sure, senior civil and military leaders also know the gravity of the situation. For more than two decades, the National Academies of Sciences, Engineering, and Medicine have conducted deeply alarming studies of US cyber vulnerabilities and quite clearly conveyed the grave nature of the threat.[28] President Obama also expressed his desire to respond far more decisively to the cyber threat in Presidential Decision Directive 20. Reporting about the still-classified directive—partially "outed" first in a *Washington Post* article in 2012 and by Edward Snowden's revelations in 2013—suggests that the directive takes an expansive view of cybersecurity, even to the point of taking preemptive action against cyber threats.[29]

All this implies clear awareness of the problem, but the proactive recommendation to seek out and attack the attackers may prove problematic, given how well hidden so many of them remain. All these years after the Code Red and Nimda computer viruses were unleashed—shortly after the attacks of September 11, 2001—the identities of those perpetrators are still unknown. As is true of many—or perhaps most—cyber attacks, digital warriors and terrorists today hide in the virtual ocean of cyberspace as well as the U-boat skippers did during their happy time along the Atlantic seaboard seventy-five years ago. Efforts to track them in advance of their attacks, to hearken yet again to the harbor lights analogy, will be as fruitless as the US Navy's first strategy in 1942 of sending out hunter-killer squadrons to search the ocean for the U-boats.[30]

In 1942 the right answer from the start was to black out coastal cities at night and to have ships evasively routed and escorted by antisubmarine vessels when they sailed. Losses still occurred after adopting these strategies but soon fell to acceptable levels. This is the lesson of the harbor lights. In cyberspace, the analogous way to embrace this approach would consist of far greater use of strong encryption and evasive routing of data via the Cloud, making it much harder for the virtual U-boat wolf packs that stalk them to find their targets.

We need not forget Pearl Harbor when thinking about cybersecurity. But surely we also need to remember the harbor lights. This is as true for the increasingly interconnected world community as it is for the United States.

Notes

1. Both quotes are from P. W. Singer and Allan Friedman, *Cybersecurity and Cyberwar: What Everyone Needs to Know* (Oxford: Oxford University Press, 2014), 8.

2. Jennifer Steinhauer, "Senate Rejects Measure to Strengthen Cybersecurity," *New York Times,* June 11, 2015, brings the record of failure up to the near present.

3. See Patrick G. Eddington and Sascha Meinrath, "Why the Information Sharing Bill Is Anti-Cybersecurity," *Christian Science Monitor,* July 22, 2015.

4. See the McAfee-Intel Security report, *Net Losses: Estimating the Global Cost of Cybercrime* (Washington, DC: Center for Strategic and International Studies, 2014), which estimates the global cost of hacking at an amount ranging as high as $500 billion, with $100 billion lost in the United States alone.

5. Spencer Ackerman, "FBI Chief Wants 'Backdoor Access' to Encrypted Communications," *The Guardian,* July 8, 2015. This is a long-standing attitude, going back to the time when it was illegal for individuals to possess strong encryption technology. See the story of the "code rebels" who disseminated these tools, despite government opposition, in Steven Levy, *Crypto: How the Code Rebels Beat the Government—Saving Privacy in the Digital Age* (New York: Penguin, 2002). See also Nicole Perlroth and David E. Sanger, "Obama Won't Seek Access to Encrypted User Data," *New York Times,* October 10, 2015.

6. See especially Richard A. Clarke and Robert A. Knake, *Cyber War: The Next Threat to National Security and What to Do about It* (New York: HarperCollins, 2010) for a comprehensive exposition of the challenges to achieving a truly robust cyber security system in the United States.

7. Thomas Bailey, *A Diplomatic History of the American People* (New York: Appleton-Century-Crofts, 1955), 798. Gallup statistics cited from the same page.

8. Axis submarines sank nearly five million tons of Britain-bound merchant shipping from 1939 to 1941. See statistics in John Ellis, *Brute Force: Allied Strategy and Tactics in the Second World War* (New York: Viking, 1990), 138–46.

9. Harald Busch, *U-Boats at War: German Submarines in Action, 1939–1945,* trans. L. P. R. Wilson (New York: Ballantine Books, 1955), 44.

10. The quotes are taken from Samuel Eliot Morison, *The Two-Ocean War: A Short History of the United States Navy in the Second World War* (Boston: Little, Brown, 1963), 109.

11. Henry H. Adams, *1942: The Year That Doomed the Axis* (New York: David McKay, 1967), 77.

12. Cited in Homer H. Hickam Jr., *Torpedo Junction: U-Boat War off America's East Coast, 1942* (Annapolis: US Naval Institute Press, 1989), 116.

13. Michael Gannon, *Operation Drumbeat: The Dramatic True Story of Germany's First U-Boat Attacks along the American Coast in World War II* (New York: Harper & Row, 1990), 385.

14. Statistics are from official German war records, cited in the appendix to Hickam, *Torpedo Junction,* 296–305.

15. Wolfgang Frank, *The Sea Wolves: The Story of German U-Boats at War,* trans. R. O. B. Long (New York: Rinehart & Company, 1955), 173.

16. Clarke and Knake, *Cyber War,* 113 for the quote, 148 for the defense rating.

17. Cited in Lolita C. Baldor, "Obama Announces U.S. Cyber Security Plan," NBC News, May 29, 2009, http://www.nbcnews.com/id/30998004/ns/technology_and_science -security/t/obama-announces-us-cyber-security-plan/#.WPJvEbGZOf4.

18. Howard Schmidt, "Defending Cyberspace: The View from Washington," *The Brown Journal of World Affairs* 18, no. 1 (Fall/Winter 2011): 50. His article was part of a debate in that same issue, to which my contribution was "The Computer Mouse That Roared: Cyberwar in the Twenty-First Century," 39–48.

19. See, for example, "Obama's 'Cybersecurity Czar' Is MIA as Hackers Run Wild," *Investor's Business Daily,* June 5, 2015. The piece reflects the opinion of the influential journal's editors.

20. On cloud security, see Vic Winkler, *Securing the Cloud: Cloud Security Techniques and Tactics* (Waltham, MA: Elsevier, 2011); and Madhin Srinavasin et al., "State-of-the-art Cloud Security Taxonomies," in *Proceedings of the International Conference on Advances in Computing, Communications and Informatics,* ed. Sabu M. Thampi, El-Sayed El-Afry, and Javier Aguiar (New York: ACM, 2012), 470–76.

21. One of the best accounts of the factors leading to Doenitz's defeat can be found in Michael Gannon, *Black May: The Epic Story of the Allies' Defeat of the German U-Boats in May 1943* (New York: HarperCollins, 1998). On breaking the Enigma codes, see David Kahn, *Seizing the Enigma: The Race to Break the German U-Boat Codes* (Boston: Houghton Mifflin, 1991).

22. Clarke and Knake, in *Cyber War,* 148, rate US cybersecurity relative to others' cybersecurity; but the authors also make clear throughout their study that they consider the absolute level of American cyber defenses to be poor as well.

23. See my critique of the Obama administration's central focus on this analogy, "Panetta's Wrong about a Cyber Pearl Harbor," *Foreign Policy,* November 20, 2012.

24. The Japanese were under no illusions about the consequences of their defeat at Midway. See Mitsuo Fuchida and Masatake Okumiya, *Midway: The Battle that Doomed Japan* (Annapolis: US Naval Institute Press, 1955). Fuchida was the flight leader in the attack on Pearl Harbor.

25. The first public demonstration of the ARPANET took place in October 1972. One of its founding fathers, Jacques Vallee, in his memoir *The Heart of the Internet: An Insider's View of the Origin and Promise of the On-Line Revolution* (Charlottesville, VA: Hampton Roads Publishing, 2003), notes that resilience was crucial to the ability of cyberspace to "grow organically" (77).

26. "Fog" is a short poem that fits the harbor lights analogy well: "The fog comes / on little cat feet. / It sits looking / over harbor and city / on silent haunches / and then moves on." A good link to and analysis of the poem is found at "Fog (poem)," Wikipedia, last modified March 31, 2017, https://en.wikipedia.org/wiki/Fog_(poem).

27. Samuel Kernell, *Going Public: New Strategies of Presidential Leadership,* 4th ed. (Washington, DC: Congressional Quarterly Press, 2006); and Richard E. Neustadt, *Presidential Power and the Modern Presidents* (New York: Free Press, 1991).

28. See especially three studies from National Academies Press (Washington, DC) that have addressed key issues raised in this chapter: Kenneth W. Dam and Herbert S. Lin, eds., *Cryptography's Role in Securing the Information Society* (1996); *Cybersecurity Today and Tomorrow: Pay Now or Pay Later* (2002); and *At the Nexus of Cybersecurity and Public Policy: Some Basic Concepts and Issues* (2014).

29. Ellen Nakashima, "Obama Signs Secret Directive to Help Thwart Cyberattacks," *Washington Post,* November 14, 2012. In January 2013 the Obama White House also issued a condensed "fact sheet" about the directive. See also Office of the Press Secretary, "Fact Sheet: Presidential Policy Directive on Critical Infrastructure Security and Resilience," the White House, February 12, 2013, https://obamawhitehouse.archives.gov/the-press-office/2013/02/12/fact-sheet-presidential-policy-directive-critical-infrastructure-securit.

30. Hickam, *Torpedo Junction,* provides a full narrative of this initial effort.

12 Active Cyber Defense

APPLYING AIR DEFENSE TO THE CYBER DOMAIN

DOROTHY E. DENNING AND BRADLEY J. STRAWSER

In the domain of cyber defense, the concept of active defense is often taken to mean aggressive actions against the source of an attack. It is given such names as "attack back" and "hack back" and is equated to offensive cyber strikes. It is considered dangerous and potentially harmful, in part because the apparent source of an attack may be an innocent party whose computer has been compromised and exploited by the attacker; so hacking back could be reckless and unfair.

But active cyber defense is a much richer concept. When properly understood, it is neither offensive nor necessarily dangerous. Our approach is to draw on concepts and examples from air defense to define and analyze cyber defenses. We show that many common cyber defenses—such as intrusion prevention—have active elements, and we examine two case studies that employed active defenses effectively and without harming innocent parties. We examine the ethics of active cyber defenses along four dimensions: scope of effects, degree of cooperation, types of effects, and degree of automation. Throughout, we use analogies from air defense to shed light on the nature of cyber defense and to demonstrate that active cyber defense is properly understood as a legitimate form of defense that can be executed according to well-established ethical principles.

Other authors have ably addressed the ethics of active defense. D. Dittrich and K. E. Himma, for example, contributed substantially to initial thinking in this area.[1] This chapter seeks to advance analysis by applying air defense principles to the cyber domain and by exploring the moral and strategic issues raised by active cyber defense.

Defining Active and Passive Cyber Defense

Because our definitions of active and passive cyber defense are derived from those for air defense, we begin by reviewing active and passive air and missile defense.

Active and Passive Air and Missile Defense

For the United States, Joint Publication 3-01, *Countering Air and Missile Threats*, defines *active air and missile defense* (AMD) as a "direct defensive action taken to destroy, nullify, or reduce the effectiveness of air and missile threats against friendly forces and assets." The definition goes on to say that active AMD "includes the use of aircraft, AD [air defense] weapons, missile defense weapons, electronic warfare (EW), multiple sensors, and other available weapons/capabilities."[2] Active AMD describes such actions as shooting down or diverting incoming missiles and jamming hostile radar or communications.

The Patriot surface-to-air missile system is an example of an active defense system. It uses an advanced aerial-interceptor missile and high-performance radar system to detect and shoot down hostile aircraft and tactical ballistic missiles.[3] Patriots were first deployed in Operation Desert Storm in 1991 to counter Iraqi Scud missiles. Israel's Iron Dome anti-rocket interceptor system has a similar objective of defending against incoming air threats. According to reports, the system intercepted more than three hundred rockets that Hamas fired from Gaza into Israel during the November 2012 conflict, with a success rate of 80 to 90 percent.[4] At the time, Israel was also under cyber assault, and Prime Minister Benjamin Netanyahu said the country needed to develop a cyber defense system similar to Iron Dome.[5]

Another example of an active air defense system is the United States' Operation Noble Eagle.[6] Launched the morning of September 11, 2001, minutes after terrorists hijacked the first aircraft, the operation has become a major element of homeland air defense, which includes combat air patrols, air cover support for special events, and sorties in response to possible air threats. Noble Eagle pilots can potentially shoot down hostile aircraft, although so far none have done so. However, over the years, they have intercepted and escorted numerous planes to airfields.

In contrast to active defense, *passive air and missile defense* is defined as "all measures, other than active AMD, taken to minimize the effectiveness of hostile air and missile threats against friendly forces and assets . . . [noting that] these measures include detection, warning, camouflage, concealment, deception, dispersion, and the use of protective construction. Passive AMD improves survivability by reducing the likelihood of detection and targeting of friendly assets and thereby minimizing the potential effects of adversary reconnaissance, surveillance, and attack."[7] Passive AMD includes such actions as concealing aircraft with stealth technology. It also covers monitoring the airspace for adversary aircraft and missiles but not actions that destroy or divert them.

Active and Passive Cyber Defense
We adapt the definitions of active and passive air defense to the cyber domain by replacing the term "air and missile" with "cyber." This gives us the basic definitions: *active cyber defense* is a direct defensive action taken to destroy, nullify, or reduce the effectiveness of cyber threats against friendly forces and assets, and *passive cyber defense* is all measures, other than active cyber defense, taken to minimize the effectiveness of cyber threats against friendly forces and assets. Put another way, active defenses are direct actions taken against specific threats,

while passive defenses focus more on protecting cyber assets from a variety of possible threats.

Using these definitions, we now examine various cyber defenses to see whether they are active or passive. We begin with encryption, which is clearly a passive defense. It is designed to ensure that information is effectively inaccessible to adversaries that intercept encrypted communications or download encrypted files, but it takes no action to prevent such interceptions or downloads. Steganography is similarly passive. By hiding the very existence of information within a cover such as a photo, it serves as a form of camouflage in the cyber domain. Other passive defenses include security engineering; configuration monitoring and management; vulnerability assessment and mitigation; application whitelisting (to prevent unauthorized programs from running); limits on administrator access; logging, backup, and recovery of lost data; and education and training of users. None of these involve direct actions against a hostile threat.

User authentication mechanisms can be active or passive. For example, consider a login mechanism based on usernames and passwords that denies access when either the username or password fails to match a registered user. We consider this passive if no further action is taken against an adversary attempting to gain access by this means. Indeed, the person might try again and again, perhaps eventually succeeding. If the mechanism locks the account after three tries, then it has an active element insofar as this particular adversary will be unable to gain entry through that account, at least temporarily. However, it does not stop the adversary from trying other accounts or from trying to gain access through other means such as a malware attack. Nor does it prevent an attacker who stole an account and password from gaining access to the system.

Now consider the Defense Advanced Research Project Agency's Active Authentication program, which seeks to validate users continuously using a wide range of physical and behavioral biometrics such as mouse and typing patterns and how messages and documents are crafted.[8] If at any time a user's actions are inconsistent with their normal biometric patterns (called their cognitive fingerprint), access could be terminated. Such a mechanism would be more active than the password mechanism, as it could keep the adversary from entering and then exploiting any legitimate account on the system. It might even thwart a malware attack, as the malware's behavior would not match that of the account under which it is running.

Consider next a simple firewall access control list (ACL) that blocks all incoming packets to a particular port on the grounds that because the system does not support any services on that port, it would be an open door for attackers. We consider this passive, as it serves more to eliminate a vulnerability than to address a particular threat. However, the ACL would become an element of an active defense if an intrusion prevention system detected hostile traffic and then revised the ACL to block the offending traffic. An intrusion detection system alone is considered more passive, as it serves primarily as a means of detection and warning.

Anti-malware (or antivirus) tools have much in common with intrusion prevention systems. They detect malicious software, including viruses, worms, and Trojans, and then (optionally) block the code from entering or executing on a protected system. Typically these tools are regularly updated to include signatures for new forms and variants of malware that are detected across the Internet. In this sense, the active defenses are applied globally over the Internet. After new malware is discovered, security vendors create and distribute new signatures to the customers of their anti-malware products.

Intrusion prevention can likewise be performed on a broader scale than a single network or even an enterprise. For example, the Internet protocol (IP) addresses of machines that are spewing hostile packets can be shared widely through blacklists and then blocked by Internet service providers. Indeed, victims of massive denial of service attacks frequently ask upstream service providers to drop packets coming from the originating IP addresses.

Anti-malware and intrusion prevention systems can be integrated to form powerful active defenses. In many respects, the combined defenses would resemble an active air and missile defense system that detects hostile air threats and takes such actions as shooting them down or jamming their communications, except that in cyberspace the defenses are applied to hostile cyber threats such as malicious packets and malware. Rather than targeting incoming ballistic missiles, cyber defenses take their aim at packets that act like cyber missiles.

Honeypots, which lure or deflect attackers into isolated systems where they can be monitored, are another form of active defense. They are similar to the decoys used in air defense that deflect missiles from their intended targets.

In addition to playing a role in network security, active cyber defenses have been used to take down botnets (networks of compromised computers) and counter other cyber threats. The following two examples illustrate.

COREFLOOD TAKEDOWN

In April 2011 the Federal Bureau of Investigation (FBI), Department of Justice, and the nonprofit Internet Systems Consortium (ISC) deployed active defenses to take down the seven-year-old Coreflood botnet.[9] At the time, the botnet comprised over two million infected computers, all under the helm of a set of command-and-control (C2) servers. The bot malware installed on the machines was used to harvest usernames and passwords, as well as financial information to steal funds. One C2 server alone held about 190 gigabytes of data stolen from more than 400,000 victims.

The active defense included several steps. First, the US District Court of Connecticut issued a temporary restraining order that allowed ISC to swap out Coreflood's C2 servers for its own servers. The order also allowed the government to take over domain names used by the botnet. When the infected machines reached out to the new C2 servers for instructions, the bots were commanded to stop. The malware reactivated following a reboot, but each time it contacted a C2 server, it was instructed to stop. The effect was to neutralize, but not eliminate, the malware installed on the compromised machines. To help victims remove the malware, the FBI provided the IP addresses of infected machines to Internet service providers (ISPs) so they could notify their customers. In addition, Microsoft issued an update to its Malicious Software Removal Tool so victims could get rid of the code.

Using the air defense analogy, the Coreflood takedown can be likened to an active defense against hijacked aircraft, where the hijackers were acting on instructions transmitted from a C2 center. In this situation, the air defense might jam the signals sent from the center and replace them with signals that command the hijackers to land at specified airports. The airports would also be given information to identify the hijacked planes so that when they landed, the hijackers could be removed.

This approach of neutralizing the damaging effects of botnets by commandeering their C2 servers has been used in several other cases. Microsoft, for example, received a court order in November 2012 to continue its control of the C2 servers for two Zeus botnets. Because Zeus had been widely used to raid bank accounts, the operation has no doubt prevented considerable harm.[10]

GEORGIAN OUTING OF RUSSIA-BASED HACKER

In October 2012 Network World reported that the Georgian government had posted photos of a Russia-based hacker who had waged a persistent, months-long campaign to steal confidential information from Georgian government ministries, Parliament, banks, and nongovernmental organizations.[11] The photos, taken by the hacker's own webcam, came after a lengthy investigation that began in March 2011 when a file on a government computer was flagged by an antivirus program. After looking into the incident, government officials determined that three hundred to four hundred computers in key government agencies had been infected with the malware and that they had acquired it by visiting infected Georgian news sites that had pages with headlines such as "NATO Delegation Visit in Georgia" and "US-Georgian Agreements and Meetings." Once installed, the malware searched for documents using keywords such as "USA," "Russia," "NATO," and "CIA" and then transmitted the documents to a drop server where the spy could retrieve them.

Georgia's initial response included blocking connections to the drop server and removing the malware from the infected websites and personal computers. However, the spy did not give up and began sending the malware out as a portable document format file attachment in a deceptive email allegedly from admin@president.gov.ge.

The Georgian government then let the hacker infect one of its computers on purpose. On that computer, it hid its own spying program in a .ZIP archive titled "Georgian-NATO Agreement." The hacker took the bait, downloaded the archive, and unwittingly launched the government's code. The spyware turned on the hacker's webcam and began sending images to the government. It also mined the hacker's computers for documents, finding one that contained instructions in Russian from the hacker's handler about whom to target and how, as well as circumstantial evidence suggesting the Russian government's involvement.

Again, using the air defense analogy, the steps taken to block the exfiltration of files from compromised computers to the drop servers could be likened to jamming the transmission of sensitive data acquired with a stolen reconnaissance plane to the thieves' drop center. The steps taken to bait the hacker into unwittingly stealing and installing spyware might be likened to a command intentionally permitting the theft of a rigged reconnaissance plane with hidden surveillance equipment that sends the data it collects about the thieves back to the command.

Characteristics and Ethical Issues in Active Cyber Defense

In this section, we offer a set of distinctions for characterizing the different types of active defense described in the preceding section and discuss some of the ethical issues raised by each.

Scope of Effects

The first set of distinctions pertains to the scope of effects of an active defense. An active defense is said to be internal if the effects are limited to an organization's own internal network. If it affects outside networks, it is said to be external.

Drawing on the air defense analogy, an internal cyber defense is similar to an air defense system that takes actions against an incoming missile or hostile aircraft after it has entered a country's airspace, while an external cyber defense is similar to an air defense system that operates in someone else's airspace or attacks the base in a foreign country where the missile is being launched or the hostile aircraft is departing. Antiballistic missile defenses that operate against warheads during their boost phase are generally external, taking place in hostile territory, while those that operate during the terminal phase are likely to be internal.

We consider defenses that involve sharing threat information with outside parties to be external. An example is the Enhanced Security Services (ECS) program operated by the Department of Homeland Security (DHS). Under the program, DHS shares with commercial service providers indicators of cyber threats. The providers, in turn, use this information to better protect their customers.[12] Defenses that involve collecting intelligence from outside sources—say, by installing early warning sensors on their networks—are also considered external. Most of the effects in the Coreflood takedown were external. The C2 servers themselves were external, and when ISC took them over, they instructed bots in outside networks to stop. In contrast, most of the effects in the Georgian case were internal. Connections to the drop server were blocked on internal networks, and internal machines were cleaned of the malware. However, the case also had external effects—for example, the infection of the hacker's own computer with spyware.

Ethical Issues

In general, most of the ethical issues regarding active defenses concern external active defenses. They are discussed in the next section when we distinguish cooperative external defenses from noncooperative ones. However, even internal defenses can raise ethical issues. For example, inside users might complain that their rights to free speech were violated if internal defenses blocked their communications with outside parties. In addition, internal defenses do nothing to mitigate threats across cyberspace. By not even sharing threat information with outsiders, internal defenses expose external networks to continued harm that might be avoided if the defenses were applied to them as well. Arguably, at least in terms of national cyber defense, a better moral choice would be to help mitigate cyber threats more broadly. As discussed in the next section, the federal government has taken several steps to promote sharing of threat data, including the ECS program.

Returning to the air defense analogy, a missile defense system that only shot down missiles headed to military bases would not be as "just" as one that also shot down missiles headed to civilian targets such as cities and malls. However, it would be unreasonable to expect that missile defense system to protect the airspace of other countries, at least absent an agreement to do so.

Degree of Cooperation

The second set of distinctions pertains to the degree of cooperation in an active defense. If all effects against a particular network are performed with the knowledge and consent of the network owner, they are said to be cooperative. Otherwise, they are classified as noncooperative. For this discussion, we assume that network owners are authorized to conduct most defensive operations on their own networks, at least as long as they do not violate any laws or contractual agreements with their customers or users. Thus, the distinction applies mainly to active defenses with external effects.

Using the air defense analogy, a cooperative cyber defense is similar to an air defense system that shoots down missiles or hostile aircraft in the airspace of an ally that has requested help, and a noncooperative cyber defense is akin to an air defense system that shoots them down in the adversary's own airspace.

Antiviral tools are cooperative defenses. Security vendors distribute new signatures to their customers, but the signatures are installed only with the customers' permission. Similarly, sharing blacklists of hostile IP addresses is cooperative. In general, any active defense that does nothing more than share threat information is cooperative.

Defenses become noncooperative when they involve actions taken against external computers without the permission of the user or network owner. In the case of Coreflood, the actions taken against the individual bots were noncooperative. Neither the users of those machines nor the owners of the networks on which they resided agreed to have the bot code stopped. But neither had they agreed to the initial malware infection and subsequent theft of their data. Arguably, any user would prefer that the malware be stopped than be allowed to continue its harmful actions. Further, even though the action was noncooperative, it was deployed under legal authorities, enabled in part by the temporary restraining order. Moreover, the actual elimination of the malware from the infected machines was a cooperative action involving the machine owners.

Noncooperative defenses include what is sometimes called an attack back, a hack back, or a counterstrike. This defense uses hacking or exploit tools directly against the source of an attack or gets the attacker to unwittingly install software, say, by planting it in a decoy file on a computer the attacker has compromised. The goal might be to collect information about the source of the attack, to block attack packets, or to neutralize the source. Noncooperative defenses also include court-ordered seizures of computers.

Although the Coreflood takedown did not include any sort of hack back, the Georgian case did. In particular, the actions taken to plant spyware on the hacker's computer constituted a noncooperative counterstrike. However, one could argue that the hacker would never have acquired the spyware had he not knowingly and willfully first infected the computer hosting it and, second, downloaded the .ZIP archive containing it. Thus, he was at least complicit in his own infection and ultimate outing.

Ethical Issues

As a rule, noncooperative defenses, particularly those involving some sort of hack back, raise more ethical and legal issues than cooperative ones. In part, this is because most cyber attacks are launched through machines that themselves have been attacked, making it hard to know whether the immediate source of an attack is itself a victim rather than the actual source of malice. They may be hacked servers or bots on a botnet. Thus, any actions taken against the computers could harm parties who are not directly responsible for the attacks. In addition, cyber attacks in general violate computer crime statutes, at least when conducted by private sector entities.

While the argument can be made that some hack backs should be permissible under the law, not everyone agrees, and the topic has been hotly debated.[13] The Department of Justice has advised victims to refrain from any "attempt to access, damage, or impair another system that may appear to be involved in the intrusion or attack." The advice contends that "doing so is likely illegal, under U.S. and some foreign laws, and could result in civil and/or criminal liability."[14] However, government entities—in particular, the military, law enforcement, and intelligence agencies—have or can acquire the authorities needed to perform actions that might be characterized as hacking under certain prescribed conditions.

One might argue that if the government cannot or will not defend private organizations from cyber attacks, then these organizations should be able to come to their own defense even if that includes hacking back. The problem with this argument is that cyber attacks can be stopped without invading the attacker's system—for example, by blocking packets, by removing malware, and by fixing vulnerabilities. For purely defensive purposes, hacking back is not usually necessary. While the primary benefit of hacking back is to identify the attackers and then possibly prosecute or neutralize them, there are well-established ethical reasons for leaving these actions in the hands of governments and avoiding vigilantism.

If we assume that noncooperative defenses are conducted by or jointly with government entities with the necessary legal authorities, then the primary concern is that innocent parties may be harmed. Then we can draw on the long tradition of just war theory to determine the conditions under which active cyber defenses that pose risks to noncombatants can be ethically justified.

Most just war theorists hold that noncombatant immunity is a key lynchpin to all our moral thinking in war.[15] Thus, noncombatants are never to be intentionally targeted for harm as any part of a justified military action. Traditional just war theory does hold, however, that some actions that will foreseeably but unintentionally harm noncombatants may be permissible so long as that harm is truly unintentional, is proportionate to the good goal achieved by the act, and is not the means itself to achieve the good goal. Grouped together, these principles are known as the doctrine of double effect. The doctrine has come under heavy scholarly debate, with many critics doubting that its principles can hold true for all cases.[16] Meanwhile, others have argued that some revised or narrowed version of the doctrine can still be defended and applied to war.[17] We cannot engage this larger debate here, but we assume that at least some narrow version of the doctrine of double effect is applicable and, as such, is critical for our moral conclusions regarding harm to noncombatants from active cyber defense.

Whether noncombatants' property can be targeted is another matter. Generally, noncombatant property is similarly considered immune from direct and intentional harm since harming a person's property also harms that person. However, as with physical harm, unintended harm of noncombatant property can be permissible in some instances. Moreover, traditional just war theory and the laws of armed conflict can allow for some level of intentional harm to civilian property if it is necessary to block a particularly severe enemy military action and the civilians in question are later compensated. Thus, the ethical restrictions on harm to civilian property are far less strict than for physical harm to civilian persons. This is true for unintentional harms of both kinds and can even allow for some intentional harm to property when necessary if the stakes are high enough and recompense can be made.

In the case of active air defense, systems like Iron Dome are not without risk to civilians. If they happen to be under an incoming rocket's flight path when it is hit, they could be harmed by fallout from the explosion. However, Israel has limited its counterstrikes primarily to rockets aimed at densely populated urban areas. In that situation, any fallout is likely to be substantially less harmful than the effects produced by the rockets themselves if they are allowed to strike. We argue that such a risk imposition can be morally warranted. Note, however, that if Iron Dome created large amounts of dangerous and lethal fallout disproportionate to the lives saved, then its use would not be permissible.

In general, if an air defense system distributes some small risk of harm to civilians under an incoming missile's flight path to protect a much larger number of civilians from even greater harm, then the present conditions make such defense morally permissible. This is precisely what we find in the case of real-world air defense systems such as Iron Dome. Further, whether the risk of harm is imposed on noncombatants from one's own state or another state is irrelevant. What matters are the moral rights of all noncombatants, including, of course, noncombatants on any side of a given conflict. The point is to minimize collateral harm to all noncombatants.

The same principles should apply to active cyber defense; that is, it should be morally permissible for a state to take an action against a cyber threat if the unjust harm prevented exceeds and is proportionate to any foreseen harm imposed on noncombatants. Indeed, in the cyber domain meeting this demand will often be easy because it is frequently possible to effectively shoot down the cyber missiles without causing any fallout whatsoever. Instead, packets are simply deleted or diverted to a log file. Nobody is harmed.

In some cases, however, an active defense could have a negative impact on innocent parties. To illustrate, suppose that an action to shut down the source of an attack has the effect of shutting down an innocent person's computer that had been compromised and used to facilitate the attack. In this case, the action might still be morally permissible for two reasons. First, the harm induced might be temporary in nature, affecting the computer for a short time until the attack is contained. Second, the harm itself might be relatively minor, affecting only the noncombatant's property and not his or her person. While such effects could possibly further impede other rights of noncombatants, such as their ability to communicate or engage in activity vital to their livelihoods, all these further harms would be temporary in nature and could even be compensated for, if appropriate, after the fact. This is not to disregard the rights of noncombatants and use of their property for furthering other rights in our moral calculus but simply recognizes that different kinds and severities of harm result in different moral permissions and restrictions.

That the harm itself is likely to be nonphysical is quite significant in our moral reasoning conclusions for active cyber defense. If it is permissible in some cases to impose the risk of physical harm on noncombatants as part of a necessary and proportionate defensive action against an incoming missile (as we argued that it could be in the air defense case), then surely there will be cases where it can be permissible to impose the risk of temporary harm to the property of noncombatants to defend against an unjust cyber attack. The point here with active cyber defense is the kind of harms that would be potentially imposed on noncombatants, in general, is the kind of reduced harms that should make such defensive actions permissible.

A caveat, however, is in order. Computers today are used for life-critical functions, such as controlling life support systems in hospitals and operating critical infrastructure such as power grids. In a worst-case scenario, an active defense that affects such a system might lead to death or significant suffering. These risks need to be considered when weighing the ethics of any noncooperative action that could affect noncombatants. In general, defensive actions that do not disrupt legitimate functions are morally preferable over those that do. If the scope of possible effects cannot be reasonably estimated or foreseen, then the action may not be permissible.

In the case of Coreflood, the takedown affected many noncombatant computers; however, the effect was simply to stop the bot code from running. No other functions were affected, and the infected computer continued to operate normally. Thus, the operation ran virtually no risk of causing any harm whatsoever, let alone serious harm. In the Georgian case, the only harm was to the attacker's own computer, and he brought it on himself by downloading the bait files, thus making himself liable to intentional defensive harm.

Although the discussion here has focused on noncooperative defenses, it is worth noting that while cooperative defenses generally raise fewer issues, they are not beyond reproach. For example, suppose that a consortium of network owners agrees to block traffic from an IP address that is the source of legitimate traffic as well as the hostile traffic they wish to stop. Depending on circumstances, a better moral choice might be to block only the hostile traffic or to work with the owner of the offending IP address to take remedial action.

Types of Effects

The third set of distinctions pertains to the effects produced. An active defense is called sharing if the effects are to distribute threat information—such as hostile IP addresses or domain names or signatures for malicious packets or software—to other parties. Sharing took place in the Coreflood takedown when the FBI provided the IP addresses of compromised machines in the United States to their US ISPs and to foreign law enforcement agencies when the machines were located outside the United States. Another example of sharing is DHS's aforementioned ECS program.

An active defense is called collecting if it takes actions to acquire more information about the threat, for example, by activating or deploying additional sensors or by serving a court order or subpoena against either the source or an ISP that is likely to have relevant information. In the Coreflood takedown, the replaced C2 servers were set up to collect the IP addresses of the bots so that eventually their owners could be notified. The servers did not, however, acquire the contents of the victims' computers. In the Georgian case, spyware was used to activate a webcam and collect information from the attacker's computer.

An active defense is called blocking if the effects are to deny activity deemed hostile—for example, the traffic from a particular IP address or the execution of a particular program. The Coreflood takedown had the effect of breaking the communications channel from the persons who had been operating the botnet to the C2 servers controlling it. As a result, they could no longer send commands to the bots or download stolen data from the servers. In the Georgian case, connections to the drop servers were blocked to prevent further exfiltration of sensitive data.

Finally, an active defense is called preemptive if the effects are to neutralize or eliminate a source used in the attacks. It can be done, for example, by seizing the computer of a person initiating the attacks or by taking down the C2 servers for a botnet. In the Coreflood takedown, the hostile C2 servers were put out of commission and the bots neutralized. With further action on the part of the victims, the malware could also be removed.

Using the air defense analogy, the cyber defense of sharing is similar to a missile defense system that reports new missile threats to allies so that they can shoot them down. The cyber defense of collecting is comparable to a missile defense system that installs or activates additional radars or other sensors in response to an increased threat level or that sends out sorties to investigate suspicious aircraft. The cyber defense of blocking is akin to a missile defense system that shoots down incoming missiles or jams their radars and seekers. Finally, the cyber defense of preemption is similar to launching an offensive strike against the air or ground platform launching the missiles.

Some authors regard retaliation or retribution as a form of active defense. However, we consider these operations to be offensive in nature, as they serve primarily to harm the source of a past attack rather than mitigate, stop, or preempt a current one.

Ethical Issues

All four types of cyber operations raise ethical issues. The act of sharing raises issues of privacy and security, particularly if any sensitive information is shared along with the threat information—for example, secret or personal data stolen by an attacker or embedded within the attack traffic. The act of collecting also raises issues about privacy and security, but in this case they relate to the new information that is acquired rather than the dissemination of existing information. One might conclude from this discussion that it is better not to share, but there are equally compelling ethical reasons for sharing threat information. By informing other victims, or potential victims, they can effectively respond to or prevent cyber attacks, and by contacting law enforcement personnel, they can investigate and prosecute those responsible for the attack, thereby preventing further attacks. This ethical dilemma has led to approaches that promote sharing while minimizing the security and privacy risks—for example, by removing sensitive and personally identifiable data.

The US government has taken several steps to encourage the sharing of threat information. With its ECS program, for instance, the government supplies known threat information to the private sector. In early 2015 President Obama issued an executive order promoting the formation of private sector information-sharing and analysis organizations for information exchange and collaboration within the private sector and with the federal government.[18] Then, at the end of the year, Congress passed legislation designed to encourage businesses to share cyber threat information with the government. Although the law requires the removal of personally identifiable information, civil liberties groups were not satisfied with the privacy provisions.[19] The Department of Justice has advised organizations that have been hit by a cyber attack to notify other potential victims, law enforcement, and the Department of Homeland Security.[20] DHS, in turn, may share this information with other potential victims (e.g., through the

ECS program) and provide the notifying organization with additional information and assistance in mitigating the attack.

The act of collecting could also lead to harm if, for example, the sensors and other tools used to collect data on a network have or introduce backdoors or vulnerabilities that other parties could exploit. Even the installation of these tools could cause harm if, in the process, other components of the network are broken. An attempt to install surveillance code in a core router of Syria's main service provider may have taken down Syria's Internet in 2012.[21]

The act of blocking communications raises ethical issues relating to free speech, loss of commerce, and over-blocking. In a worst-case scenario, traffic might be blocked that is important for operating a life support system or critical infrastructure such as power generation and distribution. Likewise, the act of preemption raises ethical issues relating to disabling software or systems. Again, in a worst-case scenario, shutting down a life support system could cause serious harm. Any possible damage would need to be considered when applying any noncooperative cyber defense as discussed in the previous section. Concern over harm should drive technical and policy efforts to limit the effects of defenses, say, by disabling only traffic and software involved in an attack rather than shutting down all traffic and complete systems.

In the Coreflood takedown, it is important to note that the government did not attempt to remove the bot code from infected machines. It only neutralized it by issuing the stop command. Part of its reason for not removing the code was a concern for unanticipated side effects that might damage an infected computer.

Because active cyber defense should not be misconstrued as a form of offense, it is worth explaining why the distinction between offensive retaliation versus legitimate defensive action is so crucial in the ethical dimensions of killing and war. Defensive harm has the lowest ethical barrier to overcome among all possible justifiable harms. That is, if one is being wrongly attacked, then the moral restrictions against using force of some kind to block that wrongful attack are (relatively) few. All people have a right not to be harmed unjustly. If one side is attempting to harm another unjustly, then the former has made itself morally liable to suffer defensive harm as part of an act taken to thwart its unjust act. The side being wrongly attacked may permissibly harm its attacker to block or thwart the attack against it so long as the defensive action meets two criteria. First, inflicting the defensive harm must be necessary to block the unjust attack. If the defensive harm in question does nothing to block the liable party's unjust attack, then it is retributive punishment, or something else, but not properly an act of defense. Second, the defensive harm must be proportionate to the unjust harm to be blocked. If a foreign plane was found conducting reconnaissance over a state's territory without permission during peacetime, then the foreign state may have made itself liable to some form of defensive action such as being escorted to an airfield. However, it would be disproportionate and wrong to shoot the plane down or, even worse, to shoot down commercial planes flying under the foreign state's flag.

In general, there must be some reasonable correlation and proper "fit" between the extent of defensive response and the degree of liability of the offending party.[22] In the case of an active cyber defense, if the act is truly a defensive effort to block an unjust attack, then so long as it is necessary and proportionate, it will usually be ethically permissible. In the Georgian case, the government responded to the cyber espionage operation against it with its own espionage operation against the hacker. It did not destroy software and data on the hacker's computer.

Degree of Automation

The final set of distinctions pertains to the degree of human involvement. An active defense is said to be automatic if no human intervention is required and to be manual if key steps require the affirmative action of humans.

Most anti-malware and intrusion prevention systems have both manual and automated components. Humans determine what goes into the signature database, and they install and configure the security software, including a range of response actions. However, the processes of signature distribution, malicious code and packet detection, and initial response are automated.

In the Coreflood takedown, the execution of the stop commands was fully automated through the C2 servers. However, humans played an important role in the operational planning and decision-making, the analysis of the botnet code and the effects of issuing a stop command, the acquisition of the restraining order, and the swapping out of the C2 servers. Thus, the entire operation had both manual and automatic aspects. In the Georgian case, much of the investigation involved manual work, including analyzing the code, determining what the hacker was looking for, and setting up the bait with the spyware. But the key element in the outing—namely, the operation of the spyware—was automated. Once the hacker downloaded the .ZIP archive, the program did the rest.

Applying once again the air defense analogy, an automatic cyber defense is similar to a missile defense system that automatically shoots down anything meeting the preset criteria for being a hostile aircraft or incoming missile, whereas a manual cyber defense would act as Operation Noble Eagle, where humans play a critical role both in recognizing and responding to suspicious activity in US airspace.

Ethical Issues

In general, on the one hand, manual actions give humans a greater opportunity to contextualize their ethical decisions. Rather than configuring a system to always respond in a certain way, humans can take into account the source or likely source of a perceived threat, its nature, the broader circumstances, and the likely consequences of taking certain actions against it. This is vital to Noble Eagle, where most incidents turn out to be nonhostile and lives are at stake. On the other hand, given that manual actions take longer to execute than automated ones, they potentially allow greater damage to incur before the threat is mitigated.

In the cyber domain, where actions can take place instantaneously, automated defenses become critical. That is, the speed of some actions in the cyber domain are such that a cyber defense must be automated to have any effect against the attack. Perhaps for this reason the Defense Department has exempted some cyber actions from its recent "man in the loop" legal requirements for automated weapon systems.[23] If a hostile actor has launched an attack to cause a power generator to explode, then an automated response that successfully blocks the attack without causing unnecessary harm is morally superior to a manual one that comes too late.

However, this does not mean that all cyber defenses should be automated. To argue that all cyber actions should be exempt from the man-in-the-loop requirement would be ethically (and strategically) problematic. The nature of a defense and its potential effects—particularly the potential severity of its foreseeable harms—must be weighed in any decision to automate. The cyber case is unique in that the speed of many cyber attacks necessitates that many defenses be automated to be effective in any way. But if the effects of automating a given defense would lead to too great a risk of impermissible harm, then it should not be done, even if this decision essentially nullifies its efficacy entirely. Thankfully, given the aforementioned reasons regarding the predictable effects that most forms of active cyber defense would produce, we find that in many cases their automation could be permissible.

Conclusions

Using analogies from air defense, active cyber defense is a rich concept that, when properly understood and executed, is neither offensive nor necessarily harmful and dangerous. Rather, it can be executed in accordance with the well-established ethical principles that govern all forms of defense—namely, principles relating to harm, necessity, and proportionality. In many cases, such as with most botnet takedowns, active defenses mitigate substantial harm while imposing little or none of their own.

While active defenses can be morally justified in many cases, we do not mean to imply that they always are. All plausible effects must be considered to determine what, if any, harms can follow. If harms cannot be estimated or are unnecessary or disproportionate to potential benefits gained, an active defense cannot be morally justified.

In considering active defenses, we have assumed that they would be executed under appropriate legal authorities. In particular, they would be conducted by authorized government entities or by private companies operating under judicial orders or otherwise within the law. We leave open the question of how far companies can go in areas where the law is unclear or untested. While such active defenses as sharing attack signatures and hostile IP addresses and domain names have raised few legal questions, an active defense that deleted code or data on the attacker's machine would raise more. No doubt, this area will likely continue to inspire lively discussions and debates.

Notes

The views expressed in this document are those of the authors and do not reflect the official policy or position of the Department of Defense or the US government. An earlier version of this chapter appeared in Emily O. Goldman and John Arquilla, eds., *Cyber Analogies* (Monterey, CA: Naval Postgraduate School, 2014).

1. D. Dittrich and K. E. Himma, "Active Response to Computer Intrusions," in *The Handbook of Information Security*, ed. H. Bidgoli (Somerset, NJ: John Wiley & Sons, 2005).

2. Department of Defense, Joint Publication 3–01, *Countering Air and Missile Threats* (Washington, DC: Department of Defense, March 23, 2012), http://www.dtic.mil/doctrine/new_pubs/jp3_01.pdf.

3. "MIM-104 Patriot," Wikipedia, no date, accessed November 6, 2012, https://en.wikipedia.org/wiki/MIM-104_Patriot.

4. I. Kershner, "Israeli Iron Dome Stops a Rocket with a Rocket," *New York Times*, November 18, 2012, http://www.nytimes.com/2012/11/19/world/middleeast/israeli-iron-dome-stops-a-rocket-with-a-rocket.html?_r=0.

5. G. Ackerman and S. A. Ramadan, "Israel Wages Cyber War with Hamas as Civilians Take up Computers," *Bloomberg,* November 19, 2012, http://www.bloomberg.com/news/2012-11-19/israel-wages-cyber-war-with-hamas-as-civilians-take-up-computers.html.

6. Air Force, Operation Noble Eagle, Air Force Historical Studies Office, posted September 6, 2012, http://www.afhistory.af.mil/FAQs/Fact-Sheets/Article/458956/2001-operation-noble-eagle.

7. Department of Defense, "Countering Air and Missile Threats."

8. A. Keromytis, "Active Authentication," Defense Advanced Research Projects Agency, no date, accessed January 13, 2016, http://www.darpa.mil/program/active-authentication.

9. K. Zetter, "With Court Order, FBI Hijacks 'Coreflood' Botnet, Sends Kill Signal," *Wired*, April 13, 2011, https://www.wired.com/threatlevel/2011/04/coreflood/; K. Zetter, "FBI vs. Coreflood Botnet: Round 1 Goes to the Feds," *Wired*, April 26, 2011, https://www.wired.com/threatlevel/2011/04/coreflood_results/; and K. J. Higgins, "Coreflood Botnet an Attractive Target for Takedown for Many Reasons," *Dark Reading*, April 14, 2011, http://www.darkreading.com/risk/coreflood-botnet-an-attractive-target-for-takedown-for-many-reasons/d/d-id/1135557?.

10. R. Lemos, "Microsoft Can Retain Control of Zeus Botnet under Federal Court Order," eWeek, December 1, 2012, http://www.eweek.com/security/microsoft-can-retain-control-of-zeus-botnet-under-federal-court-order/.

11. J. Kirk, "Irked by Cyberspying, Georgia Outs Russia-Based Hacker—with Photos," *Computerworld*, October 30, 2010, http://www.computerworld.com/article/2493051/cybercrime-hacking/irked-by-cyberspying—georgia-outs-russia-based-hacker——with-photos.html.

12. Department of Homeland Security, "Enhanced Cybersecurity Services (ECS)," October 14, 2015, http://www.dhs.gov/enhanced-cybersecurity-services.

13. D. E. Denning, "The Ethics of Cyber Conflict," in *The Handbook of Information and Computer Ethics*, ed. K. E. Himma and H. T. Tavani (Somerset, NJ: John Wiley & Sons, 2012), 407–28; E. Messmer, "Hitting Back at Cyberattackers: Experts Discuss Pros and Cons," *Network World*, November 1, 2012, http://www.networkworld.com/article/2161144/security/hitting-back-at-cyberattackers—experts-discuss-pros-and-cons.html; and Steptoe, "The

Hackback Debate," Steptoe Cyberblog, November 2, 2012, http://www.steptoecyberblog .com/2012/11/02/the-hackback-debate/.

14. Department of Justice, Cybersecurity Unit, "Best Practices for Victim Response and Reporting of Cyber Incidents," April 2015, http://www.justice.gov/sites/default /files/criminal-ccips/legacy/2015/04/30/04272015reporting-cyber-incidents-final.pdf.

15. M. Walzer, *Just and Unjust Wars: A Moral Argument with Historical Illustrations* (New York: Basic Books, 1977); T. Nagel, "War and Massacre," *Philosophy and Public Affairs* 1, no. 2 (1972): 123–44; D. Rodin, *War and Self-Defense* (New York: Oxford University Press, 2003); and B. Orend, *The Morality of War* (Peterborough, ON: Broadview Press, 2006).

16. N. Davis, "The Doctrine of Double Effect: Problems of Interpretation," *Pacific Philosophical Quarterly* 65 (1984): 107–23; F. Kamm, "Failures of Just War Theory: Terror, Harm, and Justice," *Ethics* 114 (2004): 650–92; A. McIntyre, "Doing Away with Double Effect," *Ethics* 111 (2001): 219–55; and U. Steinhoff, *On the Ethics of War and Terrorism* (Oxford: Oxford University Press, 2007).

17. J. McMahan, "Revising the Doctrine of Double Effect," *Journal of Applied Philosophy* 11, no. 2 (1994): 201–12; W. S. Quinn, "Actions, Intentions, and Consequences: The Doctrine of Double Effect," *Philosophy and Public Affairs* 18 (1989): 334–51; and D. Nelkin and S. Rickless, "Three Cheers for Double Effect," *Philosophy and Phenomenological Research*, December 2012, http://onlinelibrary.wiley.com/doi/10.1111/phpr.12002/full.

18. Barack Obama, "Executive Order—Promoting Private Sector Cybersecurity Information Sharing," The White House, February 13, 2015, https://www.whitehouse.gov /the-press-office/2015/02/13/executive-order-promoting-private-sector-cyber security-information-shari.

19. E. Chabrow, "Obama Signs Cyberthreat Information Sharing Bill," Gov Info Security, December 18, 2015, http://www.govinfosecurity.com/congress-approves-cyberthreat -information-sharing-bill-a-8762.

20. Department of Justice, Cybersecurity Unit, "Best Practices."

21. J. Bamford, "The Most Wanted Man in the World," *Wired*, August 2014, 2016, http://www.wired.com/2014/08/edward-snowden/.

22. J. McMahan, "The Basis of Moral Liability to Defensive Harm," *Philosophical Issues* 15 (2005): 386–405; and J. Quong, "Liability to Defensive Harm," *Philosophy & Public Affairs* 40, no. 1 (2012): 45–77.

23. Ashton Carter, "Autonomy in Weapon Systems," Department of Defense Directive Number 3000.09, November 21, 2012, www.dtic.mil/whs/directives/corres/pdf/300009p .pdf; and S. Gallagher, "US Cyber-weapons Exempt from 'Human Judgment' Requirement," *Ars Technica*, November 29, 2012, http://arstechnica.com/tech-policy/2012/11/us-cyber -weapons-exempt-from-human-judgment-requirement/.

13 "When the Urgency of Time and Circumstances Clearly Does Not Permit . . ."

PRE-DELEGATION IN NUCLEAR AND CYBER SCENARIOS

PETER FEAVER AND KENNETH GEERS

In a formerly top-secret document titled "Instructions for the Expenditure of Nuclear Weapons in Accordance with the Presidential Authorization Dated May 22, 1957," the US military was notified that "when the urgency of time and circumstances clearly does not permit a specific decision by the President, or other person empowered to act in his stead, the Armed Forces of the United States are authorized by the President to expend nuclear weapons in the following circumstances in conformity with these instructions."[1]

The significance of this directive was underlined by the fact that President Dwight Eisenhower informed Secretary of Defense Thomas Gates that the president himself had written parts of it. Furthermore, Eisenhower told Gates, "I cannot overemphasize the need for the utmost discretion and understanding in exercising the authority set forth in these documents. Accordingly, I would like you to find some way to brief the various Authorizing Commanders on this subject to ensure that all are of one mind as to the letter and the spirit of these instructions."[2]

Eisenhower's memo shows US national command authority wrestling with the thorniest of national security concerns—how to preserve political control when evolving technology and threats are pushing for a faster and faster response. Today the national command authority is facing similar issues in the cyber domain, and policymakers can learn from the efforts of earlier generations to adapt to the nuclear age. Cyber conflict does not constitute the same kind of civilization-ending threat that global thermonuclear war poses, but it may demand changes to the way American leaders manage national security affairs that will rival the changes wrought by the advent of nuclear weapons in the 1940s. Nuclear weapons, for example, imposed unusually dramatic constraints on traditional command-and-control (C2) arrangements; for its part, cyber conflict appears certain to strain these arrangements in new and unpredictable ways.

In this chapter, the authors examine one specific parallel, pre-delegation policy, which grants lower-level commanders the authority to use special weapons under carefully prescribed conditions. Three features of nuclear war drove policymakers to consider and, in some cases, to adopt pre-delegation: the speed with which a nuclear attack could occur, the surprise that could be achieved, and the specialized nature of the technology (that meant only certain cadres could receive sufficient training to be battle competent).

Each of these features has an obvious cyber analogue. In both the nuclear and cyber domains, defenders are under a great deal of pressure to act quickly, they may be faced with conflict scenarios no one could have imagined, and they require a high level of training and technical expertise. As a result, and in both the nuclear and cyber war cases, defenders may need some level of pre-delegated authority to act quickly and capably in defense of the nation.

Thus, the "letter and spirit" of Eisenhower's memorandum is also the topic of this chapter as it addresses the possibility that certain national security threat scenarios may oblige the national command authority to do something it would much prefer not to do—that is, to authorize military action in advance, without knowing exactly when and how it will be used.

Nuclear Pre-delegation

Early in the nuclear age, policymakers recognized a trilemma inherent in the nuclear revolution.[3] The first two horns of the trilemma constituted the "always-never dilemma": political authorities demanded that nuclear weapons always be available for use, even under the most extreme conditions (e.g., after a surprise attack), while at the same time stipulating that they would never be used accidentally or without proper authorization. Many measures designed to assure the "always" side of the dilemma posed risks for the "never" side, and vice versa. The third horn of the trilemma was that nuclear weapons should have the highest level of civilian control, far in excess of what was required for conventional military weapons and operations. Here, some measures designed to ensure strict civilian control tended to exacerbate the always-never dilemma. What happened in practice? In fact, the evolution of the US nuclear C2 system reflected an ongoing set of compromises that balanced myriad risks against these three desiderata.

As the Soviet nuclear arsenal grew in size and lethality, the challenges of this trilemma became more acute. What if a sudden illness, a natural disaster, or a surprise military attack killed or incapacitated the president, and perhaps other senior leadership figures, before he or she could even begin to manage a war? What if tactical commanders received warning of an attack or actually came under attack but political authorities delayed in responding? For certain weapons, this could create a "use them or lose them" scenario. What should US nuclear commanders do in these dire scenarios, and how could we ensure that they would not violate the principles of always, never, and civilian control?

One controversial measure designed to address these concerns was the pre-delegation of use authority (hereafter, pre-delegation), in which the president spelled out carefully delineated procedures in advance that would authorize when and how nuclear weapons could be used by tactical commanders. Of course, some form of pre-delegation is as old as warfare itself. As Martin van Creveld observed, even Stone Age chieftains wrestled with the challenges of command in war, and part of their solution likely involved explaining to the other warriors what they should do under certain anticipatable circumstances.[4] For centuries, and before technological advances solved the problem of communicating at great distances, ground and especially naval commanders departed on their missions with orders that spelled out in greater or lesser detail what political authorities expected the commanders to do while out of communication range. Indeed, some form of pre-delegation is inherent in the president's function as chief executive officer; unless the president can delegate certain of his or her powers and duties, little in the country would ever get done.[5] In 2014 Lt. Gen. Dave Deptula, USAF (Ret.), responded to a question on micromanagement in this way: "It's absolutely easy . . . trust your tactical level commanders . . . delegate engagement authority to the lowest possible level . . . give engagement authority to the people who are closest to the problem and who can observe what's going on."[6] In 2015 a paper from the Naval War College argued that the dynamic and rapidly evolving nature of the cyber domain demands that US Cyber Command (CYBERCOM) adopt the decentralized C2 doctrine of maneuver warfare to maximize the effectiveness of military cyberspace operations.[7]

Faced with the trilemma of always, never, and civilian control, US national command authorities updated the familiar tool of pre-delegation to the unfamiliar constraints of the nuclear age. It has long been known that between the Eisenhower and Gerald Ford administrations, up to seven unified and specified commanders, at the three- and four-star levels, possessed the authority to launch nuclear weapons.[8] In 1950 Commander in Chief of Strategic Air Command (CINCSAC) Gen. Curtis LeMay argued that senior officers must be able to act in the event Washington were destroyed by a surprise Soviet attack. Later he believed that he had gained this de facto authority.[9] In 1957 LeMay informed a presidential commission: "If I see that the Russians are amassing their planes for an attack, I'm going to knock the shit out of them before they take off the ground."[10] His successor, CINCSAC Gen. Thomas Power, informed Congress that he possessed "conditional authority" to use nuclear weapons. During the 1962 Cuban Missile Crisis, Supreme Allied Commander Europe Gen. Lauris Norstad was given prior authority to use nuclear weapons if Russia attacked Western Europe.[11]

The nature and scope of nuclear pre-delegation have been highly classified information in the US nuclear establishment, so the public record is murky and filled with holes. However, since 1998, a number of documents were declassified that have filled in some gaps.[12] The most dramatic revelation was the declassification of new information on "Project Furtherance," a plan that, under certain circumstances, provided for "a full nuclear response against both the Soviet Union and China," specifically "in the event the President has been killed or cannot be found."[13] In the memo dated October 14, 1968, President Lyndon Johnson's advisers recommended changes to the existing authorities: to allow the response to be tailored either to the Soviet Union or to China, to limit the response to a conventional attack at the nonnuclear level, and to outline these instructions in two documents rather than one. These revelations indicate that pre-delegation extended well beyond the use of nuclear weapons in a defensive role.

In 1976 the United States reportedly planned to revoke some, if not all, of the provisions for nuclear pre-delegation that it had established in the 1950s.[14] Currently, it is not publicly known whether any pre-delegation of authority to launch nuclear weapons continues to exist and, if so, under what constraints. However, based on recently declassified documents, into the 1980s American war planners clearly still were addressing the threat of decapitation and the difficulty of maintaining connectivity with national command authorities during a nuclear war, and pre-delegation was at least one of the options under debate.[15]

The Pros: Why Nuclear Pre-delegation

The primary benefit of pre-delegation is that it reliably circumvents the threat that an enemy could interdict communications between national command authorities and nuclear operators, decapitate the nuclear arsenal, and render it impotent. Moreover, pre-delegation accomplishes this while simultaneously reinforcing the legal chain of command. The pre-delegated instructions take the place of the orders that the national command authority presumably would have given in the scenario if it had been possible to do so; thus, pre-delegation makes the actions legal.

Pre-delegation is preferable to presidential succession, which transfers all presidential authority to subordinate officials. The Constitution and the Presidential Succession Act of 1947 prescribe a cumbersome process of succession from the president to the vice president, to the Speaker of the House, to the president pro tempore of the Senate, and finally to the cabinet officers (in the order of when the department was established). But a nuclear war could kill many if not all of these civilians suddenly or at least render them incommunicado. Given the secrecy and complexity of nuclear war planning, it is doubtful that more than a handful of these officials would be ready to manage a war, especially a nuclear war. In short, national security planners have good reason to fear that the constitutional line of succession would move too slowly during an extreme national security crisis.

A crisis-oriented alternative to succession is the "devolution" of military command, in which the president as commander in chief is replaced by the secretary of defense, who would be immediately followed by the next highest-ranking military officer, and so on. However, it is highly likely that any practicable system of de facto devolution of command would quickly diverge from the de jure line of succession. Furthermore, devolution as a national plan would seem to rest on shaky political and legal ground. It is doubtful that US civilian leadership would ever agree to cede so much power to the US military automatically, and the Supreme Court may not uphold it as constitutional. Finally, devolution of command creates the problem of "multiple presidents" if communications links with one or more of the officials in the chain of command are reconstituted and then lost again as a crisis evolved.

Pre-delegation is on much stronger legal ground and is thus preferable to devolution of command. Pre-delegation gives conditional, de facto authority to certain trusted commanders while keeping de jure authority with elected civilian leadership. Moreover, pre-delegation allows for fine-tuned civilian control since the pre-delegated authority can be as restrictive or permissive as desired. Thus, pre-delegation appears to reinforce civilian control of nuclear weapons. Last, pre-delegation allows the president to reassert command and control if communications are restored.

It is not enough, however, to have policies and doctrine aimed at mitigating the trilemma. Political authorities must also understand doctrine and actively support the policies. Military doctrine without political buy-in cannot be sustained indefinitely. Over time, gaps will emerge between what political leaders think military doctrine is and what military officers understand it to be. During a crisis, this lack of mutual understanding could lead to response failures or other breakdowns in command and control, proving disastrous for the nation.

In sum, compared with the alternatives, and provided that political authorities fully comprehend what they are doing, pre-delegation is simple and easy to implement. Building hardened command, control, and communications (C3) networks to withstand every possible worst-case scenario would be prohibitively expensive, even if it were technologically feasible in the first place. Pre-delegation offers a ready stopgap for unforeseen circumstances that could defeat C3 networks in the United States and is, by comparison, essentially free.

The Cons: Why Not Nuclear Pre-delegation

The pre-delegation of nuclear authority has an age-old Achilles' heel, human nature. For the system to work in the extreme scenarios when it would be needed, it couldn't be stymied by technical measures that physically block its use (such as a permissive action link or other coded systems that separate possession from usability[16]). Pre-delegation was intended as the solution for cases in which all communication with political authority would be broken. Therefore, a military commander possessing pre-delegation authority must also have everything that he or she would need to give a legitimate launch order. Logically, a commander with pre-delegated authority must be able to make an unauthorized use look authorized to anyone downstream in the chain of command. Thus, pre-delegation favors the "always" side of the trilemma at the expense of the "never" and the "civilian control" sides. These risks are tolerable provided that commanders honor the terms of their pre-delegated authority—that is, they must operate with complete integrity. Of course, the nuclear establishment invests extensive resources to ensure such integrity, but this risk is not inconsequential.

Pre-delegation seems to imply that de jure political control would give way very quickly to de facto military control and that there would be some level of automaticity to nuclear retaliation akin to the interlocking mobilizations of World War I or to the Soviet Union's "Dead Hand" system.[17] In short, pre-delegation poses a strain on civil-military relations. As personified by General LeMay and parodied in *Dr. Strangelove*, in war as in peacetime, civilian and military leaders may have different tendencies. On the one hand, military officers may want to use nuclear weapons in preemptive or retaliatory action to protect assets, forces, or territory, even if the bombs explode over domestic or allied territory. They may feel a certain pressure to "use them or lose them." On the other hand, civilians might prefer to absorb tactical military losses for other perceived strategic gains, such as to prevent an escalation of the conflict.

As a concept, pre-delegation is simple, but in practice it must be a highly complex mechanism. For example, how far down the chain of command should nuclear authority go? How wide should the latitude be and how specific the instructions? It is hard to anticipate in advance what would be the preferred course of action under scenarios that can only dimly be imagined. In practice, for pre-delegation to be effective, prescribed conditionality would have to be balanced with implied flexibility, yet having too much interpretive latitude with nuclear weapons is undesirable. And how public should pre-delegation policy be? Revealing some information helps deterrence, but revealing too much gives the enemy opportunities to figure out how to defeat the system.[18] It is worth noting that presidential delegations of authority should be published in the *Federal Register*, but this never happened with nuclear authorities.[19] Finally, how should nuclear authority revert to civilian control? In theory, it should happen as soon as reliable communication with the president or his or her successor is restored, but in practice it would be difficult to accomplish during a rapidly unfolding crisis.

Theoretically, pre-delegation could apply to both offensive and defensive weapons. However, the case for nuclear pre-delegation is much stronger for defensive weapons, such as air defense missiles tipped with nuclear warheads. Defensive weapons have a very short operational window to be effective, and the consequences of an unauthorized defensive use may be less severe than for an unauthorized offensive use. Defensive nuclear weapons would explode primarily over US and Canadian airspace. By contrast, offensive weapons would detonate on enemy territory, greatly increasing pressure to escalate the crisis.

However, even defensive pre-delegation scenarios threatened the territory of other states, and this proved to be one of the most sensitive and difficult aspects of the policy. The declassified record shows that President Eisenhower reluctantly acquiesced to pre-delegation policy; however, he became personally involved in the acute political challenge of pre-delegating nuclear activity that directly threatened our closest allies. In the declassified notes from a top-secret meeting held on June 27, 1958, "The President stressed the weakness of coalitions as bearing on this matter [referring to the pre-delegation of authority to fire nuclear air defense weapons]. He recalled that this was largely the secret of Napoleon's success, which was not seen until Clausewitz wrote about it. He recalled that Clausewitz had stressed that war is a political act—we must expect the civil authorities to seek control."[20]

Cyber Pre-delegation

While the nuclear revolution began with a massive explosion in the New Mexico desert, the cyber revolution has quietly sneaked up on us. The Internet has provided innumerable benefits to civilization, but a looming downside is that we may have grown too dependent on a range of networking technologies that are quite vulnerable to attack. Although we are still at the dawn of the Internet era, almost every kind of network-connected critical infrastructure has been targeted by hackers: air traffic control, financial sector, elections, water, and electricity.[21] Over time, this problem may only get worse, as formerly closed, custom information technology systems are replaced with less expensive commercial technologies that are both easier to use and easier to hack.[22] National security thinkers rightly worry that militaries, intelligence agencies, terrorists, insiders, and even lone hackers will target such systems in the future.

Cyber weapons do not pose an immediate, apocalyptic threat on the scale of nuclear weapons. For the foreseeable future, the always-never dilemma will not apply in the cyber domain quite like it applies in the nuclear domain. Indeed, in the nuclear era, apart from the bombs dropped on Hiroshima and Nagasaki, the US military always prepared for nuclear war but never fought it. By contrast, the US national security establishment (as well as the private sector) is almost always under some form of cyber attack even though many victims (and other key stakeholders) have scarcely begun to prepare for it. The United States may have a low tolerance for the kind of catastrophic cyber attack envisioned in worst-case scenarios, but it manifestly has a high tolerance for the low-level cyber attacks that its citizens endure every day.

Still, as the infamous Morris worm of 1988 and the more recent Stuxnet computer worm illustrate, there are reasons to worry about the intended and unintended effects of authorized and unauthorized use of cyber weapons.[23] And we do not know how damaging a cyber attack could be. In mid-February 2016, the *New York Times* reported that Operation Olympic Games (the alleged cyber attack on Iran's nuclear program) may actually have been dwarfed by Nitro Zeus, a proposed cyber attack that would have disabled Iran's air defenses, communications systems, and crucial parts of its power grid.[24] Moreover, cyber has a novel dimension that naturally generates concern about political control: the line between the military-intelligence and civilian-commercial domains is unclear, and activities in one domain will almost certainly seep over into the other, raising sensitive privacy and civil liberty concerns. As a result, on the notional spectrum from bayonets to ballistic missiles, cyber weapons are often considered to be closer to the ballistic missile end and, thus, much like nuclear weapons, require extraordinary C2 arrangements. After all, even conventional weapons have rules of engagement. In the cyber domain, we can expect politicians to act more conservatively not least because of uncertainty over impacts on civilian systems and challenges of attribution.

The cyber battlefield is new and evolving quickly. Iran may have waited two years to retaliate for Stuxnet, eventually hitting targets in three countries: Saudi Aramco, Qatari RasGas, and multiple US banks.[25] North Korea conducted a pre-emptive cyber attack against Sony Pictures Entertainment in a vain attempt to prevent the release of a satirical Hollywood movie about a Central Intelligence Agency (CIA) plot to assassinate North Korean leader Kim Jong-un.[26] In such cases, the nature and timing of a national response will usually be a complex and time-consuming process. In the former case, the Federal Bureau of Investigation (FBI) recently indicted seven Iranians; in the latter, President Obama announced that the United States would "respond proportionally . . . in a place and time and manner that we choose."[27]

There are three important analogues between nuclear attacks and cyber attacks: malicious code can travel across computer networks at lightning speed, successful cyber attacks are often based on novel ideas (the archetype here is the zero-day vulnerability plus exploit, which only the attacker knows about), and computer security is a complex, highly technical discipline that many decision makers do not understand. These three characteristics—speed, surprise, and specialization—may force national civilian leadership to give tactical military commanders a pre-delegated authority to operate in cyberspace so that they are able to competently and successfully defend US computer networks.

Yet the cyber challenge differs from the nuclear one in two key aspects—attribution and impact. Together, they point to the need for caution in adopting the nuclear era "fix" of pre-delegation. In cyberspace, it is often difficult to know with certainty who is attacking you, at least until a full-scope investigation is complete. This poses a significant obstacle to quick retaliation. There are analogous concerns in the area of nuclear terrorism, but for most of the Cold War, the attribution concern from state-based attacks was a secondary consideration. Ballistic missiles have a return address. In addition, if cyber attacks do not pose an existential threat to American society, they also do not pose the always-never dilemma. Therefore, it is politically fraught to assume the risks inherent in pre-delegation because the benefits and requirements are more open to debate. Pre-delegation was controversial during the nuclear era, when the C2 exigencies made it seem necessary. By contrast, cyber commanders should have more difficulty than their nuclear predecessors did in convincing political leaders on the wisdom of pre-delegation.

The Pros: Why Cyber Pre-delegation

First, planning a cyber attack may take months or even years, but once an attacker pulls the trigger, electrons move far more quickly than ballistic missiles—at close to the speed of light. In fact, even layered cyber attacks may unfold at such a high rate that pre-delegation alone is insufficient. For nuclear war, pre-delegation was deemed necessary to eliminate cumbersome interactions between national command authorities and tactical commanders; however, under most scenarios, tactical commanders would likely have enough warning to make their own deliberative response. With cyber attacks, the damage is often done before tactical commanders have a chance to collect evidence, evaluate data, and prepare a response. The cyber analogue therefore might not be the pre-delegated authority to respond but the automated authority to respond. One of the primary fears of nuclear pre-delegation was that there would be an automatic response, but with cyber attacks, the minuscule time windows involved will make some level of automation inevitable, especially to defend networks, and has led to increased discussions regarding the importance of developing autonomous systems.[28] This should be easy to pre-delegate, as long as the actions are defensive in nature. By contrast, an aggressive counterattack may not need pre-delegation because the technical challenges would require significant human intervention and deliberative planning.

Second, nuclear pre-delegation hedged against surprise attacks and unforeseen scenarios. Cyber attacks are also characterized by a high level of surprise. Information technology and cyber attacks are evolving at a blinding rate; thus, it is impossible to be familiar with every hacker tool and technique. Antivirus companies routinely gather over 100,000 unique samples of malicious code in a day, and still many cyber attacks pass undetected.[29] The most advanced attacks, which exploit so-called zero-day vulnerabilities, epitomize this challenge; such attacks are almost impossible to defend because they use a novel attack method for which there is no signature. Thus, security experts today are forced to defend against broad categories of cyber attacks instead of focusing on individual threats because it is hard to say exactly what the next cyber attack will look like.[30] The wide variety of possible attack vectors means that a cyber C2 system that restricted use authority narrowly to the topmost national command authority would likely be impotent, for by the time policymakers had figured out what was happening, and how they wished to respond, the damage would be done. Indeed, the attack may have migrated to new and unanticipated forms, always leaving policymakers several steps behind. Of course, the near-inevitability of surprise could mean that policymakers will be hard pressed to develop the carefully prescribed pre-delegation conditions of the nuclear era. Therefore, pre-delegation in the cyber domain may need to be more permissive and flexible than what was likely adopted for nuclear C2 purposes.

Third, like nuclear war, cyber war involves highly technical considerations that even dedicated policymakers are unlikely to master. The cyber sophistication of political leaders can improve with their participation in cyber exercises and their deeper familiarization with the cyber C2 system. But the rapid evolution of information technology makes it challenging even for technical professionals to keep pace, so there will likely always be a gulf in understanding between the operators and policymakers. Whereas an inability to understand the finer points of aerodynamics may not limit the quality of political guidance regarding air strikes, confusion over the nature of computer hacking could materially degrade decision-making on cyber responses. In a 2010 Black Hat conference keynote address, former CIA director Michael Hayden stated that conventional operations such as air strikes are discrete events that can be easier than cyber attacks for decision makers to manage. The president, he argued, could choose to bomb a factory at any time, but sophisticated cyber attacks take months, if not years, of painstaking, multifaceted technical subversion. Cyber pre-delegation, which would allow policymakers to develop guidance focused on desired outcomes in a deliberate manner and well before a crisis, may be the best way for political authorities to get what they want and not merely what they ask for.

Above and beyond these factors, national security decision makers around the world cannot ignore the official statements of other governments. The US military claims to employ "integrated electronic warfare, information and cyberspace operations as authorized, or directed, to ensure freedom of action in and through cyberspace and the information environment, and to deny the same to our adversaries."[31] The Israeli military claims it uses cyber attacks "relentlessly" to thwart the enemy "at all fronts and in every kind of conflict," and in peacetime it uses them to maintain Israel's qualitative military advantage over its enemies, including by influencing "public opinion."[32] The French Ministry of Defense has written that all modern military operations have a "cyber component" similar to "earth, sea, air, and space," and that the "strategic" nature of cyberspace means that operations there fall under the "highest level" of decision-making in Paris.[33] In Russia Vladimir Putin stated that "information attacks" are being used to achieve political and military goals and that their impact can be "higher" than that of conventional weapons. Anatoly Tsyganok, the director of the Center for Military Forecasting and a lecturer at Moscow State University's Global Policy Department, opined that cyber attacks are now "second in importance only to nuclear arms."[34] Given the prevalence of such high-level rhetoric, skeptics may have a point when they say the threat of cyber war is sometimes overstated, but they are living in denial if they say the threat simply does not exist.

The Cons: Why Not Cyber Pre-delegation

Cyber pre-delegation involves many of the same risks that policymakers wrestled with in the nuclear era. Pre-delegation would require trusting the cyber operators with decisions that political leaders might prefer to retain for themselves. With cyber weapons, the level to which authority would need to be delegated should be even lower in the chain of command than was needed for nuclear pre-delegation. The complexity and uncertainty of cyber mean that pre-delegation procedures could be especially fraught; specifying in advance the conditions under which certain actions would be taken might be very cumbersome. Moreover, the cyber-nuclear analogy breaks down in two ways that cut against the desirability of pre-delegation.

First, the attribution problem is much more acute in the cyber domain than in the Cold War nuclear domain. The most vexing challenge for cyber defense today is that of the anonymous hacker. Attackers hide within the international, maze-like architecture of the Internet, leaving a tenuous trail of evidence that often runs through countries with which a victim's government has poor diplomatic relations or no law enforcement cooperation. Most cyber investigations end at a hacked, abandoned computer, after which the trail goes cold. Moonlight Maze, a multiyear investigation to find a hacker group that had successfully stolen US technical research, encryption techniques, and war-planning data, discovered "disturbingly few clues" about its true origin.[35]

Vint Cerf, one of the Internet's inventors, has acknowledged that security was not an important consideration in the Internet's original design. If given the chance to start over, he maintains, "I would have put a much stronger focus on authenticity or authentication."[36] From a technical perspective, solving the attribution problem is theoretically possible. For example, the language of computer networks is now shifting from Internet Protocol version 4 (IPv4) to IPv6, which will raise the number of computer addresses from 4 billion to—for all practical purposes—infinity. Everyone and everything on planet Earth could be tagged and traced with a permanently associated number. IPv6 also supports (but does not require) Internet Protocol Security, which can be used to authenticate Internet traffic. For example, in 2006, this future capability allowed the Internet Society of China chairwoman Hu Qiheng to announce that "there is now anonymity for criminals on the Internet in China. . . . With the China Next Generation Internet project, we will give everyone a unique identity on the Internet."[37]

However, the future of cyber attribution, even in a next-generation network environment, is far from certain. Technologies such as IPv6 may be used to mitigate the threat of anonymous cyber attacks, but human rights groups fear that governments will use this new capability to quash political dissent by reducing online privacy. In 2012 the South Korean Constitutional Court overturned a five-year-old law that required citizens to use their real names while surfing the Web. Stating that the rule amounted to "prior censorship," which violated privacy, it also found the rule was technically difficult to enforce and generally ineffective.[38] Although it is possible to redress some of the Internet's current technical shortcomings, connectivity will likely continue to outdistance security for many years to come. Progress in attribution will be incremental and involve a slow harmonization of national cybercrime laws, improved cyber defense methods, and a greater political will to share evidence and intelligence.

For the time being, however, the attribution problem would often limit cyber pre-delegation to a defensive role. In the absence of reliable intelligence regarding a hacker's true identity, deterring, prosecuting, or retaliating against anyone is difficult. For example, in 2008 the US military experienced its "most serious" cyber attack ever when malicious code was discovered on US Central Command's unclassified, classified, and C2 systems.[39] The attack was presumed to be directed by a foreign intelligence agency, perhaps in Russia, but the true culprit could not be determined with precision.[40] However, the Pentagon was forced to undertake a large-scale response to the attack, code-named Operation Buckshot Yankee. Because the initial attack vector had been the insertion of a removable USB flash drive into a US military laptop in the Middle East, the Pentagon decided to issue a blanket prohibition on the use of flash drives throughout the world.[41]

The second way in which the nuclear analogy breaks down concerns impact. Nuclear pre-delegation involved extreme scenarios that were unlikely—and, indeed, never came to pass—and yet whose consequences were so daunting that political leaders saw pre-delegation as an acceptable hedge. Cyber attacks, in the extreme, could reach catastrophic levels but likely not levels contemplated at the middle range, let alone the extreme range, that were envisioned in global thermonuclear war. Some real-world examples have been alarming, but many credible national security thinkers are still skeptical of the risk posed by cyber warfare.[42] The effects of cyber attacks are often transient and may even sometimes be quickly reversed. Cyber operations typically do not move (like electrons) at light speed but at human speed, with numerous steps in a cyber "kill chain" that can be spotted and countered by defenders at numerous points in its life cycle. These aspects of cyber attacks should give victims more flexibility in decision-making relative to mitigation and response. Cyber would involve scenarios that are comparatively more likely—indeed, may already have happened—yet their consequences are not (yet) seen as so daunting that we should run the risks of pre-delegation.

Furthermore, some of the consequences of cyber pre-delegation might be readily felt, or at least perceived, in the civilian and political worlds through a loss of privacy and the politically sensitive blurring of civilian-military divides. Properly circumscribing any pre-delegated cyber authority would require common agreement on the likely threats, but cyber risk analysis and damage assessments are notoriously difficult and time-consuming endeavors. One 2013 think tank report concluded: "At present, neither the procedures nor the tools are sufficiently robust to merit a delegation of offensive cyber authorities beyond the very limited ways in which they have been utilized thus far. But a reasonable determination of whether the potential operational benefits outweigh the real and legitimate potential costs . . . necessitates further capability development, albeit in a very controlled context."[43]

At this stage in the evolution of cyber warfare, there are many more questions than answers, including national perceptions of what constitutes cyber attack, defense, and escalation. Cyber espionage and cyber attack, for example, have an odd relationship; the former is required to achieve the latter, but in fact, the latter may never actually take place. The victim, however, must take cyber espionage, especially if it occurs at a sensitive military location, as a precursor to cyber war. Many organizations today do not even have a good map of their own network infrastructure, let alone confidence in their network security. To date, still no legislation is in place that requires US commercial enterprises to employ best practices in cyber defense. Moreover, in stark contrast to a nuclear explosion, some major cyber attacks go absolutely unnoticed by the public and with only the direct participants being witting.[44] For example, according to the reporters who broke the story, the public was never supposed to know about Stuxnet, and it could be that a simple misconfiguration in some of the attack code betrayed the existence of Operation Olympic Games.[45]

If and when a real cyber war takes place, the attacker's identity should be clear because there will be other, circumstantial evidence.[46] However, the often intangible nature of most cyber attacks is likely to make cyber pre-delegation difficult for national security decision makers to approve. And if the odds of a catastrophic cyber attack are low, the consequences perceived to be manageable, and the national command authority assumed to be available to manage a cyber crisis, then the political stars may simply not align for cyber pre-delegation.

The Cyber Pre-delegation Sweet Spot?

No analogy works in all respects, but nuclear pre-delegation holds at least one clear lesson for cyber conflict: if cyber commanders do receive pre-delegation authority, it will likely be for defensive rather than offensive operations. In fact, defensive pre-delegation may be all that is needed and may even be more than is necessary to confront many cyber threats.

In stark contrast to a nuclear attack, most cyber attacks can be stopped—at least in a tactical sense—with purely defensive measures. There is no immediate need to know who the perpetrators are, where they are located, or what their true intention is. The urgency stems from a need to locate, isolate, and neutralize the malicious code as quickly as possible. Furthermore, blocking malicious data is far easier than shooting down a ballistic missile. In this light, cyber pre-delegation may not even be necessary because system administrators already have the authority and capability to protect their networks from what has become an incessant barrage of malware.

Some cyber threats, such as botnets, pose more complicated challenges and may require cyber defenders to go "outside the wire." Botnet mitigation can even entail the shutdown or hostile takeover of the botnet C2 server(s). But this type of intricate cyber operation, which normally involves the collection of evidence and acquisition of court orders, is unlikely to occur in real time. To some degree, this seems to obviate the need for cyber pre-delegation. For example, the celebrated Coreflood takedown in 2011 required both Department of Justice user notification and FBI user authorization before the federal government could remove malware from any infected computer.[47]

Still, there may be scenarios in which cyber commanders desire offensive or counterstrike options and in which there is simply no time to consult with a traditional chain of command. One could imagine that a fleeting window of opportunity would close during which crucial cyber evidence and intelligence could be gained. Here, cyber pre-delegation might be useful, but its parameters must be governed by the existing laws of war. For example, just as US forces in Afghanistan are authorized to return fire and even to pursue adversaries outside the boundaries of a military base, logically cyber pre-delegation should reflect these same principles. One limitation could be that, in hot pursuit, the counterstrike (or perhaps even a preemptive attack) could not deny, disrupt, degrade, or destroy adversary data or computer resources except when there is no other way to stop a grievous cyber attack on the United States.

Tactical cyber commanders are likely to have rules of engagement that are much more liberal than those given to nuclear commanders, because cyber attacks are simply not as dangerous as nuclear attacks. If malicious code is found already installed on a compromised US government computer, defensive actions may be straightforward, as in Operation Buckshot Yankee. If a cyber attack emanates from the US private sector, the FBI and Department of Homeland Security could take the lead with technical support from the National Security Agency and CYBERCOM if necessary.[48] When a cyber attack on the United States emanates from a foreign network, it is preferable to contact that nation's law enforcement and system administration personnel to help stop it. However, there will be occasions when foreign cooperation is not forthcoming or when there is no time for consultation before irreparable harm would be done to the United States. In this case, pre-delegation might authorize a preemptive strike or a counterattack against the offending computer or computers.

Due to the attribution problem, this pre-delegation policy should recognize that US computer networks must be protected even when the assailant is unknown. Positive cyber attribution should be required for significant retaliation, but simple, defensive blocking actions against an ongoing cyber attack should be permissible. As noted, ballistic missiles have a return address, but we may never know the true source of some cyber attacks, as we may be successfully deceived by a false flag operation. However, even without knowing the true identity of an attacker, CYBERCOM may still be able to target the proximate source of the attack according to the laws of war (e.g., with discretion, proportionality, and so on).[49] For some forms of cyber attack, such as a denial of service, the easiest and most passive form of defense is to use black holes, or silently discard the malicious traffic somewhere on the Internet, before it reaches its target.[50] For the most serious forms of cyber attack, such as a malicious manipulation of US critical infrastructure, CYBERCOM may be able to conduct a pinpoint cyber strike to terminate the malicious process(es) active on the attacking computer while leaving the other processes intact (if they are presumed to be legitimate).

If neither of these options is possible, the attacking computer may be completely shut down via cyber attack or, in extreme cases, a kinetic attack. This alternative is not ideal, because the attacking computer may have other legitimate processes or functions that could be associated with the national critical infrastructure of another country. Just as soldiers sometimes fire from within hospitals and operate against the laws of war, cyber attackers can also launch attacks from Internet servers that are related to public health and safety. Here, CYBERCOM would have to calculate risk versus reward and still minimize any collateral damage to the extent possible. Any pre-delegated cyber response should be conducted in legitimate self-defense and supported by as much public transparency as security and intelligence constraints allow. During the operation, CYBERCOM could notify the targeted computer's system administrator and national law enforcement of its actions and the rationale. In 2013 the United

States and Russia created a White House–Kremlin direct communications line between the US cybersecurity coordinator and the Russian deputy secretary of the Security Council to help manage potential crises stemming from future cyber attacks.[51]

Conclusion

The history of nuclear pre-delegation offers helpful insight into whether and how nation-states should grant pre-delegation in the cyber domain. In the United States, nuclear pre-delegation was an easy-to-implement work-around that seemed to avoid the potential pitfalls of presidential succession and command devolution. In a similar fashion, cyber pre-delegation may help national cyber commands defend critical infrastructure in the new and fast-evolving domain of cyberspace, which, like the nuclear domain, presents vexing challenges to reliable command and control.

Nuclear attacks and cyber attacks have several similarities, including speed, surprise, and specialization. Together, these characteristics could make some level of cyber pre-delegation inevitable. However, important differences between nuclear and cyber include impact and attribution, both of which national security leadership must consider before granting any level of cyber pre-delegation.

Unlike a nuclear holocaust, cyber attacks do not pose an apocalyptic threat to the United States, at least not yet. Therefore, they neither pose the always-never dilemma nor demand pre-delegation. Although attackers have a considerable tactical advantage on the cyber battlefield, it is not clear that they possess a strategic advantage. One recent study on the war in Ukraine suggested that while every facet of the crisis has been affected by cyber attacks, the cyber dimension of the conflict has nonetheless not played a critical role in the war.[52] When the element of surprise is gone, and especially if positive attribution is made, traditional political, military, and diplomatic might should determine the victor in a real-world conflict, a fact that already provides some degree of cyber attack deterrence.

The tactical advantages that hackers enjoy, however, must be addressed, and a national dialogue on cyber pre-delegation could be the right opportunity. The Internet is worth protecting as it offers a higher level of efficiency, transparency, accountability, and responsibility in government, civil society, and the marketplace. Therefore, the public should support a national effort to give cyber defenders clear rules of engagement that, in turn, would notify malicious actors (and cyber defenders) of red lines they may not cross. Finally, if some level of cyber pre-delegation already exists, it should be possible to make this policy more transparent while at the same time boosting deterrence.[53]

As with nuclear pre-delegation, a stronger case can be made for using defensive cyber weapons, especially if their impact is limited to domestic networks. However, we now know that pre-delegation during the Cold War extended beyond the use of nuclear weapons in a defensive role. It is even more likely that this would happen with cyber weapons, given their less destructive nature.

In the United States, civilian leaders demanded that they retain positive control over nuclear weapons. In the cyber domain, this will also likely be the case, as the public now spends the majority of its time connected to the World Wide Web. President Eisenhower understood that any nuclear war might take place over allied territory, and he personally took measures to address that risk. The cyber analogy is that future military conflicts will be fought on the same terrain that we use for banking, email, games, and news, all of which may come under enemy or friendly fire.

Some aspects of nuclear pre-delegation and cyber pre-delegation are similar—how far down the chain of command to go, how much latitude to give commanders for interpretation, and so on—but some characteristics of cyber conflict are unique. Information technology convergence now sends practically all communications through the same wires, so unintended damage may be difficult to avoid.[54] If any cyber attack, even in self-defense, leads to the disruption of Internet sites related to public health and safety, war crimes charges could follow. Finally, information technology is evolving so rapidly that rules for cyber pre-delegation granted today may not be valid tomorrow. That said, some aspects of modern communications could mitigate the need for pre-delegation altogether. For example, President Obama was able to sit and watch in real time the raid on Osama bin Laden's hideout in Abbottabad, Pakistan. While no real-time orders were given from the White House (so far as we know), there would have been no technological barrier to such activity. Future national decision makers will likely want to monitor operations at a similar level of intrusiveness and may choose to be more involved than the president was.

In summary, political leaders may be forced to authorize some level of pre-delegation to the military to defend national sovereignty in cyberspace, but they are also likely to be every bit as skittish about its risks. At a minimum, they will want to preserve most of the form and substance of political control.

The US experience wrestling through nuclear pre-delegation questions and scenarios during the Cold War can inform its policymakers today. Given the rapid proliferation of interest in cyber war, those same lessons learned in the United States may shed light on how other states' leaders are confronting the same challenges and opportunities. A priority for future research is to compare and contrast the US experience and interpretation of that experience with those of other relevant international actors.

Notes

Approved for public release; distribution is unlimited. The views expressed in this document are those of the authors and do not reflect the official policy or position of the Department of Defense or the US government. An earlier version of this chapter appeared in Emily O. Goldman and John Arquilla, eds., *Cyber Analogies* (Monterey, CA: Naval Postgraduate School, 2014).

    1. "Document 3: Instructions for the Expenditure of Nuclear Weapons in Accordance with the Presidential Authorization Dated May 22, 1957," declassified on April 4, 2001,

and available at the National Security Archive, Gelman Library, George Washington University, Washington, DC (hereafter National Security Archive), http://nsarchive.gwu.edu/NSAEBB/NSAEBB45/doc3.pdf.

2. Letter from President Dwight Eisenhower to Deputy Secretary of Defense Thomas Gates, November 2, 1959, declassified on January 18, 2000, National Security Archive, http://nsarchive.gwu.edu/NSAEBB/NSAEBB45/doc2.pdf.

3. Peter Feaver, *Guarding the Guardians: Civilian Control of Nuclear Weapons in the United States* (Ithaca: Cornell University Press, 1992).

4. Martin van Creveld, *Command in War* (Cambridge, MA: Harvard University Press, 1985).

5. This authority is spelled out in 3 U.S.C. § 301.

6. Lt. Gen. Russell Handy et al., "C2 Battle Management," Panel at AFA—Air & Space Conference and Technology Exposition, Washington, DC, September 15, 2014.

7. Maj. Wilson McGraw, USMC, *Beyond Mission Command: Maneuver Warfare for Cyber Command and Control* (Newport, RI: Naval War College Press, 2015).

8. Paul Bracken, *The Command and Control of Nuclear Forces* (New Haven, CT: Yale University Press, 1983). See also Scott Sagan, *Moving Targets: Nuclear Strategy and National Security* (Princeton: Princeton University Press, 1989); and Bruce Blair, *Strategic Command and Control: Redefining the Nuclear Threat* (Washington, DC: Brookings Institution Press, 1985).

9. Feaver, *Guarding the Guardians*.

10. Fred Kaplan, *The Wizards of Armageddon* (New York: Simon & Schuster, 1983).

11. Feaver, *Guarding the Guardians*.

12. The first tranche of sixteen documents was declassified and published in 1998 and summarized here: "First Documented Evidence that U.S. Presidents Predelegated Nuclear Weapons Release Authority to the Military," March 20, 1998, National Security Archive, http://www.gwu.edu/~nsarchiv/news/19980319.htm. The original declassified documents are available here: "Documents on Predelegation of Authority for Nuclear Weapons Use," National Security Archive, http://www.gwu.edu/~nsarchiv/news/predelegation/predel.htm. See also Christopher Bright, "Cold War Air Defense Relied on Widespread Dispersal of Nuclear Weapons, Documents Show," November 16, 2010, National Security Archive, http://www.gwu.edu/~nsarchiv/nukevault/ebb332/index.htm. For a good summary of more recently declassified documents, see William Burr, ed., "U.S. Had Plans for 'Full Nuclear Response' in Event President Killed or Disappeared in an Attack on the United States," December 12, 2012, National Security Archive, http://www.gwu.edu/~nsarchiv/nukevault/ebb406/. See also Marc Trachtenberg, David Rosenberg, and Stephen Van Evera, "An Interview with Carl Kaysen," MIT Security Studies Program, 1986, http://web.mit.edu/SSP/publications/working_papers/Kaysen%20working%20paper.pdf.

13. "Notes of the President's Meeting," October 14, 1968, declassified, and available at the National Security Archive, http://www.gwu.edu/~nsarchiv/nukevault/ebb406/docs/Doc%205A%20Furtherance%20document%20Oct%201968.pdf.

14. Feaver, *Guarding the Guardians*.

15. As a 1978 Defense Science Board study put it, if the attack came while the president was in Washington, DC, then "it would be possible . . . for the President either to command the forces until the attack hit Washington and he was killed or to try to escape and survive, but not both." Quoted in Joint Secretariat, "A Historical Study of Strategic Connectivity, 1950–1981," Joint Chiefs of Staff (Historical Division), July 1982, declassified September 21, 2012, and available at the National Security Archive, http://www.gwu.edu/~nsarchiv/nukevault/ebb403/docs/Doc%201%20-%20connectivity%20study%201982.pdf.

16. A permissive action link is a security device for nuclear weapons, whose purpose is to prevent unauthorized arming or detonation of the nuclear weapon.

17. Keir Lieber argues that the new historiography on World War I casts doubt on the "automaticity" of the mobilization plans. On the Soviet's Dead Hand system, which provided for a nuclear response if the system detected physical signs of a nuclear strike, see Keir Lieber, "The New History of World War I and What It Means for International Relations Theory," *International Security* 32, no. 2 (Fall 2007); and David Hoffman, *The Dead Hand: The Untold Story of the Cold War Arms Race and Its Dangerous Legacy* (New York: Random House, 2014).

18. The British resolved this public-private question with a "Letter of Last Resort," a handwritten note from the prime minister to a submarine commander that was normally kept locked in a safe and presumably never read (and then destroyed upon completion of a tour). It provided instructions for what to do in the event of a nuclear war. *See* Ron Rosenbaum, "The Letter of Last Resort," *Slate*, January 9, 2009.

19. Feaver, *Guarding the Guardians*.

20. Gen. Andrew J. Goodpaster, "Memorandum of Conference with the President, June 27, 1958—11:05 AM," June 30, 1958, declassified on April 4, 2001, available at the National Security Archive, http://nsarchive.gwu.edu/NSAEBB/NSAEBB45/doc1.pdf.

21. See Siobhan Gorman, "FAA's Air-Traffic Networks Breached by Hackers," *Wall Street Journal*, May 7, 2009. Regarding the financial sector, after the Dow Jones surprisingly plunged almost a thousand points, White House adviser John Brennan stated that officials had considered but found no evidence of a malicious cyber attack. For issues with elections, see Daniel Wagner, "White House Sees No Cyber Attack on Wall Street," Associated Press, May 9, 2010. In 2007 California held a hearing on the security of its touch-screen voting machines, in which a Red Team leader testified that the voting system was vulnerable to attack. See R. Orr, "Computer Voting Machines on Trial," *Knight Ridder Tribune Business News*, August 2, 2007. In 2006 the Sandia National Laboratories' Red Team conducted a network vulnerability assessment of US water distribution plants. See Chris Preimesberger, "Plugging Holes," *eWeek* 23, no. 35 (2006): 22. Regarding electricity, Department of Homeland Security officials briefed CNN that Idaho National Laboratory researchers had hacked into a replica of a power plant's control system and changed the operating cycle of a generator, causing it to self-destruct. See Evan Perez, "US Official Blames Russia for Power Grid Attack in Ukraine," CNN, February 11, 2016.

22. Preimesberger, "Plugging Holes."

23. William Broad, John Markoff, and David Sanger, "Israeli Test on Worm Called Crucial in Iran Nuclear Delay," *New York Times*, January 15, 2011.

24. David Sanger and Mark Mazzetti, "US Had Cyberattack Plan if Iran Nuclear Dispute Led to Conflict," *New York Times*, February 16, 2016.

25. Nicole Perlroth, "In Cyberattack on Saudi Firm, US Sees Iran Firing Back," *New York Times*, October 23, 2012.

26. David Sanger and Martin Facklerjan, "NSA Breached North Korean Networks before Sony Attack, Officials Say," *New York Times*, January 18, 2015.

27. FBI, "International Cyber Crime: Iranians Charged with Hacking U.S. Financial Sector," FBI.gov, March 24, 2016, https://www.fbi.gov/news/stories/2016/march/iranians -charged-with-hacking-us-financial-sector; and Steve Holland and Matt Spetalnick, "Obama Vows US Response to North Korea over Sony Cyber Attack," Reuters, December 19, 2014.

28. Office of the Chief Scientist, "Autonomous Horizons: System Autonomy in the Air Force—a Path to the Future," vol. 1, "Human-Autonomy Teaming," AF/ST TR 15-01 (Washington, DC: US Air Force, June 2015).

29. Author interview with Mikko Hyppönen, chief research officer for F-Secure, November 11, 2011.

30. For example, there are myriad types of SQL injection, which can be impossible to predict individually and are best defended conceptually.

31. US Army Cyber Command, "Our Mission," April 26, 2016, http://arcyber.army .mil/.

32. Rotem Pesso, "IDF in Cyber Space: Intelligence Gathering and Clandestine Operations," Israel Defense Forces, June 3, 2012, http://www.idf.il/1283-16122-en/Dover.aspx.

33. Ministère de la Défense de France, "La cyberdéfense," January 19, 2015, http:// www.defense.gov.fr/portail-defense/enjeux2/cyberdefense/la-cyberdefense.

34. Anastasia Petrova, "Russia to Get Cyber Troops," *Vzglyad*, July 16, 2013.

35. James Adams, "Virtual Defense," *Foreign Affairs* 80, no. 3 (May/June 2001).

36. Joseph Menn, "Founding Father Wants Secure 'Internet 2,'" *Financial Times*, October 11, 2011, http://www.ft.com/cms/s/2/9b28f1ec-eaa9-11e0-aeca-00144feab49a.html #axzz42YZ8qj2L.

37. Thomas Crampton, "Innovation May Lower Net Users' Privacy," *New York Times*, March 19, 2006.

38. Evan Ramstad, "South Korea Court Knocks Down Online Real-Name Rule," *Wall Street Journal*, August 24, 2012.

39. William Lynn, "Defending a New Domain: The Pentagon's New Cyberstrategy," *Foreign Affairs* 89, no. 5 (Fall 2010).

40. Noah Shachtman, "Insiders Doubt 2008 Pentagon Hack Was Foreign Spy Attack (Updated)," *Wired*, August 25, 2010.

41. Ellen Nakashima, "Defense Official Discloses Cyberattack," *Washington Post*, August 25, 2010.

42. Persuasive skeptics include Cambridge University professor Ross Anderson, former hacker Kevin Poulsen, author Evgeny Morozov, cryptographer Bruce Schneier, Professor Thomas Rid, and even the man who wrote "Cyber War Is Coming" in 1993, Naval Postgraduate School professor John Arquilla.

43. Maren Leed, *Offensive Cyber Capabilities at the Operational Level: The Way Ahead* (Washington, DC: Center for Strategic and International Studies and Georgia Tech Research Institute, 2013).

44. Martin Libicki, *Sub Rosa Cyber War* (Amsterdam: IOS Press, 2009).

45. David Sanger and Mark Mazzetti, "US Had Cyberattack Plan if Iran Nuclear Dispute Led to Conflict," *New York Times*, February 16, 2016.

46. This was the case in Estonia in 2007, for example, when even chocolate shipments to Russia were canceled.

47. Greg Keizer, "Feds to Remotely Uninstall Coreflood Bot from Some PCs," *Computer World*, April 27, 2011.

48. The private sector organization may not be ultimately responsible for the attack; rather, a hacker may be using a compromised computer in the organization's network from which to launch the operation.

49. Michael Schmitt, *Tallinn Manual on the International Law Applicable to Cyber Warfare* (Cambridge: Cambridge University Press, 2013).

50. Denial of service is an attempt to make a computer or network resource unavailable to its intended users, usually by sending it so much bogus traffic that it cannot respond to legitimate requests. For many networks, setting up a black hole can be done easily enough with a configuration change at an organization's external router, disposing of the unwanted network traffic.

51. Office of the Press Secretary, "Fact Sheet: U.S.-Russian Cooperation on Information and Communications Technology Security," White House, June 17, 2013.

52. Nation-state cyber espionage may be an exception to this rule. By 1999 the US Energy Department had determined that cyber attacks from abroad, particularly from China, posed an "acute" intelligence threat to US nuclear weapons laboratories. See Jeff Gerth and James Risen, "1998 Report Told of Lab Breaches and China Threat," *New York Times*, May 2, 1999. As stated earlier, all things nuclear may have a strategic character. See also Kenneth Geers, *Cyber War in Perspective: Russian Aggression against Ukraine* (Tallinn, Estonia: NATO Cooperative Cyber Defence Centre of Excellence, 2015).

53. Here is one existing US patent that specifically references pre-delegation: Craig Cassidy and Christopher Coriale, "Pre-Delegation of Defined User Roles for Guiding User in Incident Response," US Patent 20150242625 A1, filed February 24, 2015, issued August 27, 2015.

54. Ross Dawson, "The Flow Economy: Opportunities and Risks in the New Convergence," in *Living Networks: Leading Your Company, Customers, and Partners in the Hyper-Connected Economy* (Upper Saddle River, NJ: Financial Times/Prentice Hall, 2003).

14 Cybersecurity and the Age of Privateering

FLORIAN EGLOFF

The seas around the world are, much like the cyber domain, not governed by one single nation. We have created maritime norms and have to do the same in the cyber space to ensure a flow of information and ideas.

ADM. MIKE ROGERS

Between the thirteenth and mid-nineteenth centuries, privateering was an established state practice. Privateers (privately owned vessels that operated against an enemy with the license or commission of the government in times of war) would be used to attack the enemy's trade. In peacetime the practice of reprisal represented the means to seek redress against the harm suffered by another nation's ships at sea. A letter of marque allowed merchants to attack any ship of the offending nation until they found something of equal value to their loss.

Two months after the 2007 cyber attacks on the small Baltic country of Estonia, Defense Minister Jaak Aaviksoo used the analogy to privateering in a speech, pointing to the 1856 Declaration Respecting Maritime Law that abolished privateering.[1] He suggested that similar norms of the maritime environment were needed in cyberspace. At the North Atlantic Treaty Organization's Cooperative Cyber Defence Centre of Excellence in Estonia in 2015, Adm. Mike Rogers also referred to maritime norms when thinking about norm development for cyberspace. Policymakers' hopefulness about the analogy to the seas is understandable; maritime trade is relatively peaceful after all. However, the historical record indicates that such norms did not develop quickly nor was the process of attaining them a peaceful one. On the contrary, once a private system of force was created, states were not able to control the use of force completely. This chapter argues that the study of the historical evolution of the private system of force in maritime history offers important lessons for analyzing and shaping the evolution of cybersecurity. Scholars have used the analogy to privateering to recommend, or dismiss, the issuance of letters of marque to private companies in cyberspace.[2] At the same time, various experts have used the analogy to describe the collusion between attackers and states.[3] Thus, it may be important to explore what can be learned from the rich history of privateering, in this instance mainly from British maritime history.[4]

A longitudinal view of history is necessary to understand the development of norms against privateering. Privateering evolved from an institution that profited merchants and the Crown to one posing a threat to English naval dominance. A similar struggle is taking place today in cybersecurity. Protection from threats propagating through cyberspace has been treated as a predominantly private undertaking. At the same time, non-state actors are exploiting the insecurities of cyberspace, with the potential disregard of state versus state normative frameworks. The institution of privateering can shed light on the aligned and conflicting incentives involved for both state and non-state parties, when defensive and offensive regimes are in place and where the responsibilities of both actors are blurred. Thus, the opportunities and risks of using privateers can be explained with the aid of historical examples, and the information can then be applied to the modern-day problems of the cyber realm.

The analogy is both historical and conceptual. It makes recourse to an older world in which states were weak players when it came to the exploitation of the seas. Conceptually, the analogy points to the differing degrees of involvement and control that states can have with actors who exploit largely ungoverned spaces, such as the cyber domain. By examining the historical trajectory of privateering, we can learn from the intended and unintended consequences that the presence of such actors produced.

The analogy can shed light on specific aspects of the cyber challenge. First, it gives an insight into a system in which lines between state and non-state actors are blurred. Next, it focuses on a key aspect of the mercantilist system—that is, the economic and political realms are not differentiated. Thus, it captures two of the most important peacetime cyber challenges—cybercrime and cyber-enabled economic espionage. Finally, the analogy improves the understanding of security dynamics in a system in which capabilities are distributed among various actors. The comparison to a time in which semi-state actors (such as privateers and mercantile companies) and non-state actors (such as pirates) were abundant provides key insights into the conflicting objectives between the competition for advantage and the stability of and reliance on a system of trade.

This chapter begins with a short history of privateering and identifies the analogies found in the cybersecurity challenges. It then unravels the similarities and differences, sets up conceptual frameworks, and points to policy implications. The chapter closes with identifying the advantages and disadvantages of the analogy to privateering.

Historical Background

In the fifteenth and sixteenth centuries, several developments concurrently led to an increase in European exploitation of the seas. Shipping technology advanced so that long-distance sailing and war-fighting possibilities became more viable. At the same time a will to explore, proselytize, and conquer led seafarers into new territories.[5] Financed by investing parties who expected lucrative returns, and backed by their respective sovereigns to attack both locals and rivals, privateers represented the early means of colonial expansion. The era of the mercantile companies had begun. Mercantile companies operated by their own international policies. Merchants had to provide their own protection outside of territorial waters. They made deals with other companies or states, or were at war with them, and engaged in open warfare, piracy, and privateering, sometimes independently and against the interests of their home states.

In English history, privateering is best known through the acts of the Elizabethan sea dogs. The voyages of Sir John Hawkins, Sir Francis Drake, and Sir Walter Raleigh not only brought wealth to themselves and their investors but also inspired subsequent generations of English singers and playwrights. Besides their voyages against the Spanish in the New World, the English privateers formed a key part in the still fledgling Royal Navy. The English thus also used the skills and experience of the privateers, gained in attacking commerce abroad, for the defense of the home country.[6] For example, Sir Francis Drake and Sir John Hawkins served in the Royal Navy to fight against the Spanish Armada. Thus, privateering was used to augment national strength both militarily and through its cultural contribution to national identity.

Privateering also brought disadvantages. For one, it was a lucrative undertaking for the sailors. As a privateer also enjoyed better food and took a higher share in the prizes than he would in the Royal Navy, many of the ablest seamen served as privateers, not as sailors in the navy. Over time, the Royal Navy addressed the competition for skilled labor by forcing sailors to join the navy (impressment) and by improving working conditions on royal vessels.

The state also tried to regulate the number of sailors involved in privateering and the targets that would be attacked by issuing privateering licenses; however, effective control was not guaranteed. For example, after being knighted for his services to the court, the notorious privateer Raleigh did not stop looting, even after the peace treaty between James I and King Philip III of Spain.[7] Finally, James I had Raleigh executed. This episode illustrates one of the problems that eventually contributed to the abolition of privateering—that is, the difficulty of controlling privateers.[8] The longer wars lasted, the more privateering was professionalized and institutionalized. At the end of wars, privateers were integrated into the navy, worked on merchant ships, or became pirates.[9] The line between privateering and pirating was blurred. As Fernand Braudel noted, pirates could serve as a "substitute for declared war."[10]

Privateering as a strategy of war could distract from the more formal naval efforts of building a battle fleet. During the late seventeenth century, French privateers (corsairs and *filibustiers*) became increasingly active. While English privateers were used as a tool of influence alongside the growing navy, the corsairs were used as a primary tool of naval warfare (*guerre de course*).[11] For France they provided an ideal weapon against the English, who, comparatively, relied more on foreign trade.[12] However, this emphasis on the *guerre de course*, which was supported by the profiting investment circles, shifted (limited) funds and efforts away from building a more formal naval capacity.[13]

By the end of the eighteenth century, mostly the United States (in the War for Independence) and France (in the French Revolutionary Wars and later in the Napoleonic Wars) employed privateers against Britain. Privateering had "evolved into a weapon of the weak against the strong." However, "it was invented and encouraged by the 'strong' states of Europe, whose naval power was largely an outgrowth of privateering."[14]

The Congress of Paris decided to abolish privateering at its meeting for a settlement of the Crimean War in 1856. In the deal Britain, the dominant sea power, committed to protect neutral commerce, and, in return, the other powers relinquished the right to privateering. The settlement also represented a move against the United States, which still relied on turning its large merchant cruisers into privateers in case of naval conflict.[15] When the declaration passed, it was widely circulated so that as many powers as possible would accede. The parties of the declaration agreed that no port could receive privateers. Thus, privateering was made practically impossible also for non-signatories of the agreement, as a privateer would have to return to his home state to sell his prizes. During the US Civil War, the Northern states considered signing the Declaration of Paris to prevent the Southern states from using privateers against commerce. At that time, though, the two parties were already in a state of belligerency, thereby losing the right of signing away rights for the other party.[16] Hence, the United States never acceded to the declaration.

The Cyber Analogy

Looking at more recent technological development, the invention of the World Wide Web and the subsequent commercialization and expansion of cyberspace have rendered societies increasingly dependent on networked functionalities. Similar to the sailors' early expansionary years of activity on the seas, the users of cyberspace are largely left to protect themselves. Absent a state capacity providing redress, users must rely on their own abilities to withstand threats propagating through cyberspace. In such an environment, defensive and offensive skills are sought by a variety of actors. Just as mercantile companies could not rely on the Royal Navy to protect their trade and hence armed their merchant navy and sometimes sought protection from private men-of-war, large companies today seek to attract some of the most skilled cybersecurity experts.

Thus, a policy debate has arisen about the extent to which companies can protect themselves against state-directed attacks and about whether private actors should be engaged in hacking back.[17] Whether a private company can defend itself from a state-directed attack depends on the intent and capacity of the attacking state and the defensive capabilities of the company. If a company with a high cybersecurity maturity is a generic target, then it may be able to dissuade an attacker by making itself a hard target. However, when private companies are the direct target of a motivated, well-resourced state attacker, their defensive capabilities will not deter the attacker. Companies need additional backup capabilities, which are traditionally reserved for states. The debate on hacking back often fails to explain what the aims of such an action might be for a private company. Is it to impose costs on the attacker? Is it to help determine the attribution of the attack to a particular actor? Is it to research the motivation of the attack? Given the uncertainty about the ramifications of any offensive or retaliatory actions against an attacker, it is unclear to what extent private actors would deem such actions to be in their interests. Nevertheless, the US government dissuades corporate hacking back by claiming it is illegal under the Computer Fraud and Abuse Act and by highlighting the danger of escalation against unknown adversaries.[18]

Many states are currently building their capacities to conduct offensive and defensive cyber operations. The growing state capacities in defending against and carrying out cyber attacks may be augmented by the experience of private actors. The interests of skilled personnel and governments can overlap in various ways. First, instead of recruiting personnel for governmental positions, governments rely on the support of private personnel in several countries. Countries depend on a form of national service (e.g., formalized cyber militias), the use of contractors to buy key capacities (e.g., zero-day exploits), or the use of a range of services offered in the cyber criminal underground as part of the tool set for state exploitation of the cyber realm.

Second, there is the phenomenon of so-called patriotic hackers. Working in the political and economic interests of a country, patriotic hackers have been active in many highly visible cases ranging from the Russian hackers' attacks on Estonia in 2007 and on Georgia in 2008 to the Chinese and US hackers' attacks after the Chinese Embassy's bombing in 1999 and the Hainan spy plane incident in 2001.

Third, and less evident than the highly visible and clearly politically motivated attacks, groups have also mounted criminal intelligence-collection efforts.[19] For example, allegations have been made of close alignment between Russian and eastern European cyber criminal networks and Russian state interests. The influence and direction of criminal activity are multilayered, ranging from discretionary enforcement based on the targets selected to the way in which cyber criminals have become active in Russian political interests.[20] Empirical evidence, however, is usually incomplete and open to interpretation. For example, Ronald Deibert, Rafal Rohozinski, and Masashi Crete-Nishihata found no direct evidence linking the Russian government to the electronic attacks in Georgia in 2008, but they did not rule out the possibility that Russia quietly encouraged "malicious actions by seeding instructions on Russian hacker and nationalist forums and through other channels."[21] Tacit support can be inferred when governments do not cooperate and prosecute identified criminals in the presence of a mutual legal assistance treaty.

Reports also indicate an increase in attacks toward economic targets, focusing on economic espionage and intellectual property theft. The cultivation and utilization of private talent to effect economic wealth transfer comes closest to a modern version of privateering. Thus, at their own risk, companies, hacker groups, and some cyber criminals engage to fulfill state-sponsored goals against the interests of other commercial and noncommercial entities. The profit motives for both the state and hacker groups can differ from those of privateers. In cyberspace, states may profit indirectly by gaining plausible deniability for their own activities by hiding behind criminal hacker groups in return for tolerating their criminal activity, whereas in the case of privateering, states directly encouraged the profit-generating criminal activity.

Similarity of Regulatory Challenges

Having identified the analogical structure of the two domains, the focus now shifts to the similarities of the challenges states have faced on the sea and in cyberspace. Neither domain was controllable by a single actor, skills for deploying force mainly rested with semi-state actors, and the state alone lacked the capability or will to protect private entities. A broad range of actors was engaged in exploiting the new possibilities offered by transoceanic trade, with some having more legitimacy than others. As trade on the sea and dependence on cyberspace increased, the respective attack surfaces increased also. Therefore, states built dedicated capabilities to project force through navies and their cyber equivalents. These capabilities, however, had to coexist and compete with their private equivalents.

In the maritime space, one important factor for a navy's mobilization potential was the total number of its able seamen. Thus, countries with larger merchant fleets could draw on a larger number of able seamen. Wartime demand usually exceeded peacetime supply.[22] As a conflict began, therefore, the speed of mobilization determined who could project naval power quickly. For example, the French "système des classes . . . could recruit men up to a certain level of manpower, faster than the British practice of bounties backed by the press"; hence, France had an advantage at the beginning of the mobilization.[23] However, due to the larger number of total able seamen, the British would enjoy an advantage in the later stages of mobilization.

At this time, how this issue applies to cyber capacities is unclear. The focus on human capital is analogous, for cyber capacities are predominantly reflected in skilled manpower; however, it is not clear which specific skills a cyber operator would need to be considered a quickly mobilized capacity (in analogy to the able seamen). One reason is that offensive and defensive skills may be more distinguishable in cybersecurity than they were in naval warfare. Hence, while one can assume that a country with a large information technology sector would have an advantage in recruiting people with the necessary skills, the time lags and transformation potentials remain unexplored.

Unintended consequences of using privateers continued to create difficulties for the states that employed them, thus leading states to regulate the practice over time. As states build their own cyber capacities, their acceptance of the unintended consequences of private activities (be they commercial or criminal) may decrease. As all states grow more dependent on cyberspace, a window of opportunity for some agreement and cooperation in the cyber-criminal space may arise. However, as the history of piracy has shown, the levels of protection for cybercriminals may ebb and flow along with the polit-ical tensions of the time.

On the seas, the involvement of private force continued to play a role up to the early nineteenth century. Privateering was abolished only when the domi-nant naval force, Britain, decided that, due to its reliance on global trade, main-taining the option of private attacks against commerce was strategically and ideologically against its interests.[24] This deal did not include the United States, a rising power that was more reliant on its merchant cruisers in wartime. How-ever, in return for the smaller powers' agreement, the dominant sea power struck a deal whereby it committed to a more protected space for neutral trade.

The cybersecurity space is far removed from any international legal agree-ment settling cyber conflict. While the analogy would suggest that states learn from unintended consequences over time, they do so slowly. Thus, for some years, cyberspace may stay a relatively chaotic environment with an abundance of players present. If there is to be some sort of agreement, the analogy indicates that it does not necessarily have to include all the major powers. Leaving out one and building an effective regime constraining the usefulness of cyber attacks may suffice.

In 2015 Presidents Xi Jinping and Barack Obama agreed that their respective governments would not engage or knowingly support commercial espionage.[25] Commercial espionage was undertaken by different hackers, including some from the People's Liberation Army (PLA). Curtailing commercial espionage would then hurt the hackers' private income. Thus, given the regional power structure in the PLA, even if Xi wanted to halt this practice, he would likely face resistance.[26] As this practice has persisted for a number of years, the livelihoods and constituencies connected to the income streams need to be considered. There are strong similarities to stopping piracy, which involved a balancing act between building an alternative future for pirate communities and curbing their resistance. One theory holds that China will increasingly crack down on the work of freelancers while at the same time professionalizing the operational security of the state-conducted commercial espionage. This would mirror the British policies of the 1750s that raised the entry barriers for privateers and rendered the practice more regulated.

Differences between the Oceanic and Cyber Challenges

Having discussed the similarities between the two security spaces, we now address their differences. First, the pace of technological innovation seems faster in cyberspace than in shipbuilding. This observation has to be analyzed in conjunction with the more rapid diffusion of information and knowledge in the contemporary age. While the advancement in shipbuilding (e.g., the dread-naught) gave the English an advantage for many years, a new cyber capability, once discovered, may be repurposed by many actors within a very short time frame.[27]

Second, on the seas, human attackers expose themselves to physical risks. When an attack fails, the privateers face retribution. With remote attacks through cyberspace, this is not the case. Even if an attack is successfully traced to the responsible individuals, they may have the protection of their home state and may therefore be unreachable for prosecution.[28] This difference increases the prospects of the problem being more persistent in cybersecurity than on the seas.

Third, cyberspace widely differs from the sea because its topography is artificial; hence, it is malleable by human practice. Both technological and social changes manifest themselves in cyberspace and can change the "environment" in many, not always predictable, ways. Introducing new security-oriented technical protocols, hardware, and software for defensive purposes is a theoretical possibility. Research in networking has proposed models for new types of Internet routing; many of these proposals use security properties as guiding principles for their designs.[29] If implemented, they could contribute to a more secure environment, offering users a more explicit way of making decisions about whom to trust.

However, its malleability also means that the characteristics of cyberspace will significantly change over the coming years too. As the next two billion human users and twenty billion devices come online, the degree with which one can compare the maritime and cyber domains may change. Meanwhile, as the connectivity of societies deepens, access to security and surveillance technologies also spreads. This has already led to a balancing of the playing field in that surveillance technologies become more readily available to countries with traditionally more limited signals intelligence capabilities. Furthermore, this market is not limited to state actors, as non-state actors use some of the same technologies for defensive and offensive purposes.

Even though in many ways the cybersecurity environment seems far removed from the naval field, a lesson can still be learned. When operating in an actor-rich environment, states will not be able to control the use of cyber attacks completely. Once a system of private force is created, the institutional legacy carries forward. What state actors can do is manage the incentives, both for domestic and international actors, for using cyber attacks.

Conceptualizing the Range of Non-state and State Actors

Stepping back from the specific comparison of privateering and cybersecurity, a conceptual framework of a range of actors can capture the full potential of the analogy to the sea. In this framework, the navy, mercantile companies, privateers, and pirates are categorized according to their level of cooperation with state actors (see table 14.1).

In cyberspace, to be considered a state actor, an entity must be part of the state's organs or in direct support thereof. They are distinguished from semi-state actors that are in a close relationship with the state and sometimes advance

Table 14.1. Comparison between actors on the sea and in cyberspace

Actor type	Sea	Cyberspace
State actors	Navy (including mercenaries)	Cyber armies, intelligence, police forces, contractors, offensive security providers
Semi-state actors	Mercantile companies	Technology champions, major tele-communications companies, security vendors
	Privateers	Patriotic hackers Some cybercriminal elements
Non-state actors	Pirates	Hackers, cybercriminal elements (including organized crime)

state interests but are not organizationally integrated in state functions. Non-state actors have interests that lie outside the formal activities of a state and might reject the state's authority to govern their activities. Nevertheless, non-state actors may sometimes be in complicated relationships with states, reciprocally enabling the pursuit of respective interests. Simplifying the different relationships into three categories sufficiently captures the intuition that while some actors might be officially non-state actors, they are deeply entangled with states.[30] Overall, only those actors that directly interfere with another group's or individual's security interests are in scope.

This conceptual framework enables the analysis of different actors in cyberspace, highlighting how they are connected to the state. Importantly, the concepts do not carry an inherent moral value. The concepts of the navy, mercantile company, privateer, and pirate are understood to be by themselves morally empty. This reflects a historical understanding of them: some viewed privateers as heroes; others thought of them as criminals.

This conceptual framework enables multiple new types of analyses. It facilitates the tracing of state and semi-/non-state capabilities for deploying insecurity over time. This, along with a historical explanation of how it came about (including incentives, feedback loops, and normative changes), gives rise to a holistic analysis of the cybersecurity space. For example, figure 14.1 maps the state, semi-state, and non-state capabilities present in the international system over time and provides a richer understanding of the evolutionary development of the security dynamics. As the contemporary era witnesses a transfer from the high semi-state/non-state and low state capability quadrant to the high state and high semi-state/non-state capability quadrant, more conflicts between the different types of actors are to be anticipated. For example, as states build dedicated capabilities, they decrease their dependence on semi-state actors. This shift could be associated with a consolidation of activity in which a state cracks down on previously tolerated or sanctioned activity. One example might be China's arrest of cyber criminals after signing the Obama-Xi agreement.[31]

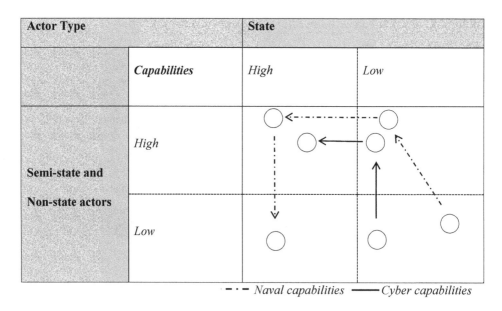

Actor Type		State	
	Capabilities	High	Low
Semi-state and Non-state actors	High		
	Low		

— · — *Naval capabilities* ——*Cyber capabilities*

Figure 14.1. State and semi-/non-state capabilities on the sea and in cyberspace over time

In a second step, the state proximity framework categories can be used to analyze the interactions between state, semi-state, and non-state actors. Lucas Kello argues that while competition between states exists, cyber insecurity has also accentuated a new state of nature involving non-state actors.[32] This global state of nature can be analyzed using the state, semi-state, and non-state framework, both for where the actors' interests collide and for where they converge. Regarding collision, the question to ask is, where is a respective actor considered the attacker and where is it the target? (See figure 14.2.)

Mapping some of the most prolific cyber attacks onto the categories identified reveals a clearer picture of the complexity of responding to cyber attacks. Each category of constellations involves different challenges for the attacked party. The framework presented reduces complexity to aid policymakers in anticipating different constellations of attackers and defenders. For example, with this framework, governments could have anticipated the Sony Pictures Entertainment and Sands Casino scenarios and considered possible policy responses. Privateering cases suggest that when the attacker has a special relationship to a state, rather than going after the attackers through the criminal prosecution system, the state must address the situation politically. For example, in the case of the attacks against the Sands Casino, allegedly conducted by hackers connected to the Iranian government, the prosecution of cybercriminals only through the legal system would not have been a fruitful response. The attack against a private corporation in this case took on a new form of signaling discontent.[33] As such, the response must address both the criminal as well as the political aspects of the actions, using the full range of policy options available.[34]

		TARGETS		
	Actor Type	*State*	*Semi-state*	*Non-state*
ATTACKERS	*State*	Stuxnet GhostNet	US → Huawei China → Google, Lockheed (e.g. Titan Rain, Operation Aurora)	PLA → Tibetan activists (GhostNet) North Korea → Sony Pictures
	Semi-state	Patriotic hackers → Estonia Iranian "hackers" → Saudi (Shamoon)		Russian "hackers" → JPMorgan Chase Iran → Sands Casino Cybercrime
	Non-state	ISIS → US Strategic Command	ISIS → AP Hacker → HackingTeam & Gamma International	Cybercrime Anonymous → Scientology Ashley Madison
	Unknown			Advanced persistent threat → German steel factory

Figure 14.2. Collision of interests between state, semi-state, and non-state actors

The framework enables the development of a better-prepared response to the next novel situation that policymakers and business leaders may encounter. It also helps the analyst to categorize the different reactions so as to keep an overview of how attacks are being treated. Thus, the semi-state category allows for a more adequate capturing of the politicized activities below the threshold of war.

When considering where the interests of the different actors converge, the question to ask is, who is seeking assistance and who is providing it? (See figure 14.3.)

		SUPPLY OF COOPERATION		
	Actor Types	*State*	*Semi-state*	*Non-state*
DEMAND FOR COOPERATION	*State*	Five Eyes	US ↔ companies in PRISM program China ↔ Huawei Russia ↔ Patriotic hackers?	Iran ↔ hackers Russia ↔ cybercrime US ↔ Hector Monsegur (Sabu)
	Semi-state	Google ↔ US (Operation Aurora) UK ↔ Huawei		
	Non-state	Sony Pictures ↔ US Cybercrime ↔ Russia		WikiLeaks ↔ Anonymous

Figure 14.3. Convergence of interests between state, semi-state, and non-state actors

Mapping some of the prolific cooperation cases onto the matrix shows various constellations that would be missed if one focused on only state-level capabilities. For example, the cooperation between technology or telecommunications service providers and states is an area that requires careful research. When US telecom providers cooperate with the US government to facilitate intelligence collection, it can violate the privacy guarantees given to the customers.[35] Similarly, the cooperation between hackers or cyber criminals and states is of interest. Examples are the aforementioned alignment of cyber criminals with Russian interests or the use of convicted hackers as informants to coordinate cyber attacks as in the case of the Federal Bureau of Investigation and Hector Xavier Monsegur (Sabu).[36]

Akin to the collision model, some of the constellations can be analyzed with the analogy to the sea. Analogizing the category of semi-state actors, or privateers and mercantile companies, and the non-state category, or the experience with pirates, allows for a richer understanding of the political and security dynamics at play in each case. In both models, constellations indicate the presence of dynamics not only of an old state-versus-state type of interaction but also of a new type of state of nature, one involving state, semi-state, and non-state actors. Through these models, the analogy can provide context and reduce complexity. It can aid policymakers in developing a strategic vision of a desirable state for the domain, including their abilities and constraints to shape the emerging normative framework.

Conclusion

The analogy to privateering has elucidated some cybersecurity challenges. Learning from four hundred years of history allows for a rich understanding of the forces giving rise to the multiplicity of actors shaping the institution of privateering and eventually leading to its abolishment. Similarly, the forces enabling and constraining the different types of actors that are active in the cybersecurity space can be identified. First, actors in cyberspace have similar proximity to the state as the mercantile companies, pirates, and privateers did in the sixteenth and seventeenth centuries. This conceptualization of actors in cyberspace captures both the expansion of transnational non-state actor activity and the devolution of responsibilities and authority to private actors.[37] The frameworks of analysis including state, semi-state, and non-state actors can reduce complexity and aid the development of a strategic vision for the domain.

Second, the levels of state capacity in cybersecurity resemble the situation in the sixteenth century, when some states transitioned from the use of privateers to professional navies. In naval warfare, this transition reduced the interest in the use of non-state actors. Judging by this process, the cyber capacities of state actors are in their infancy. The increasing dependence on cyberspace of all societies and the growth in state capability could have positive consequences for a cybercrime regime, as it could be accompanied by a decreasing interest in the use of non-state actors. However, the declining interest in the use of non-state actors is not guaranteed. Some states may still opt for a *guerre de course*.

Third, the analysis of the regime against privateering has shown that it can be traced to unintended consequences of state-sponsored and state-tolerated non-state violence, coupled with a growth of commercial opportunities for sailors. Similarly, in cyberspace one might expect unintended consequences to increase over time. Whether states will be able to coordinate their behavior to control these unintended consequences while preserving the positive effects of cyberspace remains an open question.

An awareness of the advantages and disadvantages of specific analogies is vital. It is important to clarify the type of knowledge an analogy can facilitate and where an analogy may mislead. The analogy to privateering is helpful because of the temporal distance between the two spaces, the possibility that this analogy invites long-term perspectives, and the fact that the policy innovations to redress the problems of privateering can be instructive in dealing with the cyber domain.

A comparison to a time that has long passed has a pragmatic and an analytical benefit. The pragmatic benefit is that it can depoliticize the debate and thereby focus the attention on the analytical problem at hand. The analytical benefit lies primarily in the integration of different actor types in a security space, where states are just starting to build capacities to project force.

The analogy reveals the long-term evolution of security dynamics in a space that becomes more important to the stakeholders over time. An ecosystem of security actors does not change quickly; rather, it evolves. Unintended consequences, feedback loops, and conflicting objectives influence how actors' policies change with time. In addition, the concurrent growing importance of the domain to all the actors raises the stakes and creates incentives to stabilize the domain. However, the decreasing interest in the use of non-state actors is not guaranteed.

There are some insights for more imminent policies. Particularly, the challenges in recruitment—both for states and non-state actors—can be better understood with the aid of the analogy. The analogy suggests that competition for skilled personnel is persistent and influences the way a formal state capacity can be developed. The analogy offers some appreciation for the various ways in which states have tried to work with skilled personnel (be it militias, volunteers, public-private partnerships, contractors, or army personnel). Linked to this aspect is the risk of policymakers profiting financially from cybersecurity policies. It is important to understand why some governments seem to enable economic and commercial espionage. Also, the job prospects of policymakers once they leave governmental employment have to be carefully evaluated. Parties that invested in privateering may shed some light on how governments are persuaded to sanction policies from which both officials and private corporations can profit. For example, critics of privateering argued that commerce raiding diluted the state's efforts to build an effective state-owned warfare capability.

The privateering analogy poses hazards too. Among them are the risks of further militarizing the discourse on cybersecurity, of advocating empire, and of assuming that history will predict the future.

The analogy considers cyberspace in relation to another relatively aggressive and militarized discourse. Civilian analogies may be more productive in creating opportunities for dialogue and cooperative solutions. Thus, other, more peaceful analogies could be more desirable in the context of a multilateral forum when looking to reshape the perception of the cybersecurity problem and to extend the possible range of solutions.

The analogy could be read as advocating empire, with all its oppressive and dominating aspects, as a solution to the security problems of cyberspace. After all, when privateering was abolished in 1856, the Royal Navy was the unchecked predominant naval power. The British had a very strong position from which to influence norm development, as they could assert those norms by force. It is not clear whether it would be desirable or feasible for any single power in the twenty-first century to reform the international cyber domain as Britain did the maritime domain centuries ago. Unlike the maritime case, where order was imposed by a Western power, in the cyber era China and perhaps others with different historical, cultural, and political predilections will be influential players.

Although this analogy can be used as a productive tool to enhance current thinking about cybersecurity, the general caveat that historical experience cannot guarantee a parallel course of events today must not be neglected. Just as past policymakers made decisions in the face of uncertainty, knowledge of the past should not lead the scholar into the misguided belief that history will repeat itself. While considering the lessons from the analogy to privateering, policies for the twenty-first century must take into account the idiosyncrasies of today's political landscape. The twenty-first century does offer some new opportunities, which policymakers can and should embrace. Higher degrees of international integration broaden the space for a cooperative solution between different stakeholders. Thus, as the Royal Navy guaranteed a principle of free trade in the past, a large group of stakeholders today is engaged in trying to make cyberspace a more open, transparent, interoperable, and inclusive environment.

Notes

Extracts of this chapter have previously appeared in Florian Egloff, "Cybersecurity and the Age of Privateering: A Historical Analogy," University of Oxford Cyber Studies Working Papers, no. 1 (Oxford: University of Oxford, 2015), http://www.politics.ox.ac.uk /materials/centres/cyber-studies/Working_Paper_No.1_Egloff.pdf. The publication is funded by the European Social Fund and the Estonian government. Printed with the permission of the Oxford Cyber Studies Programme. The author thanks the book editors, the participants of the author workshop in December 2015, and the Oxford Cyber Studies Working Group for their constructive feedback and comments.
Epigraph: Mike Rogers, "The Importance of Partnership in Cyberspace," keynote speech presented at the CYCON: International Conference on Cyber Conflict, Tallinn, Estonia, May 27, 2015.

1. Jaak Aaviksoo, "Cyber Defense: The Unnoticed Third World War," speech presented at the Twenty-Fourth International Workshop of the Series on Global Security, Paris, June 2007.

2. For an example discussing such a policy, see Michael Lesk, "Privateers in Cyberspace: Aargh!," *IEEE Security & Privacy* 11, no. 3 (2013). For an example of recommending the private sector in the United States should be given letters of marque, analogizing the privateers as a solution to the pirate problem, see Michael Tanji, "Buccaneer.Com: Infosec Privateering as a Solution to Cyberspace Threats," *Journal of Cyber Conflict Studies* 1, no. 1 (2007).

3. See Stewart Baker, Rafal Rohozinski, and Nigel Inkster in US Congress, Senate, Committee on Homeland Security and Governmental Affairs, 111th Cong., 1st sess., "Cyber Security: Developing a National Strategy" (Washington, DC: US Government Printing Office, 2009) (S. Hrg. 111–724); US Congress, US-China Economic and Security Review Commission, 111th Cong., 1st sess., "2009 Report to Congress" (Washington, DC: US Government Printing Office, 2009); and Peter Apps, "Analysis: Agreement Seen Distant at London Cyber Conference," Reuters, October 26, 2011.

4. Some scholarship on the lessons of privateering for cybersecurity exists, but it is underdeveloped, focuses predominantly on warfare, or centers on privateering as a policy option rather than assessing its potential for the reexamination of the public-private distinction. See J. Laprise, "Cyber-Warfare Seen through a Mariner's Spyglass," *IEEE Technology and Society Magazine* 25, no. 3 (2006); Peter W. Singer and Noah Shachtman, "The Wrong War: The Insistence on Applying Cold War Metaphors to Cybersecurity Is Misplaced and Counterproductive" (Washington, DC: Brookings Institution Press, August 15, 2011), https://www.brookings.edu/articles/the-wrong-war-the-insistence-on-applying -cold-war-metaphors-to-cybersecurity-is-misplaced-and-counterproductive/; Thomas Dullien, "Piracy, Privateering . . . and the Creation of a New Navy," keynote speech presented at the SOURCE Conference, Dublin, May 2013; B. Nathaniel Garrett, "Taming the Wild Wild Web: Twenty-First Century Prize Law and Privateers as a Solution to Combating Cyber-Attacks," *University of Cincinnati Law Review* 81, no. 2 (2013); and Peter W. Singer and Allan Friedman, *Cybersecurity and Cyberwar: What Everyone Needs to Know* (New York: Oxford University Press, 2014).

5. Paul M. Kennedy, *The Rise and Fall of British Naval Mastery* (London: Penguin, 2004).

6. However, this entailed significant risks. As Paul Kennedy points out, the privateers were "prone to alter carefully formulated plans in favour of rash enterprises and all too easily tempted by the prospect of plunder and glory into forgetting the national strategy." As an example, he points to Sir Francis Drake's abandoning the chase of the Armada to attack the *Rosario*. Ibid., 38.

7. Francis R. Stark, *The Abolition of Privateering and the Declaration of Paris* (New York: Columbia University, 1897), 66.

8. Another famous example is Captain Kidd, who was hired to go after French vessels and pirates but was eventually hanged for piracy. While some debate whether Kidd actually committed piracy or acted as a privateer, his actions upset some powerful interests and resulted in his execution.

9. Matthew S. Anderson, *War and Society in Europe of the Old Regime, 1618-1789* (Stroud: Sutton, 1998), 57; Michael Arthur Lewis, *The History of the British Navy* (Harmondsworth: Penguin Books, 1957), 74–75; and Stark, *Abolition of Privateering*, 97.

10. Fernand Braudel, *The Mediterranean and the Mediterranean World in the Age of Philip II* (Berkeley: University of California Press, 1995), 2:865.

11. Anderson, *War and Society*, 97–98, 147.

12. Kennedy, *Rise and Fall*, 79.

13. The degree of choice should not be overstated, however, as the French did not have the financial means to invest in a navy comparable with the British. In addition, there was much enthusiasm for privateering. For more details, see Halvard Leira and Benjamin de Carvalho, "Privateers of the North Sea: At Worlds End—French Privateers in Norwegian Waters," in *Mercenaries, Pirates, Bandits and Empires: Private Violence in Historical Context*, ed. Alejandro Colás and Bryan Mabee (London: C. Hurst, 2010), 60–62.

14. Janice E. Thomson, *Mercenaries, Pirates, and Sovereigns: State-Building and Extraterritorial Violence in Early Modern Europe*, Princeton Studies in International History and Politics (Princeton: Princeton University Press, 1994), 26.

15. Jan Martin Lemnitzer, *Power, Law and the End of Privateering* (Basingstoke: Palgrave Macmillan, 2014), 48–51.

16. Stark, *Abolition of Privateering*, 155–56.

17. See, for example, Paul Rosenzweig, "International Law and Private Actor Active Cyber Defensive Measures," *Stanford Journal of International Law* 103 (2014); Lucas Kello, "Private-Sector Cyberweapons: Strategic and Other Consequences," June 15, 2016, http://dx.doi.org/10.2139/ssrn.2836196. See also Florian Egloff, "Cyber Privateering: A Risky Policy Choice for the United States," Lawfare, November 17, 2016, https://www.lawfareblog.com/cyber-privateering-risky-policy-choice-united-states.

18. Leslie R. Caldwell, "Assistant Attorney General Leslie R. Caldwell Delivers Remarks at the Georgetown Cybersecurity Law Institute," Department of Justice News, Washington, DC, May 20, 2015, https://www.justice.gov/opa/speech/assistant-attorney-general-leslie-r-caldwell-delivers-remarks-georgetown-cybersecurity. For information about how the United States tries to boost cooperation with industry, see Department of Homeland Security, "Enhanced Cybersecurity Services (ECS)," Department of Homeland Security, accessed March 19, 2016, https://www.dhs.gov/enhanced-cybersecurity-services. For the background history of the precursor to the ECS program, see Milton Mueller and Andreas Kuehn, "Einstein on the Breach: Surveillance Technology, Cybersecurity and Organizational Change," paper presented at the Twelfth Workshop on the Economics of Information Security, Washington, DC, 2013.

19. See, for example, the investigation into a strain of the GameOver Zeus banking Trojan, which was configured to also collect security-related documents in Georgia, Turkey, and Ukraine pertaining to Russia's involvement in the conflict: Michael Sandee, "Gameover Zeus: Backgrounds on the Badguys and the Backends" (Delft: Fox-IT InTELL, 2015). For recent evidence of collaboration between Russia's FSB and cyber criminals, see "U.S. Charges Russian FSB Officers and Their Criminal Conspirators for Hacking Yahoo and Millions of Email Accounts," Department of Justice News, Washington, DC, March 15, 2017, https://www.justice.gov/opa/pr/us-charges-russian-fsb-officers-and-their-criminal-conspirators-hacking-yahoo-and-millions.

20. Indications for this are criminal sites, which exclude content that "can adversely affect the Russian Federation, the Ukraine, and Belorussia." Example is from Max Goncharov, "Criminal Hideouts for Lease: Bulletproof Hosting Services" (Los Angeles: Trend Micro, 2015).

21. Ronald J. Deibert, Rafal Rohozinski, and Masashi Crete-Nishihata, "Cyclones in Cyberspace: Information Shaping and Denial in the 2008 Russia-Georgia War," *Security Dialogue* 43, no. 1 (2012): 16.

22. Nicholas Rodger, "Mobilizing Seapower in the Eighteenth Century," in *Essays in Naval History, from Medieval to Modern*, ed. N. A. M. Rodger (Farnham: Ashgate Publishing, 2009), 4.

23. Ibid., 5.

24. Lemnitzer, *Power, Law*, 39–40.

25. Barack Obama and Xi Jinping, "Remarks by President Obama and President Xi of the People's Republic of China in Joint Press Conference," White House, September 15, 2015, https://obamawhitehouse.archives.gov/the-press-office/2015/09/25/remarks-president -obama-and-president-xi-peoples-republic-china-joint.

26. James A. Lewis, "Moving Forward with the Obama-Xi Cybersecurity Agreement" (Washington, DC: Center for Strategic & International Studies, October 21, 2015), https:// www.csis.org/analysis/moving-forward-obama-xi-cybersecurity-agreement.

27. See, for example, the diffusion of techniques employed in the Stuxnet attack.

28. One way the United States tried to overcome this was by issuing bounties—unsuccessfully in the case of the Zeus author Evgeniy Mikhailovich Bogachev (FBI, "Evgeniy Mikhailovich Bogachev," Most Wanted List, https://www.fbi.gov/wanted/cyber/evgeniy -mikhailovich-bogachev) and successfully in the case of the Sasser worm author, Sven Jaschan.

29. Xin Zhang et al., "SCION: Scalability, Control, and Isolation on Next-Generation Networks," paper presented at the IEEE Symposium on Security and Privacy (SP), Oakland, California, May 22–25, 2011.

30. A good heuristic for adjudicating state responsibility regarding actors operating from a state's territory is the ten-point "Spectrum of State Responsibility" developed in Jason Healey, "Beyond Attribution: Seeking National Responsibility for Cyber Attacks," Issue Brief (Washington, DC: Atlantic Council, 2011).

31. Adam Segal, "The Top Five Cyber Policy Developments of 2015: United States–China Cyber Agreement," Net Politics (Washington, DC: Council on Foreign Relations, January 4, 2016), http://blogs.cfr.org/cyber/2016/01/04/top-5-us-china-cyber-agreement/.

32. Lucas Kello, "The Meaning of the Cyber Revolution: Perils to Theory and Statecraft," *International Security* 38, no. 2 (2013).

33. This ties in with the signaling function privateering took in the mid-seventeenth century as argued by Gijs Rommelse in "Privateering as a Language of International Politics: English and French Privateering against the Dutch Republic, 1655–1665," *Journal for Maritime Research* 17, no. 2 (2015).

34. Not much is known about the actual US response in the Sands Casino case. However, in the attacks against Sony Pictures Entertainment, the United States publicly attributed the attacks to North Korea and imposed sanctions.

35. Julia Angwin et al., "AT&T Helped U.S. Spy on Internet on a Vast Scale," *New York Times*, August 15, 2015.

36. Mark Mazzetti, "F.B.I. Informant Is Tied to Cyberattacks Abroad," *New York Times*, April 23, 2014.

37. Ronald J. Deibert and Rafal Rohozinski, "Risking Security: Policies and Paradoxes of Cyberspace Security," *International Political Sociology* 4, no. 1 (2010).

Conclusions

GEORGE PERKOVICH AND ARIEL E. LEVITE

International society and governments are only beginning to understand the attributes of cyber capabilities and the possible nature of cyber conflicts. A principal value of analogies is to clarify which features of cyber capabilities and potential conflict are most pertinent to analyze and understand and which are less relevant or important to preventing conflict and to conducting cyber operations. Analogies help sharpen questions and identify dilemmas.

Readers may disagree, of course, with the observations we make here. Debate is welcome. One of the attractions of analogies is that, like art, they elicit the perspectives, experiences, and outlooks of the beholder. This invites conversation or debate among observers in which all parties may gain by appreciating a new angle, by affirming or discarding a prior assumption, by seeing or learning something entirely new. Our conclusions here are offered in this spirit. We have resisted the temptation to make policy recommendations for the US or other governments. Some of the foregoing chapters imply or suggest steps that governments could take to avert or minimize dangers of cyber conflicts, but as a general proposition we find that the cyber world is evolving so fast, with so many complicating factors (as we describe in the following sections), that policy prescriptions not made in context and in real time would be suboptimal. However, we do offer general principles and objectives later in this conclusion for policy-makers to consider.

What Are Cyber Weapons Like (Not Like)?

Drawing on all the chapters in this volume, here we attempt to summarize the defining characteristics of cyber weapons and, where helpful, how they differ from other military technologies.

Distinct, Essential Qualities of Cyber Capabilities

Information and communication technology (ICT), hereafter referred to as cyber technology, is dual use in the sense of serving benign and malign purposes. No previous dual-use technology has so thoroughly and quickly produced both peaceful and hostile applications on a global scale as cyber technology has.

Aircraft, railways, telegraph, and radio dramatically augmented civilian life, including commerce, while also boosting military potency. But the combined pace and scale of dual-use cyber technologies' dispersion and impact are unique. The growing dependence on cyber in many facets of modern life, including the Internet of things, makes managing vulnerabilities to disruption extremely challenging. We discuss implications of this in the second and third sections of this conclusion.

Cyber capabilities also are uniquely protean as instruments of intelligence gathering and coercion. The same basic tools and operators can be utilized in multiple ways to achieve a wide range of objectives. The range includes intelligence collection, political-psychological warfare, deterrence signaling, discrete sabotage, combined-arms military attacks, and campaigns of mass disruption. Other technologies also can be effective to achieve each of these objectives. Burglars could penetrate a political party's headquarters and steal and then publish sensitive files, though probably not as fulsomely as can be done by cyberespionage and by disseminating exfiltrated information via social media. Radio can transmit propaganda and other subversive information, just as the Internet does. Aircraft can drop laser-guided bombs to destroy a nascent weapons-plutonium production capability, as they did in Syria in 2007. But no other type of technology can be utilized in such diverse ways as cyber technology.

The versatility and ubiquity of cyber capabilities greatly enhance their appeal for intelligence collection, military operations, covert actions, and clandestine signaling. For powerful states, cyber can be a substitute, a precursor, and a complement to classic operations. For other actors—state and non-state—who find themselves in highly competitive security environments with technically sophisticated adversaries, cyber instruments may be uniquely attractive as power balancers.

The benefits of versatility are magnified by the relative ease of entry into the offensive cyber world. Compared to advanced kinetic and nuclear weapons and platforms, and their related reconnaissance and battle management capabilities, cyber capabilities are compellingly affordable. It is much easier and cheaper to recruit and train personnel to develop and operate cyber capabilities than it is to develop an effective conventional, nuclear, or biological war-fighting capability. For states that cannot develop their own cyber capabilities and cadres, these "goods" and services can be procured readily in the gray and black markets. The low barrier to entry is another important way in which cyber capabilities are unlike most other weapon technologies and why so many diverse actors can compete in this domain.

Still, turning cyber capabilities into effective and dependable weapons—as distinct from criminal tools and terror weapons—presents formidable challenges. In military and covert operations, the ability to ascertain impact and tailor effects is critical both to build confidence that intended results would be achieved and to avoid collateral damage and unintended secondary and tertiary consequences; however, the effects of cyber weapons may be uncertain and therefore difficult to predict with confidence. Cyber weapons may be viable only for short or at least uncertain amounts of time. Maintaining their effectiveness in the face of routine systems maintenance or specific defensive countermeasures is time and resource consuming. Constant monitoring of targeted networks is required.

Assuming concerns over the uncertainty of effects and durability can be mitigated, another distinct and attractive feature of cyber capabilities—for espionage and attack—is their stand-off potential. They can be operated from a distance to achieve global reach, sparing the conductors from friction with intermediaries and from risks of interdiction, capture, or death. (Satellites are similar in this regard, though they differ in that they are more difficult and expensive to build and deploy and do not directly carry out attacks.) The personnel protection afforded by cyber capabilities reduces the risk that an agent or soldier will be captured and used to create a spectacle that can escalate the intruded-on state's determination to retaliate and, in turn, create pressure on the captured asset's government either to escalate or to bargain to obtain his or her release. The worldwide range of cyber capabilities makes them especially attractive to states (and criminals) that cannot otherwise reach targets. For states that do have other long-range strike capabilities, the relatively low cost and personnel risks of cyber attack may be preferable. These effects cut multiple ways: they open not only a wide potential of offensive operations but also a similarly vast vulnerability to being attacked by cyber weapons from anywhere.

Cyber capabilities can be uniquely secretive, only partly due to the distances and conduits through which they can be operated. The development, the deployment, and often even the use of cyber assets are not easily observable—unlike air bases, naval forces, and drones. Secrecy provides operational advantages for intelligence gathering, covert operations, and war fighting. The low visibility of (some) cyber operations also allows potential victims to choose whether to publicize them. This creates political space for them to save face if they choose not to retaliate.

Seen from a different angle, the imperative to keep capabilities and, in most cases, operations secret makes it less likely that governments, let alone citizenries, will conduct informed debates over whether, when, and how (1) to conduct offensive cyber operations and (2) to develop effective international norms and rules for such operations. We discuss this implication further in the third section of this chapter.

Secrecy facilitates and is augmented by the distinct difficulty of confidently attributing the precise origin of attack and who actually is responsible for authorizing it. A precision-guided missile (PGM) or bomber can be readily and confidently traced to its source. Chemical explosives used in terrorist or covert operations also can often be traced to their sources, albeit with ambiguity about whether a particular state sponsored the attack. In the case of cyber weapons, the multitude of possible attackers, the potential concealment or falsification of attackers' identities, the diversity of effects, and the globally distributed vectors of attack make determining the authorial source of an observed effect difficult and time consuming. It is especially daunting to attribute cyber penetrations and attacks with the speed, resolution, and confidence needed to enable states to decide on, publicly justify, and tailor punitive responses. Forensic analysis alone rarely enables sure attribution; other sources and methods are often necessary. But states are reluctant to reveal these means, thus further complicating the challenges of attribution and the viability of deterrence by punishment.

The attribution problem may be less pressing in the context of overt warfare when combatants will reasonably presume that cyber attacks are coming from the same source as other forms of attack and the risk of misidentification may be modest. However, in covert contests short of armed conflict, attribution will complicate when and how antagonists interpret and respond to attacks. We discuss these challenges further later.

Cyber Effects and Their Implications

Turning now to specific functions, Michael Warner describes in chapter 1 how modern information and communications technologies have greatly enhanced states' and other actors' capacities to gather and analyze intelligence and to conduct covert operations. The advent of the telegraph and radio in the nineteenth and twentieth centuries also enhanced intelligence collection and guidance of military operations in radical ways. But these enhancements were not nearly as swift, widely distributed, and powerful as those provided by cyber. The dispersal of information and communications technologies and networks means that "anyone with a network connection can be a victim of espionage mounted from nearly anywhere." The growing availability of techniques to overcome the so-called air gap, and thereby siphon information from ICTs that are unconnected to the Internet, extends vulnerability. Both developments are new.

A major distinction between cyber capabilities and other technologies, including radio and satellites, is that the software and operations deployed to gather intelligence by cyber means also can be readily and directly used to conduct attacks. Satellites and related communications technologies must be linked both to physical delivery means such as missiles, planes, ships, and other platforms and to payloads to conduct attacks. By contrast, the same cyber tool and operation used to exfiltrate information also can be used as precursors as well as platforms for attack, even if some capabilities (especially payloads) would need to be added to fully weaponize them.

It is possible that intelligence gathering, if discovered by the target, could instigate a crisis or escalate a conflict. Yet, experience to date, albeit limited, suggests that states understand that cyber intelligence gathering is to be expected and managed without recourse to war, as is the historical norm. For example, Edward Snowden's revelations of US cyber intelligence gathering did not prompt targeted states to initiate conflict. Nor did the discovery that China had penetrated and exfiltrated massive amounts of data from US government files and health insurance providers cause the United States to take classical military action.

It is also possible that rather than cause conflict, the discovery of deep cyber penetration of a system, especially when it is air gapped (i.e., never connected to the Internet), may cause the penetrated state to reconsider the conduct of illicit or otherwise threatening activity. Some observers believe, for example, that the Stuxnet operation motivated Iran to accept constraints on its nuclear program, not because roughly a thousand centrifuges were damaged, but because the cyber penetration of the enrichment program alarmed Iranian leaders that they could not maintain the secrecy required to complete the development of nuclear weapons.

New challenges do arise from the proven potential of cyberespionage tools to become precursors as well as instruments of attack. If and when adversaries—say, the United States and Russia or China—are embroiled in a crisis and have increased the deployments and readiness of their military forces, the discovery of a cyber penetration into one's command-and-control systems and/or nuclear early warning systems could be unnerving and destabilizing. The state that conducted the penetration may be intending only to gather intelligence and early warning, and perhaps to communicate a deterrent signal, but the penetrated state may perceive the penetration as a harbinger of attack. Depending on the circumstances and the vulnerability that the penetrated state feels to preemptive attack, pressures could grow on that state to use its forces while it still can. Furthermore, the tight compartmentalization that intelligence organizations customarily impose on sensitive sources and methods, such as when monitoring adversary strategic command-and-control systems, greatly exacerbates the potential for cyber espionage to trigger inadvertent escalation. Clearly, the nature of cyber intelligence-gathering technology has uncharted implications for the potential conduct of cyber war and for efforts to prevent or manage it.

Moving from intelligence gathering to coercion, cyber weapons are especially—perhaps uniquely—useful in the gray zone of confrontation below armed conflict. The Russians' 2007 cyber attacks on Estonia, the US-led Stuxnet attack on Iranian centrifuges, the Iranians' destructive 2012 attack on Saudi Aramco's desktop computers and 2011–13 denial of service attacks on US banks, and the 2014 North Korean attack on Sony Pictures Entertainment—all exemplify diverse secretive coercive cyber operations short of warfare.

That said, cyber capabilities also create effects that can be vital to the conduct of war (and terrorism). Cyber capabilities now indispensably enable most of the communications, reconnaissance, command-and-control, and operational functions of modern militaries. Cyber networks do more for militaries than any other single previous technology, greatly boosting the defensive and offensive capacity of states and their militaries. At the same time, of course, it also makes them vulnerable to disruption in unprecedented ways.

Complementing their enabling potential, cyber weapons also can be disruptive or destructive in and of themselves. Like earlier forms of electronic warfare, they can blind, deceive, degrade, and even disable and destroy an opponent's communications, reconnaissance systems, navigation, command and control, and weapons' targeting and operation. Cyber weapons also can have much greater impact on infrastructure and physical systems than traditional instruments of electronic warfare can. Major powers have already used such offensive cyber capabilities against others, but they have not directly engaged each other in escalatory warfare in the cyber era. Thus, there is no empirical basis or analogy for evaluating how the dynamic competition between enabling and disabling cyber operations would play out in actual combat among near peers.

Operational Challenges

The enabling and disabling functions of cyber capabilities can serve both offensive and defensive purposes. While this is true of other intelligence-gathering tools and weapons too, the duality of cyber capabilities and their technical and operational features render them especially ambiguous. States often claim (fairly or not) that their military actions are defensive and precautionary while their opponent's activities are aggressive. This should be expected in the case of cyber operations also. The ambiguous offensive-defensive duality of cyber tools, operations, and, in some cases, organizations (like computer emergency response teams) raises challenges similar to those posed by national ballistic missile defenses in nuclear-armed states. Building effective defenses on a scale that could match an opponent's nuclear-armed missile arsenal could be a way to lower the risk of initiating offensive attacks (by blunting the victim's capacity to retaliate); yet, declarations of purely defensive intent in peacetime do not obviate the physical capacity of such capabilities to augment offensives in wartime.

The ambiguities that inhere in cyber operations are not simple to resolve. Means and procedures for clarifying intent have yet to be developed. Here the difficulty is that clarification of intent hinges at least in part on the willingness of actors to disclose the penetrations they have achieved or for the defender to reveal to a prospective attacker that its penetration has been detected. Both steps would be fraught with acute dilemmas.

In terms of target disablement and destruction, three types of weapons discussed in this volume merit comparison with potential cyber weapons. Robert Schmidle, Michael Sulmeyer, and Ben Buchanan in chapter 2 describe how experience with nonlethal weapons such as pepper spray, temporarily blinding lasers, foam guns, and road spikes to puncture vehicle tires may suggest potential applications of cyber weapons. Nonlethal weapons are meant to incapacitate rather than destroy an adversary's equipment and personnel. This incapacitation would often be localized, temporary, and possibly reversible. These effects could make such weapons relatively benign and therefore politically acceptable to use. Cyber weapons can achieve such effects via denial of service attacks or corruption of a military system's operation, as Israel reportedly did with Syrian air defenses in the 2007. Cyber attacks on cellular phone and other communications systems can disrupt the capacity of actors to coordinate illegal, hostile, or otherwise dangerous behavior without physically harming these actors and innocent people nearby. But, equally, such attacks may complicate efforts to terminate a conflict because communications between the political leadership and the field may be undermined.

On the one hand, nonlethal weapons have been and are most likely to be used in overt conflicts short of warfare. The distance from which cyber weapons can be operated, and the difficulty of attributing their use, makes them even more appealing for gray-zone operations than are other lethal and nonlethal weapons.

On the other hand, the challenge of confidently predicting and limiting the effects of cyber weapons may make them more difficult to apply than other non-lethal weapons are. As Schmidle, Sulmeyer, and Buchanan note, "With capabilities as new and complex as cyber ones, the unintended consequences of particular capabilities may cause additional or unexpected damage." The uncertainty—which includes possible over-performance or under-performance—then "greatly complicates the confidence a commander can have in the ability to achieve precise effects exactly when desired." This uncertainty may lead to self-restraint in opting to use them, as has been the case with nonlethal weapons. Yet, in overt warfare, when destruction and civilian casualties are already occurring, this inhibition presumably would decrease.

For purposes of destruction, as distinct from temporary disablement, PGMs such as guided gravity bombs and cruise missiles have been especially attractive to those states that can acquire them and support their use with adequate targeting information. The comparative advantage of these weapons lies in the precision, the exchange ratio, and the probability of destruction, especially as their operational ranges grow. Cyber weapons can have similar advantages. Unlike PGMs, cyber weapons do not require expensive, observable platforms and extensive logistical development and support infrastructure. But, as James M. Acton explores in chapter 3, cyber weapons may pose more challenges for their users than PGMs do.

First, in any missile or cyber attack, precision depends on the accuracy of the targeting coordinates and the effects of the weapon. A missile aimed at the wrong target or a malware attack on a network whose connections were not accurately mapped by the attacker can result in adverse consequences for the attacker (as well as the victim). The potential scale of unintended (imprecise) effects of cyber attacks, however, could be significantly greater than that of a cruise missile or bomb because of the lower certainty about the desirable and undesirable effects of their employment. For example, the cyber weapon could fail to neutralize the target, or malware used in an attack could spiral out of control or could be reverse engineered and proliferated by adversaries. The latter is an especially grave concern for cyber weapons, analogous perhaps only to biological weapons.

Second, the intelligence collection required for accurate, effective targeting of weapons is in some circumstances greater for cyber attacks insofar as the attacker needs to learn not only how to enter a network but also its extent, how it works, and what consequences any attempts to manipulate it might have. Moreover, targeted information and communications systems are not "stationary" as some prime targets of PGMs would be. Software is regularly updated, hardware is replaced, security protocols are changed, vulnerabilities are discovered, and so on. In the future, as offensive and defensive cyber systems become more automated, adoptive, and artificially intelligent, the challenge of attacking them will be akin to targeting mobile missiles, which is quite daunting though not impossible.

Third, as with PGMs, attackers need to be able to assess the damage they inflict on targets in a timely manner. Aviators conducting a bombing raid on a suspected nuclear plant need to know that the targeted country's air defenses have been disabled, whether by missiles or malware, before the planes come into range. As Acton suggests, the need for quick and reliable battle damage assessment may be even greater for cyber attacks in time-sensitive military operations insofar as their effects can be temporary, reversible, and less observable. A cyber attacker seeking to disable a set of adversary capabilities in order to conduct other operations needs to know not only that the degradation of their performance has indeed occurred but also when they may be fixed. Making such damage assessments is also vital to figuring if and when malware has spread unintentionally to other targets that might cause political, diplomatic, economic, or strategic harm to the attacker's position. Knowing the full extent of damage that has been or may be caused is necessary to inform efforts to neutralize or minimize unintended effects and to manage the consequences, including with parties not involved in the conflict.

Fourth, while cyber weapons might be used repeatedly over a short period, they are not likely to be usable in multiple successive campaigns against the same adversaries, unlike aircraft and missiles. Further, once adversaries (and the global information technology community) detect malware, they can develop ways to defend against it in days or months rather than the years or even decades required to develop new defenses to advanced kinetic weapons.

Finally, and importantly, PGMs with few exceptions do not achieve strategic objectives. Similarly, cyber capabilities may deter and coerce states and non-state actors or, theoretically, serve as firebreaks against escalation in crises or war, but they don't win wars, remove governments, end insurgencies, or restore order to regions beset by violence. As Acton writes, "Even if cyber attacks prove highly effective at disrupting an enemy's [buildup and] military operations, physical force will almost certainly be required to exploit this disruption." This is not to deny that governments, depending on the circumstances, can use cyber capabilities to undermine or influence the composition of other governments and to intimidate and weaken internal opponents. Opponents of governments can use cyber capabilities to do the reverse too. The larger point here is that, to take and hold power, actors likely will need other capabilities to consolidate opportunities created by cyber operations. For example, as Peter Feaver and Kenneth Geers note in chapter 13, every facet of the conflict in Ukraine has been affected by cyber attacks, yet "the cyber dimension of the conflict has nonetheless not played a critical role in the war."

Weaponized drones are an advanced form of PGM. They are especially attractive for targeting single or small numbers of people or other soft targets over long distances with an extremely short time between the decision to fire and the impact on the target. To date, drones have not been used to attack substantial matériel, infrastructure, or military targets, although their payload and lethality are rapidly increasing.

As David E. Sanger records in chapter 4, drones and cyber weapons share a number of advantages, notwithstanding the different types of targets they are directed against. They are relatively cheap to procure and operate. They do not put their operators at risk. For intelligence-collection purposes and for attack, drones can hover over targets for long times, helping reduce risks of mistaken targeting and unintended casualties. Cyber assets can reconnoiter targets for even longer periods.

While the destructiveness of missiles delivered from drones can be predicted precisely, and drone reconnaissance reduces risks of targeting errors compared with aircraft or cruise missiles, their primary liability remains mistaken targeting. A wedding party is killed because it was misidentified. A family believed to include the leader of the Islamic State is killed in a strike on a house that would have been acceptable to the leadership of the attacking state and perhaps under international law and public opinion, but due to faulty (or dated) intelligence, the leadership was not there. Cyber weapons are not equally prone to the consequences of mistaken identity, because thus far they are not (and perhaps in the future they will not be) used primarily to injure or kill people as opposed to disabling the systems the people depend on for various purposes. Nevertheless, as Sanger reports, uncertainty over effects and the related risks of collateral damage have inhibited the broad-scale use of offensive cyber weapons by the United States and probably by other countries too.

Risk of proliferation is another inhibitor of cyber attack that pertains less to drones. The software and operational techniques used to mount a cyber attack could be captured and replicated by adversaries, perhaps with attribution to the attackers. This risk of reverse engineering obtains with drones too, of course, but copying drone technology takes longer than copying and adapting malware. Further, drones are not nearly as versatile and potentially dangerous as cyber weapons are. Nor can most adversaries as easily deploy drones over long distances as they can cyber weapons.

On the other end of the war-fighting spectrum from drones, cyber weapons, in theory, can achieve large-scale effects on war-supporting facilities and infrastructure that otherwise might require extensive bombing campaigns to destroy. Norms and laws of armed conflict have evolved since World War II so that widespread indiscriminate bombing by states is now taboo. (This proposition has been tested by the destruction visited on Syria, including by Russian attacks; by the Sri Lankan government's prosecution of civil war against the Liberation Tigers of Tamil Eelam; and by the attacks of hybrid organizations such as Hezbollah, Hamas, and the Islamic State.) In any case, states engaged in overt warfare today increasingly seek to target their adversaries' war-supporting infrastructure, especially energy production and transmission resources, as well as telecommunications networks and services. The scale and intensity of operations required to achieve these results with surgical application of kinetic weapons could be daunting both for the potential target states and their populations and for the states that would conduct such attacks due to the risk of dramatic over-kill. To the extent that cyber weapons could surgically undermine the function-ality of war-supporting facilities, infrastructure, and industry with much less destruction of life and property, and lesser permanent or environmental dam-age, such weapons may be relatively desirable. Indeed, the option of first trying to hit a target with a cyber weapon and, only if and when that fails, then resort-ing to kinetic attack is appealing.

Yet three caveats must be raised. First, as noted earlier, there are significant uncertainties regarding the effects of malware attacks over time. The develop-ment of the field of operations research for offensive cyber operations is merely in its infancy. Second, if superiority can be attained in the ability both to inflict damage and to confidently limit effects, the inclination to conduct such attacks may grow. Third, if an adequate level of defensive protection exists to limit the prospects of retaliation, then incentives to conduct attacks will increase. Each of these three possibilities can be destabilizing; however, they also can be stabiliz-ing if the credibility of threats to conduct precision cyber attacks strengthens deterrence of wrongful actions. Hence, the impact of introducing offensive cyber capabilities is highly context specific, as we discuss further later.

On the ultimate end of the spectrum of destruction lies attacks on major pop-ulation centers via weapons of mass destruction. International humanitarian law and the laws of armed conflict proscribe such attacks per se. Nuclear-armed states nonetheless plan to target leadership, military command and control, and other strategic assets that are located amid or close to population centers. More-over, those actors willing to violate international norms and laws could use nuclear and perhaps biological weapons for massive destruction.

Cyber weapons raise several questions in this context. First, is massive disrup-tion meaningfully distinguishable from massive destruction? Steven Miller in chapter 10 notes that "any use of nuclear weapons will have devastating conse-quences. The same is not true of cyber." While this assertion is undoubtedly true in a political sense, it is less certain in physical terms. A single high-altitude nuclear detonation, say, in a desert or over a ship at sea clearly would not cause widespread destruction. But while a nuclear warning shot could cause an oppo-nent to eschew further escalation of conflict, in the case of a nuclear dyad it also could have the opposite effect and unleash an escalatory exchange of nuclear weapons, leading to mass destruction. There is no data on limited nuclear war-fare, only conjecture. If cyber weapons could be unleashed to massively disrupt or disable—even temporarily and reversibly—systems on which whole societies depend without directly killing large numbers of people, would such use be more likely and, if so, with what justification? How would the potential availability of achieving mass disruption by cyber operations affect the behavior of states in crises or in escalating warfare? In potential conflict among nuclear-armed states, could the use of cyber weapons to inflict massive disruption of the adversary's military and economy be a less objectionable alternative to nuclear attack? Could this cyber option, especially when intended as a political signal, act as a firebreak to nuclear escalation and thereby reduce the risks of nuclear war?

Again, nuclear war has not been experienced between adversaries who both possessed these weapons. Many argue that the unique, horrifying, and irreversible destructiveness inherent in nuclear weapons has encouraged restraint and, to date, made mutual deterrence work. In contrast, cyber attacks have become commonplace, albeit on a limited scale thus far, and clearly have not exhibited the mass-disruption potential they are widely believed to possess. Does the relative nondestructiveness of cyber weapons open the way to their potential use to achieve massive disruption? Or is the inability to confine and predict their effects (much as with biological weapons) a major cause of restraint? And would this unpredictability make deterrence as developed in the nuclear domain less tenable, as Steven Miller suggests?

Cyber weapons are more analogous to biological than to nuclear weapons in terms of their low visibility and the type of effects they could have.[1] Thankfully, experience with biological weapons is limited, and the possession and use of these weapons are now banned by the Geneva Protocol of 1925 and the Biological Weapons Convention of 1972. The latter has not been universally signed or adhered to, and it does not preclude some future use of biological weapons that could have implications similar to cyber warfare. But for now it can be said the restraints on biological warfare are far greater than those on cyber warfare. On the one hand, as with biological weapons, the development, penetrations, and attacks—for offensive and defensive purposes—of cyber weapons could go undetected, at least for a time. They could remain concealed even after indications of their existence appeared. At least for some time, it could be difficult to distinguish reliably whether an effect was due to an attack or a natural occurrence, further complicating attribution in case of attack. Biological attacks, like cyber attacks, can have minor effects and can be temporary and eradicable. On the other hand, some cyber and biological attacks could be immediately detected and identified as such, and they could cause extensive and long-lasting damage. Just as with biological weapons, their effects may hinge on numerous factors, some of which are transient, thereby creating serious uncertainty about their real-world effects and further undermining the capacity to precisely tailor the damage they inflict.

The Contingent Future of Cyber Weapons

To summarize, the versatility of cyber weapons, the unbounded distance over which cyber intelligence gathering and attack can be conducted, the safety that cyber technologies afford their operators, the secrecy and difficulty of attribution they entail, the low cost of attacks, and the potential precision and reduced violence of their effects—all make these weapons not only more tempting to use but actually more usable than other coercive instruments are. Is their overall impact, then, stabilizing or destabilizing? If the uncertainty over attribution and the reduced scope, level, and duration of damage imposed by cyber weapons

make states (and others) less inhibited to conduct such attacks, war and its escalation could become more likely. In parallel, the attractiveness of versatile cyber
capabilities could diminish states' attention to and investment in diplomacy and
other instruments of statecraft. Conversely, cyber weapons could be unusually
stabilizing in two ways—if the relative utility of cyber weapons makes them a
credible instrument of deterrence or compellence and if they can serve as an
effective firebreak against further escalation, thus reducing incidents of conflict
or capping escalation in crises. The flip side is that many, if not most, of the
states that are most capable of deploying potent cyber weapons are themselves
vulnerable to the damage such weapons can inflict. Moreover, the uncertainty of
cyber weapons' effects and the risk that malware can be replicated and proliferated relatively easily have made leading cyber states cautious in using them, at
least to date.

Overall, it is important to recognize that this new technology is not deterministic. Great uncertainty and contingency exist in this domain. Manifold potential
scenarios for war fighting, deterrence, and restraint can be imagined. They are
central to the future that states and societies must struggle to shape, as we discuss in the next two sections.

What Might Cyber Conflict Be Like (or Not)?

Having drawn on analogies to summarize some of the distinguishing qualities
and effects of cyber weapons, and some of the advantages and challenges these
qualities and effects pose, we now consider what conflicts involving these weapons might be like.

Information Warfare

Stephen Blank describes in chapter 5 how Russia has incorporated cyber technologies and operations into its long-standing approach to political warfare and
information operations to unnerve, weaken, and otherwise intimidate adversaries. In Georgia and in Ukraine, actors believed to be motivated by Russia placed
malware in electricity supply systems to deter Russia's adversaries and others
from escalating competition. In the case of Ukraine, a malware attack was
unleashed, apparently in retaliation for a physical bombing of transmission lines
feeding Russia-controlled Crimea and to signal the capacity to inflict greater
damage on Ukraine. Following the imposition of Western sanctions on Russia,
Russian cyber actors massively penetrated major US banks but were not reported
to have stolen money or corrupted data. This suggests that the purpose was to
deter the United States from pushing for more sanctions against Russia. Russia
apparently interfered in the 2016 US presidential election, presumably to dissuade the United States from sanctioning or otherwise contesting the Russian
government's behavior at home and in the "near abroad." The general point is
that cyber capabilities can be potent tools for waging political warfare and
deterring adversaries from countering such exertions in escalatory ways.

Preventive and Preemptive Use of Force

Preventive war and preventive use of force have been undertaken throughout history; likewise, cyber capabilities may be applied for both purposes. As John Arquilla describes in chapter 6, preventive war generally involves "starting a war at a most opportune moment . . . while the prospects of defeating an enemy's military, seizing territory, or toppling a regime are good—or at least before a growing threat worsens." Preventive force, by distinction, seeks to apply measured violence "in the hope of avoiding a full-blown war or to keep the strategic situation in an ongoing confrontation from deteriorating." (Preemption is different still, referring to action to defeat or weaken an adversary's *imminent* attack).

For various reasons, as Arquilla suggests, preventive force has become a common feature of international affairs in recent years and is likely to take cyber forms in the future. The Stuxnet attack is the leading example of using cyber force to prevent nuclear proliferation, to reassure allies, and thereby to avoid conventional war with Iran over its nuclear program. Other purposes could be to deter adversaries from aggression or to compel them to change their behavior and thus avoid suffering a wider-scale cyber or kinetic attack. Cyber attacks can be used against adversaries' offensive cyber capabilities and to degrade conventional military or other assets that are enabled by cyber systems and therefore are vulnerable to cyber attack.

Arquilla recounts how Britain's preventive uses of naval forces against Denmark (and indirectly France) in 1801 and 1807 succeeded in their immediate objectives of thwarting Napoleon's naval ambitions. But these successes also had the unintended effect of spurring Germany's subsequent century-long buildup of its naval forces and other defenses against British naval mastery. Analogizing to cyber preventive warfare, Arquilla posits that cyber techniques offer greater potential than kinetic weapons for protracted efforts to retard or delay an adversary's acquisition of dangerous capabilities and its conduct of aggression. In particular, "cyber prevention might also prove an ideal means for detecting and disrupting terrorist networks, for slowing their recruitment processes, and for generally undermining [their] trust and morale." The advantages of cyber stem in part from the covertness and difficulty of detecting and diagnosing correctly cyber capabilities and operations. These issues may provide both operational and political advantages to the attackers, especially the more tech savvy among them. If an attacker can initially mask the failures that are induced as a mere technical accident and subsequently plausibly deny involvement and keep personnel out of harm's way, the risks of being held publicly accountable and sparking escalatory retaliation appear much less than would be the case if other means of attack were used. However, once used, a particular cyber weapon can be detected, countered, and adapted by adversaries for their own use. For these and other reasons, the preventive use of cyber weapons can cause and intensify arms racing. Ultimately, Arquilla concludes, cyber preventive force is likely to increase, and it is impossible now to predict whether on balance the consequences will be welcome or unwelcome.

Preemptive use of force differs somewhat from its preventive use and is likely to be an attractive role for cyber weapons. Preemption occurs when conflict appears imminent or unavoidable, and preemptive attack appears to offer some prospect of significantly diminishing the adversary's capacity and will to prevail. States and non-state actors increasingly envision how they could use cyber attacks to preemptively weaken adversaries' conventional and cyber potency. At the same time, states and societies that depend heavily on ICTs feel acutely vulnerable to preemptive surprise attack. One actor's vulnerability is another actor's opportunity.

The Japanese attack on Pearl Harbor in December 1941 is, at least for Americans, emblematic of such a preemptive attack. Chapter 9 by Emily O. Goldman and Michael Warner usefully dispels the common misunderstanding that Pearl Harbor was a "bolt from the blue." Relations between the United States and Japan already were conflictual. The United States was crippling Japan's economy with sanctions to compel it to quit China. Washington expected war. In this sense, the Pearl Harbor attack was a preemptive escalation of an existing conflict. If anything, what happened at Pearl Harbor is now more germane to current situations in Europe, the South and East China Seas, the Korean Peninsula, and the Middle East, where competitors are engaged in disputes and mobilizing coercive capabilities and policies to influence each other. The Pearl Harbor analogy invites analysis of how cyber attack could be integrated into, or even stimulate, preemptive combined-arms warfare.

The purpose of the Pearl Harbor attack was to weaken and delay the capacity of the United States to prosecute a war with Japan, which hoped that it would diminish the resolve of Washington and the American people to fight and instead would motivate accommodation with Japan. In the cyber era, an analogous purpose of a preemptive attack could be to disable and disrupt a stronger power's capacity to deploy distant forces while creating time and space to achieve initial victories against extant forces. If a cyber attack could be counted on to reliably inflict relatively minimal casualties or damage to an adversary's forces, unlike the kinetic attack at Pearl Harbor, then such an attack could alter the strategic context of a confrontation by presenting the adversary with a fait accompli at minimal cost to both parties. The burden of escalation would shift to the state seeking ingress on the other's territory, advantaging the side that conducted the preemptive cyber attack. (Of course, an ingressing state also could attempt a preemptive cyber attack to weaken the receiving side's capacity to repel it, with the goal of motivating the defender to accede to the attacker's demands.)

In the case of Pearl Harbor, the attack failed both tactically and strategically in ways that may offer lessons for potential cyber preemption. The Japanese attackers missed the US Pacific Fleet's more important aircraft carriers, heavy cruisers, and submarines. The attack also did not destroy or durably incapacitate the fleet's fuel depots, dry docks, repair facilities, and undersea cable landings. By analogy, an actor contemplating a preemptive cyber attack against a powerful adversary would (should) need confidence both that intelligence on the systems being attacked is comprehensive, accurate, and timely and that the predicted physical and strategic effects of the attack are accurate. Significant overestimates of effects could mean that the adversary would be insufficiently damaged, while significant underestimates of effects could mean the adversary (and other states) would become more, rather than less, motivated to escalate in response. Relatedly, to calibrate the next steps, the cyber attacker would need accurate and comprehensive assessments of the damage caused by the attack. The attacker especially would wish to know how much time would be available before the adversary could remobilize capabilities that had been disrupted. Japanese leaders made both mistakes in the Pearl Harbor attack, underestimating the extent of the damage it would inflict and the motivation it would give the United States to fight.

Pearl Harbor also provides lessons for defenders against cyber attack. Had the Japanese attack significantly destroyed the Pacific Fleet's logistics capabilities, the attack would have achieved more lasting effects. Among other things, this highlights the importance of defending a state's logistics capabilities, and the infrastructure required to sustain it, from cyber attack. In the cyber era, military logistics are extremely dependent on ICT and the backbone of the Internet, with all the vulnerabilities this reliance entails. In many states, the entities and systems involved in logistics are controlled by private sector contractors and not exclusively by governments. Thus, the challenge of defending them is enormous, complicated, and very costly. Huge numbers of personnel and organizations, operating under diverse cultures and institutional imperatives (including cost management), must be trained repeatedly and harmonized to keep defenses effective and current. Resilience—again complicated and costly—must be built into the networks required to supply and operate military forces.

No one now has the experience of conducting or defending against a major cyber attack analogous to Pearl Harbor. Potential attackers and defenders, therefore, do not really know whether and how their capabilities to conduct and defend against a major preemptive attack would operate in real-life interactive conditions. The reconnaissance of adversaries' networks and the deployment of tools required to prepare an attack could be negated through routine actions taken by defenders to surveil, maintain, and upgrade their systems. Optimistically, states could deploy early warning sensors internationally, including in adversaries' systems, and marry them to active defenses to blunt attacks. Such defenses could be automated either as a default condition or in response to a command during crisis. All these possibilities create a premium on early action—simultaneously offensive and defensive—in a potential war situation to exploit vulnerabilities in the adversary's systems before any hostility begins ("use it or lose it") and to thwart an attack before one suffers losses. These incentives, amid the broader uncertainties of how interactive cyber war would actually unfold, highlight the potential of misunderstandings and miscalculations triggering preemptive attack and escalation, all of which can be destabilizing.

Escalatory Warfare

World War I provides a different example of war being triggered by an act whose timing and nature were unpredicted and then escalating with a scale and pace that leaders were unprepared to manage. Some fear this could be a likely course for a cyber conflict. No one foresaw that a Serbian terrorist cell, with ambiguous ties to the Serb state, would attack the heir to the Habsburg throne and thereby instigate a war involving all the world's major powers and killing seventeen million people. The governments and societies that then moved step by step into massive warfare did not imagine how it would escalate and persist. Early historical accounts of the war argued that technology—specifically, rail transport—drove the escalation process.

The question pertinent to analogizing World War I and potential cyber conflict is whether technology—railways in World War I and cyber today—importantly affects crisis stability and whether a conflict once started will escalate. The Balkans were a tense region at the beginning of the twentieth century, with lingering territorial and political disputes and competitive major powers jockeying for advantage directly and via proxies. This description also pertains to the Middle East, Russia's periphery, and South and East Asia today. Similarly, it is possible, especially in the Middle East, the Russian periphery, and the Indian subcontinent, that actors with relationships that are difficult to attribute conclusively to states could commit acts that other states would perceive as aggression.

In 1914 once a terrorist attack occurred and precipitated the crisis, Germany, France, and Russia had powerful incentives to mobilize their troops and railways first, before their adversaries did. Railways compressed space and time—key variables in military conflicts—and thereby increased pressures on decision makers during a crisis. This compression coupled with the technical challenges of reversing course (rerouting traffic) rigidified response options, resulting in a strong first-mover advantage. The space and time for managing the crisis narrowed. Thus, following the June 28 attack on Archduke Franz Ferdinand, the competing powers felt imperatives to move quickly to mobilize their forces. Once mobilization began, it created momentum toward conflict that was difficult to arrest or reverse, especially as little time was allowed for deliberations and diplomacy. Yet, even if leaders had wanted before the war to negotiate measures to regulate or limit military uses of railways, this effort would have been severely complicated by the imperative not to undermine the great economic benefits of this technology.

Francis J. Gavin, in chapter 7, carefully sketches this dynamic while taking pains to emphasize that railways did not cause World War I. Political factors did. The idea that new military capabilities, especially railways, would advantage the offensive and make war short and decisive "created a more permissive environment for states to gamble and risk war." But this did not determine events. Moreover, railways themselves were potent only insofar as they were integrated with military forces that could take and occupy territory. Similarly, cyber capabilities *alone*—even if, unlike railways, they can directly cause extensive damage—cannot by themselves fight and win large-scale wars and determine postwar outcomes. Achieving these strategic purposes requires a mix of cyber and other major military capabilities, but cyber operations could certainly compress the time and space for diplomacy and create serious incentives to undertake pre-emptive action in a crisis situation.

The conflict-hastening potential of cyber weapons could be partially miti-gated by robust defenses against cyber attack and by resilient communications and command-and-control systems. Such robust passive defensive measures were not available to counter railway mobilizations. But it remains uncertain whether and how states would have confidence in their defenses and resil-iency given the nature and rapid evolution of offensive cyber capabilities. This observation points to the importance of political institutions within states and diplomatic processes among them. Europe in 1914 was woefully under-institutionalized. The capacity to mobilize massive numbers of soldiers and military equipment grew much faster than each state's ability to accurately assess, share, and deliberate on such mobilization and then collectively to negotiate measures to control escalation. More developed institutional capac-ity would not have prevented the terrorist act in Sarajevo, but it could have averted or contained the escalation that followed it. This observation is perti-nent for Indo-Pakistani relations today as well as for regional tensions in the Middle East and perhaps East Asia.

Economic Warfare

The World War I era invites cyber analogies in part because the economy of that time was more globalized than ever before and after until the information and transportation revolution took hold in the 1990s. In the early 1900s, as Nicholas A. Lambert narrates in chapter 8, "accurate and instantaneous information relaying details of supply, demand, and prices was essential to all businesses and especially to the financial services industry that facilitated the movement of commerce with ever-increasing velocity." Connectivity created the potential to attack vital nodes of a country's economy or the global economy, as it does today. British naval strategists in the ten years leading up to August 1914 sought to leverage the combined supremacy of London's available commercial maritime information and the Royal Navy's command of the seas to develop plans to wage strategic economic warfare. Their aim was to quickly shock and derange the enemy's entire economy, thereby driving German society to abandon support for its misguided government. This differed from the "strategic" bombing cam-paign later practiced (and mislabeled) in World War II that targeted military and related industrial assets such as ball-bearing plants, aircraft manufacturers, and oil refineries. The broad analogy today to what the British had planned in 1914, obviously, is to a massive cyber attack on national and international financial institutions or other critical infrastructure that would promptly cause chaos, economic paralysis, enormous financial losses, and so on.

Britain's hegemonic position at sea and in global trade and finance enabled its officials to imagine this strategy. Their optimism was bolstered by the British economy's capacity to withstand the financial and economic shocks that a war would bring better than all other major powers could. These observations prompt analogous questions for the cyber world. Is hegemony different in cyberspace than, say, on the ground, at sea, and in the air, where armies, navies, and air forces heretofore have determined hegemonic power? Does any single power, most likely the United States, have sufficiently more capacity than any other power to combine financial intelligence and cyber superiority with physical power projection to gain a decisive capacity to influence international affairs and win conflicts? And does the power of a particular state to exert its will in cyberspace come with a corresponding vulnerability to cyber exertions of other actors against it? It took a long time for competitors to build navies to contest the British and for alternatives to British traders and financiers to rise. In cyberspace, are capacities and vulnerabilities more readily changeable?

The potential "disruptiveness" of the British economic warfare strategy inspired years of secret debate within the Admiralty and between it and other departments of the government, including the Board of Trade, the Treasury, and the Foreign Office. Officials perceived that the strategy could not only be effective but also have unprecedented and perhaps unforeseeable implications for Great Britain and for national and international law. The British process leading up to the strategy in 1914 resembles more recent accounts of US governmental deliberations over various forms of offensive cyber operations whose implications would be uncertain and unprecedented.

When hostilities in Europe started, purposeful economic strangulation compounded the natural contractions of war, as planned. But the ensuing unintended tightening of finance, trade, production, and employment was so alarming that British workers, businesses, and political representatives turned on the government. So did neutral countries that were damaged collaterally. Within three months, the British government aborted the economic warfare strategy. In a sense, it had been too effective. Its consequences were so sharp and widespread that they hurt British interests too, and political leaders could not ignore them.

Would a cyber campaign to inflict massive disruption on another state's economy yield similar blowback that could render it unsustainable and therefore strategically of little value? Would this be true for any state or only for some? Can cyber attacks be designed and conducted to achieve *massively* disruptive economic effects on a targeted state or group therein without harming others and the global economic and financial system as a whole? Would the risks of blowback depend significantly on the level of dependence on international connectivity of particular states and societies? For example, would Russia be significantly less susceptible than the United States or the United Kingdom or even China would be to blowback if Russia conducted cyber attacks against its adversaries' banks that in turn led to a crisis and major losses in the global financial system?

Strategic Context

The analogies presented in this book to explore what cyber conflict might be like are far from exhaustive. Two broad contextual possibilities for future cyber conflict deserve brief additional discussion here. The first centers on conflicts between states that clearly are unequal in aggregate capability to conduct nuclear or conventional warfare, covert operations, and cyber warfare, as distinct from conflict among near equals. The second centers on potential conflicts between nuclear-armed states.

Exemplifying the first context, the United States and Israel have used drones to attack adversaries in Afghanistan, Yemen, and Gaza that lack defenses against this technology and commensurate counteroffensive capabilities. The United States has used PGMs similarly against adversaries that cannot defend against them or retaliate long distance to threaten the US homeland. In circumstances where opponents were more evenly matched, the probability and conduct of offensive operations presumably would be different. A central question, then, is whether the accessibility of cyber war-fighting capabilities, and the global reach they provide, would significantly change the correlation of forces between adversaries that otherwise are highly unequal in coercive power. That is, do cyber capabilities offer the potential to further consolidate the stronger power's absolute advantage, or, conversely, do they enable weaker actors to shift the balance of power?

We are obviously unable to provide a definitive answer here, yet the Israeli experience may be illuminating. Notwithstanding Israel's overwhelming superiority in both unmanned aerial vehicle and cyber technologies, as well as other military capabilities, Hezbollah and at times Hamas have been able to achieve a conventional balance of terror by acquiring inferior but nonetheless meaningful retaliatory capabilities. This balance, though precarious at times, has largely confined the confrontation between the adversaries to sporadic, mostly short-duration exchanges that have denied the stronger party conclusive achievement of its aims. Offensive cyber capabilities could have similar strategic effects in confrontations among "unequals."

Finally, most of the states with active offensive cyber capabilities also possess nuclear weapons, with Iran being a notable exception. Thus, in competitions and warfare between these states and between them and other adversaries, the shadow of nuclear deterrence will be visible. Conflict, including with cyber dimensions, that could begin or threaten to escalate to major warfare will carry the potential to go nuclear. Nuclear-armed states, to date, have not conducted such warfare against each other (or allies of nuclear-armed states). Many observers attribute this to nuclear deterrence. Thus, possibly, and perhaps likely, deterrence of *major* warfare among nuclear-armed states will persist regardless of whether cyber attacks are unleashed.

Two caveats to this observation must be offered. First, cyber attacks on strategic command-and-control systems could be attempted to negate an adversary's nuclear deterrent. This possibility is especially worrisome in situations where the first use of nuclear weapons in escalating conventional war, even their tactical use, is real—for example, in Pakistan or perhaps North Korea. The threat of cyber attacks on strategic command-and-control assets could then cause that adversary to take countervailing steps, including using (or at least pre-delegating the use of) nuclear weapons earlier than otherwise expected, to avoid or minimize losses to its nuclear forces. This challenge is greatly aggravated by the similarity between cyber penetration of the nuclear command-and-control systems for purposes of intelligence gathering and early warning on the one hand and offensive operations against these systems' functionality on the other hand. Second, if cyber attacks were perceived to be attractive because a state believed they would fall below the adversary's threshold for nuclear escalation, and such attacks then had unexpected massive effects, nuclear escalation could inadvertently result. In any case, the relevance of nuclear deterrence is important in the cyber era, and it should be remembered that none of the analogies in this volume involve warfare among nuclear-armed states.

What Is Managing Cyber Conflict Like (Not Like)?

The cyber attacks on Estonia and Georgia in 2007, Iran beginning in 2007, Sony Pictures Entertainment in 2014, and the Ukrainian electricity system in 2015, as well as the hack into the US Democratic National Committee in 2016, reflect how criminal activity and interstate confrontation are now channeled to and through the cyber domain. In terms of its centrality on the international agenda, cyber confrontation is assuming the role that nuclear weapons occupied during the height of the Cold War. We invoke the term "confrontation" as a background condition to convey that the actors that may find themselves moving toward conflict cannot easily be categorized as aggressors or defenders. Some states and non-state actors, for example, dispute sovereignty over territory, with each claiming to be right. Some dispute the legitimacy of governments and feel the need to compel them to end repressive behavior; others feel the need to prevent or punish interference in their domestic affairs. The classic dichotomy of offense versus defense, however, does not capture the nature of confrontation in cyberspace. This is not merely because some of the purposes of cyberspace activities are either indistinguishable or mutually reinforcing. Many major contestants are acting, or preparing to act, offensively and defensively at the same time while using the same basic instruments for both efforts. The action is persistent.

Notwithstanding the persistence of confrontational interactions, most of the actors on the world scene need the cyber domain to function stably enough so they can operate in it both nationally and internationally. Thus, there is a tension between one's potential interest in using cyber operations to exercise control over one's population or to weaken or otherwise harm adversaries, and one's interest in preserving the functionality of the global cyber system. Each competitor wants to exert itself as fully as it can against its adversaries, but each objectively has an interest (though not necessarily of equal salience) in not causing a total breakdown in the functioning of the system itself.

The stability of the cyber system and of the broader political economy are now more connected than ever. In both cases stability is dynamic, not static. Both feature constant change, disruption, and creation. Systemically destabilizing actions would be those that exceed the normal flux of change and competition and that destroy or severely undermine the functioning of cyber communications and commerce and/or the broader political economy. Cyber conflict has the potential to threaten both systems. Acts that deeply and durably sow doubt in the reliability and integrity of the international cyber system—or otherwise disrupt its functioning, much less destroy it—would also cause international economic crisis. Major interstate warfare with cyber aspects would likely threaten both the cyber system and the broader geopolitical system.

Objectives in Preventing and Managing Cyber Conflict

To prevent or minimize activities that threaten the functioning of the global ICT system and the global political economy, states and relevant private actors should be expected to undertake a range of policies and activities to fulfill the following functions:

- enhance the capacity to detect and attribute cyber exploitations and attacks and to distinguish their purposes;
- augment various forms of defense against such activities, both to protect assets and raise the costs to potential perpetrators;
- increase the resilience of key cyber-dependent systems;
- while more difficult, pursue political and technical analogues to arms control agreements, or understandings that could inspire confidence that malware and other "weapons" will be sparingly used and will not have unintended consequences, including proliferation;
- assert state control over actors that use their territories to conduct unlawful cyber activities and over their citizens who do so abroad;
- upgrade capabilities to signal, threaten, and initiate cyber and other actions to inflict sufficient "pain" on adversaries to motivate them to eschew or desist from hostile activities; and
- develop, over time, norms to restrain the most potentially destabilizing sorts of cyber activities.

These steps would contribute to the prevention and mitigation of actions that could threaten the dynamic stability of the cyber domain and of the international political economy.

Defensive Measures

States and private actors are increasingly devoting resources to detect cyber intrusions and possible attacks. Little tension or debate affects such salutary efforts. Yet other steps or forms of defense do elicit various concerns relating to cost, legal propriety, and international stability.

Defensive capabilities and operations are, and will continue to be, central to preventing and contending with cyber attack in ways that are very different from countering nuclear attack, for example. "In stark contrast to a nuclear attack," Peter Feaver and Kenneth Geers write in chapter 13, "most cyber attacks can be stopped—at least in the tactical sense—with purely defensive measures." Moreover, many ways exist, and can be developed, to defend against cyber attacks without resorting to counterattacks that inflict damage outside of those networks being defended. Security "hygiene" within networks and organizations, anti-malware tools, and honeypots to lure or deflect attackers into isolated systems are well known. Other techniques could include analogues to emplacing "dies" in data the way banks do in cash so that if data or software are stolen, they become marked or do not "work" when located in unauthorized systems. Beacons can be embedded in data to emit signals if and when they are stolen, augmenting the capacity of authorities to trace perpetrators. Such defensive measures can raise the risks and potential costs of illicit cyber activity without eliciting dangers of unwanted or difficult-to-manage conflict.

However, as John Arquilla's "harbor lights" analogy warns, even the most seemingly obvious and risk-free defensive measures—turning the lights off to avoid being targeted at night—may not be taken due to organizational dysfunction, fear of protests by those who might be inconvenienced, market pressures on technology producers to minimize costs and operational complexity for users, and concerns of intelligence and law enforcement interests who want to maintain opportunities to penetrate systems. Here is another complicating effect of the breadth and depth of ICT diffusion throughout advanced states and societies: many prosaic factors impede implementation of otherwise uncontroversial forms of defense.

More difficult challenges arise when activities could extend beyond the defender's networks and into those of others, including perhaps neutral parties, especially if a competent legal authority did not authorize such activities. Dorothy E. Denning and Bradley J. Strawser, in chapter 12, analogize such defensive cyber measures to air defenses. Taking down hostile botnets, for example, could be akin to taking active defenses against hijacked aircraft. Such measures can range from forcing hijackers to land at specified airports to, in the extreme, shooting down such aircraft in a scenario like the September 11 attacks on the United States. Defensive attacks "out of network," Denning and Strawser argue, should follow established principles of the laws of armed conflict that require necessity, discrimination, and proportionality. Their treatment of these issues, of course, is more nuanced than summary here allows, and the challenges deriving from uncertainties in the potential effects of cyber operations in external networks must figure prominently in the planning and conduct of defensive measures. One important question that will increasingly be raised is whether and under what conditions to allow, or even encourage, private businesses and other actors to conduct defensive measures. Here, too, governments nationally and internationally will need to refine relative distinctions between passive and active defensive measures and whether they operate within or outside of the defender's own networks, under what oversight, and to what effects.

Feaver and Geers explore issues of command and control and of the pre-delegation of authority to conduct defensive cyber operations. They note that as with nuclear weapons, cyber threats and operations feature great speed, surprise, and specialization on the part of those charged with planning and conducting responses to attack. Fortunately, thus far, "cyber weapons do not pose an immediate, apocalyptic threat on the scale of nuclear weapons." This obviates the need always to be able and ready to retaliate quickly and therefore to establish pre-delegated authority. Indeed, thus far, most states, if not all, have demonstrated notable caution in authorizing even defensive cyber operations that could affect others' ICT. As we discuss later, this restraint stems in part from the difficulty of confidently attributing not only vectors of hostile action but also who authorized them, and from uncertainties over the intended and unintended effects of robust defensive actions. Moreover, the need for pre-delegation and for the automation of out-of-network defensive measures is attenuated by the likelihood that attacks on hostile botnets will require intricate sustained efforts to collect evidence and mobilize countervailing capabilities. Such time allows for the acquisition of court orders or other political-legal authority. Again, this distinguishes defensive cyber operations from nuclear defense and retaliation.

Still, cyber conflict remains a relatively new and rapidly evolving phenomenon. Artificial intelligence capabilities are developing quickly, as is the effort to leverage them into autonomous fighting systems, including in cyberspace. Most states are wrestling with how to structure and authorize command and control of cyber operations. As various military and civilian services organized on functional and regional lines face cyber threats and develop cyber capabilities, leaders struggle to define who can conduct what kinds of operations and under what conditions. Feaver and Geers highlight that the pre-delegation of some types of operations may be useful, "but its parameters must be governed by the existing laws of war." However, within this general framework, to which not all states may adhere, many important and complex issues are yet unresolved.

Nonproliferation

Complementing efforts to defend against cyber attacks, states will naturally seek ways to prevent the proliferation of software (malware) that could be particularly threatening to the maintenance of social stability. Unfortunately, the diffusion of offensive cyber capabilities is proving hard to control or restrict either by new arms control agreements or by the expansion of export control regimes, especially the Wassenaar Arrangement of 1996. This is partly because cyber tools are almost always inherently dual use, with their civilian applications being quite valuable and widely supported. Verification of arms limitations, so germane to the arms control agreements of the Cold War era, is practically infeasible with cyber technology.

Resiliency

Because defenses against cyber intrusion and attack are not perfect, and the spread of offensive capabilities cannot be blocked with confidence, states and major private enterprises must invest in resiliency. Resiliency will mean and require different things in different contexts—that is, in the military, the financial sector, the energy sector, and so on. The general aims are to decentralize potential points of failure or loss, to deploy backup capabilities and plans, to prepare users of systems for the possibility of disruption, and to plan contingencies accordingly. Of course, while resiliency may deny attackers the gains they seek, pursuing it runs counter to normal economic logic.

Controlling Proxies
Whether resiliency is embraced and implemented widely, limitations in the effectiveness of defenses and nonproliferation initiatives mean that a central challenge now is to narrow the range of threatening actors. Against the tide of globalization, states must affirm as fully as possible their position as the monopolistic controllers of the projections of force (including cyber force) from their territories and by their citizens abroad. The importance of this effort addresses the need to diminish room for cyber privateering to take place.

Florian Egloff aptly describes in chapter 14 the rise and fall of naval privateering between the fifteenth and seventeenth centuries, suggesting that we are currently at a somewhat analogous early stage in the cyber domain. Indeed, we may be in a stage reminiscent of the earlier phase of privateering, which saw a significant growth in the phenomenon's scope. There are various systemic and case-specific motivations for abetting cyber privateering, but one that deserves special attention is the rapid progress being made in attributing cyber attacks to their perpetrators. This progress, facilitated in large part by intelligence breakthroughs even more so than by cyber forensics, has the unfortunate side effect of encouraging states to privatize some of their offensive cyber operations. Herein lays the importance of codifying the sovereignty, the jurisdiction, and the responsibility of states in the cyber domain.

Deterrence and Compellence
Turning to more dynamic and fraught approaches to using cyber instruments to prevent or manage conflict, deterrence and compellence assume major importance. It should be emphasized that cyber instruments and operations could be used alone or in conjunction with other capabilities to affect a wide range of adversarial behavior beyond cyber attack, espionage, and thievery. The choice of instruments should depend on whom is to be influenced to eschew or desist from doing what. In this broader context, potential cyber operations can serve signaling purposes short of war. Indeed, such signaling is a feature of deterrence and coercion, or compellence. The aim is to signal a threat of future action and motivate the adversary to eschew or de-escalate violence or to otherwise change behavior.

Before considering what concepts, strategies, and tactics are most suitable to contest cyber and other threats today, it is helpful to recognize that the contemporary environment differs greatly from the Cold War period in which major states developed their national security strategies and instruments. The nature and frequency of confrontation in the contemporary world, especially in cyberspace, are quite different from the central challenge that deterrence addressed in the Cold War. During the Cold War, few said that illicit activities by mafias, terrorist acts by the Red Brigades of Italy, propaganda campaigns by major powers, or revolutionary conflicts in Southeast Asia, Africa, and South America were failures of deterrence. Commentators today, however, say that the Sony Pictures Entertainment attack, the hack of the US Democratic National Committee, and the takedown of the Ukrainian power grid represent failures of deterrence. Several dubious assumptions operate here. One is that deterrence is *the* appropriate lever to counter such a wide range of hostile actions. Another is that cyber capabilities and policies should be the primary tool for deterring any and all nefarious acts conducted by cyber means. As the eminent strategist Robert Jervis recently noted, "There is no such thing as cyber deterrence."[2] While this may well be an overstatement, the unique features of cyber capabilities—versatility, low cost, vast range, high speed, and difficulty of detection and attribution—can be used to support a wide range of national policies including deterrence and, more broadly, coercion to influence an extensive array of adversarial activities.

The use of cyber threats (and attacks) for deterrence and compellence of hostile cyber or other activities presents several nearly unique challenges. For a threat to be effective, the opponent needs to perceive it, but, as discussed previously, one attraction of cyber capabilities to date has been their secrecy. On the one hand, because cyber attacks remain generally anathema to much of the world, states perceive reputational risks in being exposed as active or potential attackers. On the other hand, as cyber weapons spread and cyber warfare becomes commonplace, actors may find secrecy unnecessary and unhelpful insofar as deterrence could be augmented by making their offensive capabilities better known. Further, if norms of behavior for the legitimate conduct of cyber conflict are developed, then states may become more transparent, at least regarding the types of operations that could be justified under such norms and be consistent with rules of armed conflict. Of course, operational imperatives are another motivation for secrecy. States (and others) do not want to alert potential targets of possible types and methods of operations and thereby enable adversaries to take defensive measures that would erode offensive capabilities. The Snowden revelations are an example of both motives for secrecy: the United States suffered reputational damage, and the revelations helped competitors to defend against US techniques. Whether effectiveness of deterrent threats requires an opponent to perceive in some detail what harm can be inflicted on it or, instead, whether a vague sense of possibilities is sufficient for deterrence remains speculative.

Of course, it is possible, and perhaps likely, that revelations (such as Snowden's) of capabilities and attacks that have already occurred give other states and nonstate actors a sense of what can be done and thereby create a general basis for future deterrence and compellence. Any group contesting a state with demonstrable capabilities such as the Stuxnet operation against Iran, the reported Israeli corruption of Syria's air defenses, the Chinese penetration of US government and health insurer files, and the Russian hack into the US Democratic National Committee files should anticipate that similar actions can be taken against it.

Looking ahead, for deterrence and compellence to be effective, the issuer of such threats needs to have not only credible capabilities but also plans to win, or at least not to lose, an escalatory process if the adversary does not relent. Yet, to date, there is no consensus on the meaning of "escalation" with cyber instruments. Uncertainties surround the potential effects of many possible cyber moves. The inability to distinguish computer network exploitation from computer network attack is fundamental in this regard. In crises or early stages of mobilization toward armed conflict, states that detect adversarial penetrations of their networks may find it difficult to assess whether the intentions are to gather intelligence or to conduct an attack—or both. They will prudentially assume the worst. More broadly, as Robert Jervis notes, "one could not know the physical, let alone the psychological and political impact, of exercising various cyber options; the country that is the object of the attack would assume that any effect was intended."[3] This situation can stimulate preemptive attack or other forms of escalation.

States (and other actors) will simultaneously seek to prepare for escalation and to minimize the risk. There is great tension between the two imperatives. Preparing the "battlefield" could advantage one or all parties to a potential conflict and could perhaps augment deterrence by motivating the adversary to back down. Conversely, it could lead to escalation, whether desired by one or both parties or not.

The tension between secrecy and deterrence exacerbates this conundrum. Revealing that one has penetrated or can penetrate various networks and do discrete kinds of harm can bolster deterrent or compellent threats. Such displays, however, weaken the weapon. For military operational purposes, as distinct from deterrence, one wants the adversary to *underestimate* its vulnerabilities and the threat one poses to it.

For these and other reasons, calculating whether and how deterrence and compellence by threats of cyber punishment can work is difficult. As Steven E. Miller and others note, the number and types of actors and scenarios that need to be deterred and compelled in the cyber domain are large. Moreover, attributing who is responsible for acts that would warrant retaliatory attack is fraught with technical, legal, and political difficulties, which would obtain even if the conductor of deterrence and compellence threats was entirely certain of the effects its potential attack(s) would have. Similar concerns likely would or should affect strategies to use offensive cyber attacks to limit the damage adversaries otherwise might threaten to inflict by cyber or other means—that is, deterrence by denial as distinct from deterrence by retaliation.

A number of these considerations are addressed in this volume. Three additional issues deserve mention here.

First, compellence is inherently more difficult to achieve than deterrence is. States that have committed national prestige and considerable resources to pursue fundamental objectives are highly resistant to external pressures to desist. It is easier to deter actors from doing things they are not already doing. The Stuxnet attack on Iran is perhaps the leading, if not only, example of a cyber operation that appears to have helped motivate a determined state to halt a threatening activity it was already conducting. Further research should be devoted to seeking examples wherein cyber threats and actions have compelled adversaries to abandon hostile behavior, whether such behavior is cyber related or not.

Second, the development of effective cyber defenses and hardening to cyber attack could significantly bolster cyber deterrence and compellence. A blend of robust offensive and defensive capabilities could make a state's deterrent and compellent threats against an adversary more credible by lessening the adversary's probability of retaliating effectively. It could also mitigate risks of escalation. However, if one or more adversarial groups of states achieved defensive capabilities that made their offensive threats more credible, would this enhance stability or more likely fuel the equivalent of arms racing and crisis instability?

Third, it is an open question whether offensive cyber capabilities are more or less strategically valuable for weaker parties than they are for stronger ones. States with greater capabilities are likely to be more vulnerable and sensitive to disruptions in cyberspace that could affect them directly or could emerge as blowback from their own cyber operations. Weaker actors, whose coercive capabilities, societies, and economies are less digitally dependent, may feel they have relatively less to lose in cyber conflict. The Pearl Harbor analogy suggests that weaker parties may see comparatively less risk in conducting cyber attacks, though in the end the stronger party prevailed, while Lambert's chapter on British economic warfare illustrates the risk of blowback from a stronger party's use of economic warfare.

The Potential for Restraint
Finally, as experience with cyber conflict is thus far rather limited, any state contemplating major attack—as part of a strategy of deterrence, compellence, or preemptive use of force—will create a precedent that could open the way for others to follow suit. There may be advantages to being the first mover—as, for example, the United States felt in the use of nuclear weapons against Japan and the use of Stuxnet and its predecessors—but the first move is rarely the last one. An actor's relatively easy entry into the world of cyber conflict practically ensures that once the actor gains significant advantage from a type of attack, others will be inclined to do something similar. Notwithstanding the great uncertainty about the future of cyber conflict, several dynamics currently seem clear enough: First, offensive cyber operations are on the rise in terms of both intensity and frequency. Second, capabilities to undertake far bigger attacks than have occurred to date are already in the hands of several states. Their ranks are growing. Third, the possessors of strategic cyber capabilities, even those employing offensive cyber operations regularly, still consciously refrain from exercising them to the utmost, regardless of how otherwise appealing the use of these weapons might appear. This pattern of caution on high-end uses holds true even for states that otherwise challenge the status quo.

Historical analogies and our preceding analysis prompt us to speculate what could be the reasons for such restraint, even if considerations vary among the actors that possess the pertinent capabilities. The following factors, which we do not attempt to weight or ascribe to particular actors, may help explain the apparent moderation of offensive cyber operations to date.

- Ethical considerations and legal concerns related to the laws of armed conflict clearly are one factor, even if less obviously reflected by Russia. Even in the shadowy cyber world, where few formal rules apply, most countries deem certain activities to be extremely unethical if not altogether illegal.
- Vulnerability to retaliation by cyber means seems to be another prominent concern insofar as some of the leading possessors of cyber armory are also heavily dependent on ICT for their prosperity and strength.
- Duality of intelligence and offensive cyber resources and opportunities discourages the use of offensive cyber tools for fear of compromising or burning unique intelligence assets.
- Uncertainty about the identity and affiliations of real sponsors of cyber aggression, as well as reluctance to divulge sources and methods that have made attribution possible, could explain inhibitions to undertake retaliatory cyber attacks.
- Considerations of efficacy, and the possibilities that cyber weapons could underperform or over-perform, seem to inspire restraint. This seems especially pronounced when contemplating attacks that might have strategic consequences. Some persistent and profound concerns are that cyber offensives might fail to achieve the objective assigned to them, that their perpetrators might be exposed, and that the attacks could produce unintended consequences including extensive collateral harm and damage bordering on the effects of using weapons of mass destruction. A particularly prominent concern with cyber weapons, analogous perhaps only to the massive use of nuclear weapons, is the possibility that their adverse consequences would not only affect noncombatants in adversaries' territories but could also seriously undermine the entire human habitat or its commercial lifeline, thereby also gravely harming the interests of the perpetrator.
- Growing concerns are that the employment of offensive cyber tools might not only incentivize others to use similar tools against highly sensitive targets but also provide adversaries the capability to reverse engineer these tools.
- Recognition that, once used and detected, cyber capabilities likely will not be usable again creates a "use only in extreme emergency" mentality.

Obviously these constraints are not of equal weight and effect on particular states (such as Russia). We are also at an early stage in the practice and study of offensive cyber operations and conflict. The observations, analyses, and speculations we offer here are meant to stimulate further analysis and debate, drawing on pertinent analogies from the past, and help scholars and practitioners to shape the future.

Notes

1. Gregory D. Koblentz and Brian M. Mazanec, "Viral Warfare: The Security Implications of Cyber and Biological Weapons," *Comparative Strategy* 32, no. 5 (2013): 418–34, DOI: 10.1080/01495933.2013.821845.

2. Robert Jervis, "Some Thoughts on Deterrence in the Cyber Era," *Journal of Information Warfare* 15, no. 2 (Spring 2016).

3. Ibid., 70.

CONTRIBUTORS

James M. Acton is the codirector of the Nuclear Policy Program at the Carnegie Endowment for International Peace.

John Arquilla is a professor and the chair of defense analysis at the US Naval Postgraduate School, a former analyst at the RAND Corporation, and a consultant to senior leaders during Operation Desert Storm, the Kosovo War, and other conflicts. His books include *In Athena's Camp* (RAND, 1997), *Networks and Netwars* (with David Ronfeldt, RAND, 2001), and *Afghan Endgames: Strategy and Policy Choices for America's Longest War* (edited with Hy Rothstein, Georgetown University Press, 2012).

Stephen Blank, a senior fellow at the American Foreign Policy Council, is a former professor of national security studies at the Strategic Studies Institute of the US Army War College in Carlisle, Pennsylvania.

Ben Buchanan is a postdoctoral fellow at the Cyber Security Project at the Belfer Center for Science and International Affairs in the Harvard Kennedy School of Government.

Dorothy E. Denning is a distinguished professor of defense analysis at the US Naval Postgraduate School and the author of *Information Warfare and Security* (Addison-Wesley, 1998).

Florian Egloff, a Clarendon Scholar, is a doctoral candidate at the Centre for Doctoral Training in Cyber Security and a research affiliate at the Cyber Studies Programme in the Department of Politics and International Relations at the University of Oxford.

Peter Feaver is a professor of political science and public policy at Duke University, the director of the Triangle Institute for Security Studies, and the director of the Duke Program in American Grand Strategy. He formerly was the special adviser for strategic planning and institutional reform on the National Security Council.

Francis J. Gavin is the Giovanni Agnelli Distinguished Professor and inaugural director of the Henry A. Kissinger Center for Global Affairs at the School of Advanced International Studies at Johns Hopkins University. He is also the author of *Nuclear Statecraft: History and Strategy in America's Atomic Age* (Cornell University Press, 2012).

Kenneth Geers, a senior research scientist for Comodo, serves as an ambassador with the Cooperative Cyber Defence Centre of Excellence with the North Atlantic Treaty Organization (NATO) and a senior fellow of the Atlantic Council. Geers is also an affiliate with the Digital Society Institute–Berlin and a visiting professor at Taras Shevchenko National University of Kiev. His twenty years of US government service include time in the US Army, National Security Agency, Naval Criminal Investigative Service, and NATO.

Emily O. Goldman is the senior adviser to the commander, US Cyber Command, and the director, National Security Agency. Formerly a professor of political science at the University of California–Davis, she is the author of *Power in Uncertain Times: Managing the Fog of Peace* (Stanford, 2011).

Nicholas Lambert is the Class of 1957 Chair in Naval Heritage (2016–17) in the Department of History of the US Naval Academy. His books include *Planning Armageddon: British Economic Warfare and the First World War* (Harvard, 2012).

Ariel E. Levite, the nonresident senior associate at the Nuclear Policy Program of the Carnegie Endowment for International Peace, was the former principal deputy director general for policy at the Israeli Atomic Energy Commission and a deputy national security adviser for defense policy to Prime Minister Ehud Barak.

Steven E. Miller is the director of the International Security Program at Harvard University's Belfer Center for Science and International Affairs. He also serves as the editor in chief of *International Security* and is a coeditor of *Going Nuclear* (with Michael Brown, Owen Coté, and Sean Lynn-Jones, MIT Press, 2010).

George Perkovich is the Ken Olivier and Angela Nomellini Chair and Vice President for Studies at the Carnegie Endowment for International Peace. A former foreign policy adviser to Senator Joe Biden, Perkovich is the coauthor of *Not War, Not Peace? Motivating Pakistan to Prevent Cross-Border Terrorism* (with Toby Dalton, Oxford University Press, 2016).

David E. Sanger is the national security correspondent for the *New York Times* and a senior writer for the paper. He is also an adjunct lecturer at the Kennedy School of Government at Harvard University and a Senior Fellow for National Security and the Press at the school's Belfer Center for Science and International

Affairs. His latest book is *Confront and Conceal: Obama's Secret Wars and Surprising Use of American Power* (Crown, 2012).

Lt. Gen. Robert E. Schmidle Jr. (USMC, Ret.) served as the deputy commander of US Cyber Command from 2010 to 2012.

Bradley J. Strawser is an assistant professor in the Defense Analysis Department at the US Naval Postgraduate School and a research associate at the Oxford University Institute for Ethics, Law, and Armed Conflict.

Michael Sulmeyer is the director of the Cyber Security Project at the Belfer Center for Science and International Affairs in the Harvard Kennedy School of Government. He formerly was the director for plans and operations for cyber policy at the Office of the Secretary of Defense.

Michael Warner is the command historian of US Cyber Command and an adjunct faculty member at American University and Johns Hopkins University. He is also the author of *The Rise and Fall of Intelligence: An International Security History* (Georgetown University Press, 2014).

INDEX

Figures and tables are denoted by f and t following the page number.

Wake Tech. Libraries
9101 Fayetteville Road
Raleigh, NC 27603-5696

DATE DUE

GAYLORD | PRINTED IN U.S.A.

WITHDRAWN

CPSIA information can be obtained
at www.ICGtesting.com
Printed in the USA
LVHW04s2157070518
576285LV00024B/322/P

9 781626 164987